Handbook of Technical Writing

Second Edition

Charles T. Brusaw
NCR Corporation

Gerald J. Alred
University of Wisconsin—Milwaukee

Walter E. Oliu
U.S. Nuclear Regulatory Commission

St. Martin's Press
New York

Library of Congress Catalog Card Number: 81-51836
Copyright © 1982 by St. Martin's Press, Inc.
All Rights Reserved.
Manufactured in the United States of America.
654
fedc

For information, write:
St. Martin's Press, Inc.
175 Fifth Avenue, New York, NY 10010

Interior design by Arthur Ritter.
Cover design by Claire Nelson.

ISBN 0-312-35808-3

Handbook
of Technical Writing

Second Edition

Preface

The *Handbook of Technical Writing,* Second Edition, is designed to be a comprehensive, thoroughly practical reference guide for students in technical writing courses and as a desktop reference for people already working in industry or government. In the classroom, it can be used along with a standard textbook, although instructors wishing to avoid the constraints imposed by a textbook might easily use this handbook as the basic resource.

The *Handbook* offers coverage far beyond the scope of conventional English handbooks. In addition to a comprehensive treatment of grammar, usage, style, format, and writing procedures (planning, research, outlining, methods of development, and so forth), it provides information on all kinds of technical communication—reports of various kinds (including oral reports), proposals, instructions, specifications, job descriptions, letters, and memorandums. It gives abundant examples, all drawn from technical or industrial contexts to provide the greatest possible relevance for professionally oriented readers.

We believe that the unique four-way access system in the *Handbook of Technical Writing* makes it more flexible and easier to use than other writing handbooks we have seen. The four ways of finding information are explained in detail in "How to Use This Book," beginning on page xi. The main body of the book is organized alphabetically, so that readers can usually go immediately to the topic at hand. The Index provides an exhaustive list of the topics, indexed not only by the terms actually used in the entries but also by various other terms that readers might think of instead. The "Topical Key to the Alphabetical Entries" groups the entries into categories, helping readers to pull together information on broad subjects covered by several different entries and facilitating correlation of the *Handbook* with standard textbooks or writing guides. "Five Steps to Successful Writing" offers a brief, step-by-step guide through the writing process, which is then summarized in list fashion, with page references to pertinent entries, by the "Checklist of the Writing Process." In this second edition, we have added a "List of Usage Entries." We have also incorporated directly into the book a correction chart cross-referenced to pertinent entries, which was furnished as a separate pamphlet to instructors using the first edition.

In preparing this second edition, we have carefully reexamined every entry, clarified or updated a great many, deleted a few that proved to have been seldom used, and added a substantial number of new ones. Among the added entries are those on feasibility reports, internal proposals, investigative reports, literature reviews, mathematical equations, references, test reports, and trouble reports. Other new entries discuss complaint letters, dictation, executive summaries, forms, indexing, job interviews, news releases, order letters, policies and procedures, proofreading, reference letters, and telegraphic style.

As in the first edition, the entries themselves are complete without being exhaustive; that is, they focus on solutions to specific writing problems. Therefore, the general tone is prescriptive. For example, questions about the history of a term such as *data* and the controversy surrounding its use may be interesting, but they are secondary to the advice the user seeks. So, although debate over divided usage is acknowledged, a guideline is offered to get the writer on with the business of writing. This principle holds true for the grammatical entries as well; they do not attempt to settle the debate between traditional and linguistically oriented grammarians, although current linguistic thinking about grammatical problems is presented where it contributes to the explanation of a concept. The guidance offered in the entries will solve most of the problems users of this book will encounter, but students should, of course, understand that some companies and government agencies have their own special standards governing some of the subjects dealt with in this book and that persons employed by those companies and agencies must conform to those standards.

We should mention that the *Handbook of Technical Writing* shares not only its format but a majority of the topics in the alphabetical entries with *The Business Writer's Handbook,* also published in a second edition by St. Martin's Press. Within the alphabetical entries that are common to both books, however, the examples given are often different—illustrating the same writing principles but in language that reflects a technical context in the one book and a business context in the other. In a course combining both technical and business writing, either book might be used, the choice depending upon the emphasis the instructor thinks most appropriate for the particular class.

We are deeply grateful to the many instructors, students, technical writers, and others who have helped us prepare the *Handbook of*

Technical Writing. For their contributions to the first edition, which continue to lend strength to the second, we are indebted to Conrad R. Winterhalter of the University of Dayton and NCR Corporation, who wrote the original entries on the résumé, letter of application, letter of acceptance, and letter of technical information, and to Keith Palmer of the Harnischfeger Corporation, who wrote the original entry on specifications. Both also read the manuscript of the first edition in various drafts, as did the following others, to whom we are happy to renew our thanks: James H. Chaffee, Normandale Community College; Paul Falon, University of Michigan; Wayne Losano, Rensselaer Polytechnic Institute; Brian Murphy, NCR Corporation; Lee Newcomer, University of Wisconsin–Milwaukee; Thomas Pearsall, University of Minnesota; John Taylor, University of Wisconsin–Milwaukee; and Arthur Young, Michigan Technological University. For assistance with the second edition, special thanks must go to Thomas Warren and his staff at Oklahoma State University, whose detailed advice was as valuable as it was generous. We are also deeply indebted to Edmund Dandridge, North Carolina State University; James Farrelly, University of Dayton; Patrick Kelley, New Mexico State University; Carol Mablekos, Temple University; Roger Masse, New Mexico State University; Judy McInish, Hornbake Library, University of Maryland; Wendy Osborne, Society of American Foresters; Diana Reep, University of Akron; Charles Stratton, University of Idaho; Mary Thompson, Krannert Graduate School of Management, Purdue University; and Merrill Whitburn, Rensselaer Polytechnic Institute. We also express our appreciation to the following organizations for permission to use examples of their technical writing: Biospherics, Inc., *Chemical Engineering,* Harnischfeger Corporation, Johnson Service Company, NCR Corporation, and Professional Secretaries International. Finally, we must acknowledge the superb editorial guidance we have received from St. Martin's Press, especially from Marilyn Moller and Nancy Perry.

C.T.B.
G.J.A.
W.E.O.

Contents

How to Use This Book

The purpose of this book is to help you solve specific writing problems as quickly and efficiently as possible. To get the greatest benefit from the book, you should be familiar with the four different ways you can find the advice and examples you need.

1. The Alphabetical Text Entries (pages 1–635)

To solve many problems, you will find it most convenient to go directly to the handbook entries on pages 1–635. These entries are in alphabetical order, just as in a dictionary or encyclopedia. For example, if you wish to know the difference between *activate* and *actuate*, or want to review the uses of adverbs, or need information about writing an annual report, you can go directly to the "A" section of the handbook and find the entries: **activate/actuate, adverbs, annual report.** Key words at the top of each page will help you locate any alphabetical entry quickly. Within each entry, cross-references to other entries are printed in **bold type**, so that you can easily investigate any problem to whatever depth may be necessary to solve it. For example, if you go to the entry on **conjunctions**, you will find not only brief explanations of the various kinds of conjunctions but also bold-type cross-references to more detailed entries about each kind: **coordinating conjunctions, correlative conjunctions, subordinating conjunctions**, and **conjunctive adverbs**. Within each of these entries, you will find still further cross-references, in case you wish to explore a particular problem in even more detail. This reference system enables you to get out of the book and on with the job as soon as you have enough information to solve a specific problem.

2. The Index (page 639)

If you are uncertain about what entry to look for, the Index will help you. There you will find the item you need, with its page reference, listed not only under the term used in the handbook but under various other terms that you might think of instead. For example, if you wish to know how to handle the omission of words in quoted material, you can look under "omitted words" in the Index and be directed to the page with the entry for **ellipses** (the spaced dots indicating that words have been left out). You can also find the discussion of ellipses by looking up "dots, spaced," "periods, spaced," or "quotations."

The Index is especially valuable for locating information about the usage of individual words or phrases. For example, if you want to know the difference between *diagnosis* and *prognosis*, you can look up either word in the Index and be directed to the page for the entry **diagnosis/prognosis** in the "D" section of the handbook.

3.   The Checklist of the Writing Process (page xx)

The Checklist of the Writing Process is a tool not only to help you locate information but to help you detect your specific writing problems. It presents a disciplined step-by-step approach to successful writing, with page references to the entries that discuss each step in detail. The Checklist is a guide to planning, organizing, and writing your draft, and it is equally valuable as a guide to revision, to help you pinpoint and eliminate specific problems in a systematic way. The Checklist is preceded, on page xiii, by a brief, thoroughly practical discussion of the writing process entitled "Five Steps to Successful Writing."

4.   Topical Key to the Alphabetical Entries (page xxii)

Should you wish to review a broad subject that is dealt with in a number of entries—for instance, the parts of speech, the various methods of development, the elements of style, or the types of illustrations—you will find the alphabetical entries conveniently grouped by subject in the Topical Key, which is, in effect, a topical table of contents. Further, if you are using the *Handbook of Technical Writing* along with a standard textbook, the Topical Key provides an easy way to correlate the contents of the handbook with those of your textbook. Thus, if a particular discussion within the textbook is inadequate, unclear, or lacking in examples, you can readily turn to the related entries in the handbook for additional help. Immediately following the Topical Key is the List of Usage Entries, which can guide you to the entries on the usage of particular words or phrases.

Five Steps
to Successful Writing

The main reason that many technical people have difficulty writing, no matter how proficient they may be in their technical specialties, is that they do not know how to begin. They sit down at a typewriter or with a pen in hand and hope to fill a blank page; they do not realize that skill in writing, like any other technical skill, requires a systematic approach.

Successful writing is not the product of inspiration, nor is it merely the spoken word transferred to paper—it is primarily the result of knowing how to *structure ideas* on paper. The best way to ensure that a writing project will be successful—whether it is a letter, a proposal, or a formal report—is to divide the writing process into five major steps: preparation, research, organization, writing the draft, and revision. At first, these five steps must be consciously—even self-consciously—followed. But so, at first, must the steps involved in operating a computer, designing a circuit, troubleshooting a motor, interviewing a candidate for a job, or chairing a meeting of the board of directors. With practice, the steps involved in each of these processes become nearly automatic. This is not to suggest that writing becomes easy; it does not. But the easiest way to do it—and the only way to ensure that your writing will accomplish its objective—is to do it *systematically.*

The following pages describe the five major steps of the writing process, and the Checklist that follows provides a summary that can guide you as you write. The Checklist can also help you detect any problem you may have and refer you to the specific entry that explains the cause of the problem and shows you how to solve it. In the following discussion, as elsewhere throughout this book, words and phrases printed in **bold type** refer you to specific alphabetical entries on pages 1–635.

Step 1.   Preparation

Writing, like most technical tasks, requires solid **preparation**—in fact, adequate preparation is as important as writing the draft. Preparation for writing consists of (1) establishing your objective, (2) identifying your reader, and (3) determining the scope of your coverage.

Establishing your **objective** is simply determining what you want your reader to know or be able to do when he has finished reading your report or paper. But you must be precise; it is all too common for a writer to state his objective in terms so broad that it is almost useless. An objective such as "To report on possible locations for a new manufacturing facility" is too general to be of real help. But "To present the relative advantages of Chicago, Minneapolis, and Salt Lake City as possible locations for a new manufacturing facility in such a way that top management can choose the best location" gives you an objective that can guide you throughout the writing process.

The next problem is to identify your **reader**—again, precisely. What are your reader's needs in relation to your subject? What does your reader already know about your subject? You need to know, for example, whether you must define basic terminology or whether such definitions will merely bore, or even insult, your reader. Is your reader actually several readers with differing interests and levels of technical knowledge? For the objective stated in the previous paragraph the reader was described as "top management." But *who* is included in this category? Will one of the people evaluating the report on three cities as possible plant locations be the personnel manager? If so, he is likely to have a special interest in such things as the availability of qualified personnel in each city, the presence of colleges where special training would be available for employees, housing conditions, perhaps even recreational facilities. The purchasing manager will be concerned about available sources for the materials needed by the plant. The marketing manager will give priority to such things as the plant's closeness to its primary markets and the available transport facilities for distribution of the plant's products. The financial vice-president will want to know about land and building costs and the local tax structure. The president may be interested in all of these things, and more; for example, he might be concerned about the convenience of personal travel between corporate headquarters and the new plant.

In addition to knowing the needs and interests of these readers, you should know as much as you can about their backgrounds. For example, have they visited all three cities? Have they already seen other studies of the three cities? Is this the first new plant, or have they been through the process of choosing locations for new plants before? Finally, if you have multiple readers, then the best course, once you have learned as much as possible about their needs and backgrounds, is to combine all of your readers in your mind into one composite reader and write to *that reader*. By writing as if to a single reader, you are less likely to adopt a lecturing tone (as if you were addressing a group rather than individuals) or to be intimidated by your audience.

Determining your objective and identifying your reader will help you decide what to include and what not to include in your writing. When you have distinguished the important from the unimportant on the basis of your objective and your reader, you have established the **scope** of your writing project. If you do not clearly define the scope before beginning your research (the next step), you will inevitably spend needless extra hours on research because you will not be sure what kind of information you need, or even how much. For example, given the objective and reader established in the preceding two paragraphs, the scope of your report on plant locations would include such things as land and building costs, the available labor force, transportation facilities, proximity to sources of supply, and so forth; but it probably would not include information on the history of the cities being considered or on the altitude and geological features (unless these were directly pertinent to your particular business).

Step 2. Research

The purpose of most technical writing is to explain something—usually something that is complex. This kind of writing cannot be done by someone who does not understand the subject he is writing about. The only way to be sure that you can deal adequately with a complex subject is to compile a complete set of notes during your **research** and then to create a working outline from the notes. Three sources of information are available to you: the library (see **library research**), personal **interviews** (or written **questionnaires**), and your own knowledge. Consider them all when you begin your research, and use those that fit your needs. Of course, the amount of research you need to do depends on your project; for a simple memo or letter your "research" may amount to nothing more than jotting down all your ideas before you begin to organize them.

Step 3. Organization

Without **organization**, the material gathered during your research would be incomprehensible to your reader. To provide effective organization, you must determine the sequence in which your ideas should be presented—that is, you must choose a **method of development**.

An appropriate method of development is the writer's tool for keeping things under control and the reader's means of following the writer's presentation. Your subject may obviously lend itself to a particular method of development. For example, if you were giving instructions for starting an engine, you would naturally present the steps of the process in the order of their occurrence. This is the sequential method of development. If you were writing about the history of the computer, your account would go from the beginning to the present. This is the chronological method of development. If your subject naturally lends itself to a certain method of development, use that method—don't attempt to impose another method on it. Many different methods of development are available to you; this book includes all of those that are likely to be used by technical people: chronological, sequential, spatial, increasing- or decreasing-order-of-importance, comparison, division-and-classification, analysis, general-to-specific, specific-to-general, and cause-and-effect. As the writer, you must choose the method of development that best suits your subject, your reader, and your objective. You are then ready to prepare your outline.

Outlining makes large or complex subjects easier to handle by breaking them into manageable parts, and it ensures that your writing will move logically from idea to idea without omitting anything important. It also enables you to emphasize your key points by placing them in the positions of greatest importance. Finally, by forcing you to structure your thinking at an early stage, creating a good outline releases you to concentrate exclusively on writing when you begin the rough draft. Even if the task is only a letter or short memo, successful writing needs the logic and structure that a method of development and an outline provide, although for such simple projects the method of development and outline may be in your head rather than on paper.

If you intend to include **illustrations** with your writing, a good time to think about these is when you have completed your outline—especially if the illustrations need to be prepared by someone else while you are writing and revising the draft. If your outline is reasonably detailed, you should be able to determine which ideas wll require graphic support in order to be clear.

Step 4.   Writing the Draft

When you have established your objective, reader, and scope, when you have done adequate research, and when you have chosen a method of development and created a good outline, **writing the draft** is relatively easy. Writing the draft is simply the process of transcribing and expanding the notes from your outline into **topic sentences** and then into **paragraphs**. Write the draft quickly, concentrating entirely on converting your outline to sentences and paragraphs. Don't worry about a good **opening** or **introduction** unless it comes easily. Concentrate on ideas. Don't attempt to polish or revise. Don't let concern about grammatical rules or spelling get in your way, for these are of little importance in the rough draft. Do, above all, keep your reader's needs and knowledge in mind.

Step 5.   Revision

Chances are that the clearer a piece of writing seems to the reader, the more effort the writer has put into **revision**. If you have followed the steps of the writing process to this point, you have a very rough draft that could hardly be considered a finished product. Revision is the obvious final step. A different frame of mind is required for revising than for writing the draft. Read and evaluate the draft from the point of view of the reader. Be anxious to find and correct faults, and be honest. Be hard on yourself for the reader's convenience; never be easy on yourself at the reader's expense.

Do not try to do all your revision at once. Read through your rough draft several times, each time looking for and correcting a different set of problems or errors.

Check your draft for accuracy and completeness. Your draft should give the reader exactly what he needs, but it should not burden him with unnecessary information or get sidetracked into loosely related subjects. If you have not yet written an **opening** or **introduction**, this is the time to do it. Your introduction should serve as a frame into which your reader can fit the information that follows in the body of your draft; check your introduction to see that it does, in fact, provide the reader with such assistance. Your opening should both suggest the subject and capture the reader's attention.

Check your draft for unity, coherence, and transition. If it has **unity**, all of the sentences in each paragraph contribute to the development of that paragraph's central idea (expressed in the topic sentence), and all of the paragraphs contribute to the development of the main topic. If the draft has **coherence,** the sentences, and also the paragraphs, flow smoothly from one to the next; the relationship of each sentence or paragraph to the one before is clear. Coherence is accomplished in many ways, but especially by the careful use of **transition** devices and by the maintenance of a consistent **point of view**. Also check your draft for proper **emphasis** and **subordination** of your ideas. This is also a good time to adjust the **pace**: if you find places where too many ideas are jammed together, space the ideas out and slow the pace; conversely, if you find a series of simple ideas expressed in a series of short, choppy sentences, combine them into fewer sentences and speed the pace.

Check your draft for **clarity**. Much of the revision you have done has already contributed to greater clarity. But check now for any terms that need to be defined or explained for your reader (see **defining terms**), and also check for **ambiguity**. Is your writing free of **affectation**? Is it free of **jargon** that your reader (or some of your readers) may not understand? Are there **abstract words** that could be replaced by concrete words? Check your entire draft for appropriate **word choice**.

Check your draft for **style**. By now you have already done much to improve the style, but there is more you can do. Check for **conciseness**: can you eliminate useless words or phrases—what some writers call deadwood? It is almost always possible to do so, and your writing (and your reader) will benefit greatly. Get rid of **clichés** and other **trite language**. Is your writing active? Especially in technical writing there is a great temptation to use the passive **voice** when the active would be far stronger (and also more concise). Replace negative writing with **positive writing**. Check, too, for **parallel structure**. Examine your **sentence construction** and look for ways to achieve more interesting **sentence variety**.

Check your draft for **awkwardness** and for departures from the appropriate **tone**. Awkwardness and tone are hard to define because they are the result of a great many different things, most of which you have already dealt with in your revision. But try reading your draft aloud (better still, have someone read it to you) while you listen as if you were the reader. You will recognize awkwardness because it sounds forced or clumsy—like something you would never say if you were *talking* to your reader because it would sound unnatural to both of you. You will recognize departures from the proper tone because they are words, phrases, or statements which are inappropriate to the relationship that exists between you and your reader. For example, in a memo to a superior in your organization you would want to avoid phrases or statements that sounded like lecturing. If your reader is serious about your subject (and it is best to assume that he is), don't try to be witty. Correcting the tone is frequently a matter of replacing a word with another that has more appropriate **connotations**. For example, consider the difference in tone between "I'm puzzled by your *stubbornness* in opposing the plan" and "I'm puzzled by the *firmness* of your opposition to the plan."

Finally, check slowly and carefully for problems of **grammar, punctuation**, mechanics (**spelling, abbreviations, capital letters**, etc.), and **format**.

The following Checklist of the Writing Process is designed to guide you as you follow the five steps to successful writing. It can help you to diagnose any problem you may have, and it also refers you to the specific entry that explains the causes of the problem and shows you how to correct it.

Checklist of the Writing Process

This checklist arranges key entries from pages 1–635 of the handbook into a recommended sequence of steps for any writing project. The exact titles of the entries are in **bold type** for quick reference, and the page numbers are given. When you turn to the entries themselves, you will find bold-type cross-references to further entries that may be helpful. (You may also wish to refer to the Topical Key to the Alphabetical Entries on page xxii and to the Index on page 639.)

Topical Key to the Alphabetical Entries

This tabulation arranges the alphabetical entries on pages 1–635 into subject categories. Within each category the entries are listed mainly, but not strictly, in alphabetical order. Excluded from this tabulation are the numerous entries concerned with the usage of particular words or phrases (*imply/infer, already/all ready,* etc.); a separate List of Usage Entries follows the Topical Key. For a listing of key entries in a recommended sequence of steps for any writing project, see the Checklist of the Writing Process.

Types of Technical Writing

Planning and Research

Paragraphs, Sentences, Clauses, and Phrases

Parts of Speech, Inflection, and Agreement

Punctuation and Mechanics

Mechanics (*see also* Format)

List of Usage Entries

This list contains the entries that are concerned with the usage of particular words and phrases.

A

a/an

A and *an* are indefinite **articles;** "indefinite" implies that the **noun** designated by the article is not a specific person, place, or thing but is one of a group.

> EXAMPLE   The computer operator ran *a* program. (Not a specific program, but an unnamed program.)

Use *a* before words beginning with a consonant or consonant sound (including *y* or *w* sounds).

> EXAMPLES   The year's activities are summarized in *a* one-page report. (Although the *o* in *one* is a vowel, the first *sound* is *w*, a consonant sound. Hence the article *a* precedes the word.)
> *A* manual has been written on that subject.
> It was *a* historic event for the laboratory.
> The project manager felt that it was *a* unique situation.

Use *an* before words beginning with a vowel or vowel sound.

> EXAMPLES   The report is *an* overview of the year's activities.
> He seems *an* unlikely candidate for the job.
> The interviewer arrived *an* hour early. (Although the *h* in *hour* is a consonant, the word begins on a vowel *sound;* hence it is preceded by *an.*)

Be careful not to use unnecessary indefinite articles in a **sentence.**

> CHANGE   There is no more difficult *a* subject than chemistry.
> TO   There is no more difficult subject than chemistry.

> CHANGE   I will meet you in *a* half *an* hour.
> TO   I will meet you in *a* half hour.
> OR   I will meet you in half *an* hour.

a lot/alot

A lot is often incorrectly written as one word *(alot).* Write the **phrase** as two words: *a lot.* The phrase *a lot* is very informal, however, and should not normally be used in technical writing.

> CHANGE   The peer review group had *a lot* of objections.
> TO   The peer review group had many objections.

abbreviations

An abbreviation is a shortened form of a word, formed by omitting some of its letters. Most abbreviations are written with a period; abbreviations of measurements are an exception.

EXAMPLES   company/co.
boulevard/blvd.
electromotive force/emf
horsepower/hp

Abbreviations, like **symbols,** can be important space savers in technical writing, where it is often necessary to provide the maximum amount of information in limited space. Use abbreviations, however, only if you are certain that your **reader** will understand them as readily as he would the terms for which they stand. Remember also that a **memorandum** or **report** addressed to a specific person may be read by others, and you must consider those readers as well. Use common sense to decide when abbreviations are appropriate. Do not use them if they might become an inconvenience to the reader. A good rule of thumb: When in doubt, spell it out.

In general, use abbreviations only in charts, **tables, graphs,** and other **illustrations,** where space is at a premium. Normally you should not make up your own abbreviations, for they will probably confuse your reader. Except for commonly used abbreviations (U.S.A., a.m.), a term to be abbreviated should be spelled out the first time it is used, with the abbreviation enclosed in **parentheses** following the term. Thereafter, the abbreviation may be used alone. (See also **acronyms and initialisms.**)

EXAMPLE   The National Aeronautics and Space Administration (NASA) has accomplished much in its short history. NASA will now develop a permanent orbiting space laboratory.

FORMING ABBREVIATIONS

Measurements.   When you abbreviate terms that refer to measurement, be sure your reader is familiar with the abbreviated form. The following list contains some common abbreviations used with units of measurement. Notice that, except for *in.* (inch), *tan.* (tangent), and others that form words, abbreviations of measurements do not require periods.

amp, ampere	Hz, hertz
atm, atmosphere	in., inch
bbl, barrel	kc, kilocycle

bhp, brake horsepower
Btu, British thermal unit

kg, kilogram
km, kilometer

bu, bushel
cal, calorie
cd, candela
cm, centimeter
cos, cosine

lb, pound
lm, lumen
min, minute
oz, ounce
pa, pascal

ctn, cotangent
doz or dz, dozen
emf or EMF, electromotive force
F, Fahrenheit
fig., figure (illustration)

pt, pint
qt, quart
rad, radian
rev, revolution
sec, second or secant

ft, foot (or feet)
gal., gallon
hp, horsepower
hr, hour

sr, steradian
tan., tangent
yd, yard
yr, year

Abbreviations of units of measure are identical in the singular and plural: one *cm* and three *cm* (not three *cms*).

Personal Names and Titles. Personal names should generally not be abbreviated.

CHANGE Chas., Thos., Wm., Geo.
TO Charles, Thomas, William, George

An academic, civil, religious, or military title should be spelled out when it does not precede a name.

EXAMPLE The *doctor* asked that the report be finished.

When it precedes a name, the title may be abbreviated.

EXAMPLES Dr. Smith, Mr. Mills, Capt. Hughes

Reverend and *Honorable* are abbreviated only if the surname is preceded by a first name.

EXAMPLES The *Reverend* Smith, *Rev.* John Smith (but not *Rev.* Smith)
The *Honorable* Commissioner Holt, *Hon.* Mary J. Holt

An abbreviation of a title may follow the name; however, be certain that it does not duplicate a title before the name.

CHANGE *Dr.* William Smith, *Ph.D.*
TO *Dr.* William Smith
OR William Smith, *Ph.D.*

The following is a list of common abbreviations for personal and professional titles.

Atty.	Attorney
B.A. or A.B.	Bachelor of Arts
B.S.	Bachelor of Science
B.S.E.E.	Bachelor of Science in Electrical Engineering
D.D.	Doctor of Divinity
D.D.S.	Doctor of Dental Science (or Surgery)
Dr.	Doctor (used with any doctor's degree)
Ed.D.	Doctor of Education
Esq.	Esquire (*Esquire* is a nearly obsolete title following a name. It never precedes the name. Equivalent to *mister,* it is used primarily in legal notices.)
Hon.	Honorable
Jr.	Junior (used when a father with the same name is living)
LL.B.	Bachelor of Law
LL.D.	Doctor of Law
M.A. or A.M.	Master of Arts
M.B.A.	Master of Business Administration
M.D.	Doctor of Medicine
Messrs.	Plural of Mr.
Mr.	Mister (spelled out only in the most formal contexts)
Mrs.	Married woman
Ms.	Woman of unspecified marital status
M.S.	Master of Science
Ph.D.	Doctor of Philosophy
Rev.	Reverend
Sr.	Senior (used when a son with the same name is living)

Company Names. Many companies include in their names such terms as *Brothers, Incorporated, Corporation,* and *Company.* If these terms appear as abbreviations in the official company names, they should always be used in their abbreviated forms: *Bros., Inc., Corp.,* and *Co.* If not abbreviated in the official names, such terms should be spelled out in most writing other than addresses, **footnotes, bibliographies,** and **lists,** where abbreviations may be used. A similar guideline applies for use of an **ampersand** (&); this symbol should be used only if it appears in an official company name.

Geographical Locations. It is generally better to spell out geographical locations, including the names of streets, cities, states, regions, and countries (except for long multiword names, such as United States of America, Union of Soviet Socialist Republics, and United Kingdom). Spelling out such names avoids possible misunderstanding and makes reading much easier. The following list includes some common geographical abbreviations.

N.E.	Northeast	Ave.	Avenue
N.H.	New Hampshire	Blvd.	Boulevard
N.Y.	New York	Ct.	Court
U.K.	United Kingdom	Dr.	Drive
U.S.A.	United States of America	Ln.	Lane
		St.	Street

The U.S. Postal Service recommends the following abbreviations for states and protectorates on envelope addresses. Notice that they are written in capital letters without punctuation.

Alabama	AL	Michigan	MI
Alaska	AK	Minnesota	MN
Arizona	AZ	Mississippi	MS
Arkansas	AR	Missouri	MO
California	CA	Montana	MT
Colorado	CO	Nebraska	NB
Connecticut	CT	Nevada	NV
Delaware	DE	New Hampshire	NH
District of Columbia	DC	New Jersey	NJ
Florida	FL	New Mexico	NM
Georgia	GA	New York	NY
Guam	GU	North Carolina	NC
Hawaii	HI	North Dakota	ND
Idaho	ID	Ohio	OH
Illinois	IL	Oklahoma	OK
Indiana	IN	Oregon	OR
Iowa	IA	Pennsylvania	PA
Kansas	KS	Puerto Rico	PR
Kentucky	KY	Rhode Island	RI
Louisiana	LA	South Carolina	SC
Maine	ME	South Dakota	SD
Maryland	MD	Tennessee	TN
Massachusetts	MA	Texas	TX

Utah	UT	Washington	WA
Vermont	VT	West Virginia	WV
Virgin Islands	VI	Wisconsin	WI
Virginia	VA	Wyoming	WY

Dates and Times. The following are common abbreviations.

A.D.	*anno Domini* (beginning of calendar time—A.D. 1790)	Jan.	January
B.C.	before Christ (before the beginning of calendar time —647 B.C.)	Feb.	February
		Mar.	March
		Apr.	April
		Aug.	August
A.M.	*ante meridiem* (before noon)	Sept.	September
P.M.	*post meridiem* (after noon)	Dec.	December

Months should always be spelled out when only the month and year are given. The proper forms for standard abbreviations are in your **dictionary,** either in regular alphabetical order (by the letters of the abbreviation) or in a separate index.

above

Avoid using *above* to refer to a preceding passage, **illustration,** or **table** in your writing. Its reference is often vague, and, if the passage referred to is far from the current one, the use of *above* is distracting because your **reader** must go back to the earlier passage to understand your reference. The same is true of *aforesaid, aforementioned, above mentioned, the former,* and *the latter.* In addition to distracting the reader, these words also contribute to a heavy, wooden **style.** If you must refer to something previously mentioned, either repeat the **noun** or **pronoun** or construct your **paragraph** so that your reference is obvious. (See also **former/latter.**)

absolute words

Absolute words (such as *round, unique, exact* and *perfect*) are not logically subject to **comparison** *(rounder, roundest)*; however, language is not always logical, and these words are sometimes used comparatively.

> EXAMPLE　Phase-locked loop circuits make the FM tuner performance *more exact* by decreasing tuner distortion.

Absolute words should be used comparatively only with the greatest caution in technical writing, where accuracy and precision are crucial.

absolutely

Absolutely means "definitely," "entirely," "completely," or "unquestionable." It is not an **intensifier** and should not be used to mean "very" or "much." When used as an intensifier, it can be deleted.

CHANGE   A new analysis is *absolutely* impossible.
TO   A new analysis is impossible.

abstracts

An abstract is a condensed version of a longer piece of writing that summarizes and highlights the major points, enabling the prospective **reader** to decide whether it will be worthwhile to read the work in full. Abstracts appear in many **formal reports** and most dissertations, as well as in many other long works. Usually two hundred words or less, abstracts must make sense independently of the longer works they summarize. They should contain no direct references to the **reports** they are describing, nor should they include **illustrations, tables,** or bibliographic **references.** Depending on the kind of information they contain, abstracts are usually classified as either descriptive or informative.

DESCRIPTIVE ABSTRACTS

A descriptive abstract tells what topics a document discusses but does not reveal its conclusions. It is almost an expanded **table of contents** in sentence form. The essence of brevity, a descriptive abstract is usually no more than a few **sentences.**

EXAMPLE   Water discharges from cooling towers constructed with asbestos fill are examined for concentrations and types of fibers present. The source of asbestos fibers and suggested mechanisms for their release are presented. The current literature on the effects of drinking water containing asbestos fiber is reviewed, and recommendations on continued asbestos use and monitoring of fiber levels are given.

INFORMATIVE ABSTRACTS

An informative abstract almost always accompanies a formal technical report or **journal article.** Unlike the descriptive abstract, the informative abstract contains some of the information from the report. It includes the most important facts on which **conclusions** are based, the conclusions themselves, and any recommendations made. The **tone** and the basic scope of the report are retained in the informative abstract, but the details are omitted.

EXAMPLE   Water discharges from cooling towers constructed with asbestos fill were found to contain chrysotile-asbestos fibers at concentrations ranging from "none detected" to 10^8 fibers/liter (mass concentrations ranged to 37 μg/liter). The major source of these fibers in nonurban areas appears to be the components of the towers rather than the air drawn through the towers or the makeup water taken into the towers from natural surface waters. Suggested mechanisms for the release of chrysotile fibers from cooling-tower fill include freeze-thaw cycles and the dissolution of the cement due to acidic components of the circulating water. Ash- and other material-settling ponds were found to reduce asbestos-fiber concentrations in cooling-tower effluent. The literature reviewed does not suggest a causal relationship between adverse human health effects and drinking water containing on the order of 10^6 chrysotile-asbestos fibers/liter; for this reason, and because of ignorance of the health risks and costs for the manufacture and use of other types of cooling-tower fill, it is not presently suggested that the use of asbestos fill be discontinued. However, it is strongly recommended that monitoring programs for asbestos fibers in cooling-tower waters be carried out at sites where asbestos fill is used; data from such programs can be used to determine whether any mitigative measures should be taken. On the basis of estimates made in this study using drift-deposition models, monitoring for asbestos in drift from cooling towers does not presently appear to be warranted.

TIPS FOR WRITING AN ABSTRACT

1. If writing an abstract for your own report, you should write it after the report has been completed.
2. Summarize the report's most important topics (remember that these should be found in the major and minor **heads** of your outline).
3. Retain the basic ideas, but omit as many details as possible.
4. Begin with a **topic sentence.**
5. Avoid overloading the abstract with technical terms, since your readers may not all be familiar with the subject matter.
6. Be careful not to make direct references to the report that the abstract condenses; the abstract must be able to stand alone.
7. Condense the information as concisely as you can, eliminating unnecessary words and ideas. Do not, however, delete **articles** (*a,*

an, the) or transitional words or **phrases** (*however, therefore, for example, in summary,* etc.) or your abstract may become too brusque and choppy. (See also **telegraphic style.**)

8. Do not include illustrations or tables.
9. Do not include bibliographic references.

There are certain journals, such as *Chemical Abstracts, Journal of Economic Abstracts,* and *Highway Research Abstracts,* that are devoted exclusively to abstracts, articles, and reports in specific fields of study. These journals enable researchers to scan the **literature** in a specialized field in a short time. (See also **library research.**)

abstract nouns

An abstract noun refers to something that is intangible, something that cannot be discerned by the five senses. Abstract nouns are those that are used to name qualities, concepts, and actions, such as *kindness, fantasy, love, idealism, sportsmanship,* or *hate.* (See also **abstract words/concrete words,** and **concrete nouns.**)

abstract words/concrete words

Abstract words refer to general ideas, qualities, conditions, acts, or relationships.

EXAMPLES   work, courage, crime

Abstract words must frequently be qualified by other words

CHANGE   What the members of the Research and Development Department need is *freedom.*

TO   What the members of the Research and Development Department need is *freedom to explore the problem further.*

Concrete words refer to specific persons, places, objects, and acts that can be perceived by the senses.

EXAMPLES   soldier, sawing, auto theft
Skiing (concrete) is a strenuous *sport* (abstract).

Concrete words are easier to understand, for they create images in the mind of your **reader.** Still, you cannot express ideas without using some abstract words. Actually, the two kinds of words are usually used best together, in support of each other. For example, the abstract idea of *transportation* is made clearer with the use of specific concrete words, such as *subways, jets,* or *automobiles.*

In fact, a word that is concrete in one context can be less concrete in another. The same word may even be abstract in one context and concrete in another. The choice between abstract and concrete terms always depends on the context. Just how concrete a particular context might require you to be is shown in the following **illustration,** which goes from the most abstract, on the left, to the most concrete, on the right.

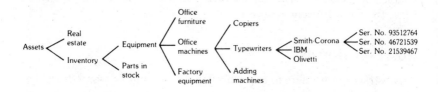

The illustration represents seven levels of abstraction; which one would be appropriate depends on your purpose in writing and on the context in which you are using the word. For example, a company's **annual report** might logically use the most abstract term, *assets,* to refer to all the property and goods the firm owns; shareholders would probably not require a less abstract term. Interoffice **memorandums** between the company's accounting and legal departments would appropriately call the firm's holdings *real estate* and *inventory.* To the company's inventory control department, however, the word *inventory* is much too abstract to be useful, and a report on inventory might contain the more concrete terms *equipment* and *parts in stock.* But to the assistant inventory control manager in charge of equipment, that term is still too abstract; he or she would speak of several particular kinds of equipment: *office furniture, factory equipment,* and *office machines.* The breakdown of the types of office machines for which the inventory control assistant is responsible might include *copiers, adding machines,* and *typewriters.* But even this classification would not be concrete enough to enable the company's purchasing department to obtain service contracts for the normal maintenance of its typewriters. Because the department must deal with different typewriter manufacturers, *typewriters* would have to be listed by brand name: *Olivetti, IBM,* and *Smith-Corona.* And the Smith-Corona technician who performs the maintenance must go one step further and identify each Smith-Corona typewriter by serial number. As the illustration shows, then, a term that is sufficiently concrete in one context may be too abstract in another.

accept/except

Accept is a **verb** meaning "consent to," "agree to take," or "admit willingly."

> EXAMPLE   I *accept* the responsibility that goes with the appointment.

Except is normally a **preposition** meaning "other than" or "excluding."

> EXAMPLE   We agreed on everything *except* the schedule.

acceptance letters

When you have received an offer of a job that you want to accept, reply as soon as possible—certainly within a week. The **format** for such a letter

2647 Patterson Road
Beechwood, OH 45432
March 6, 19--

Mr. F. E. Cummins
Personnel Manager
Calcutex Industries, Inc.
3275 Commercial Park Drive
Bintonville, MI 49474

Dear Mr. Cummins:

State the job and salary you are accepting. →

I am pleased to accept your offer of a position as a junior design engineer in the Calcutex Servocontrol Group at a salary of $X,XXX per month.

Specify the date on which you will report to work. →

Since graduation is August 30, I plan to leave Dayton on Tuesday, September 2. I should be able to locate suitable living accomodations within a few days and be ready to report for work on the following Monday, September 8. Please let me know if this date is satisfactory to you.

State your pleasure at joining the firm. →

I look forward to a rewarding future with Calcutex.

Sincerely,

Craig Adderly

Letter 1

is simple: Begin by accepting, with pleasure, the job you have been offered. Identify the job you are accepting, and state the salary so there is no confusion on these two important points.

The second **paragraph** might go into detail about moving dates and reporting for work. The details will vary depending on what occurred during your **job interview.** Complete the letter with a statement that you are looking forward to working for your new employer. Letter 1 is a typical acceptance letter. (See also **correspondence.**)

accumulative/cumulative

Accumulative and *cumulative* are **synonyms** that mean "amassed" or "added up over a period of time." *Accumulative* is rarely used except with reference to accumulated property or wealth.

EXAMPLES   Last month's x-ray exposures were acceptable, but we must be careful so that *cumulative* doses do not exceed our guidelines.
The corporation's *accumulative* holdings include nineteen real-estate properties.

acknowledgment letters

It is sometimes necessary (and always considerate) to let someone know that you have received something sent to you. An acknowledgment letter serves such a function. It should usually be a short, polite note that mentions when the parcel arrived and that expresses thanks. See **correspondence** for general advice on letter writing; study the following example for an idea of how to phrase a typical acknowledgment letter.

Dear Ms. Evans:

I received your report today; it appears to be complete and well done. Thank you for sending it so promptly.

When I finish studying it thoroughly, I will send you our cost estimate for the installation of the Mark II Energy-Saving System.

Thank you again for your effort.

Sincerely,

acronyms and initialisms

An acronym is an abbreviation that is formed by combining the first letter or letters of several words. Acronyms are pronounced as words and are written without periods.

EXAMPLES   *r*adio *d*etecting *a*nd *r*anging/radar
*Co*mmon *B*usiness-*O*riented *L*anguage/COBOL
*s*elf-*c*ontained *u*nderwater *b*reathing *a*pparatus/scuba
*N*orth *A*tlantic *T*reaty *O*rganization/NATO

An *initialism* is an abbreviation that is formed by combining the initial letter of each word in a multiword term. Initialisms are pronounced as separate letters. When written lowercase, they generally require periods; when written uppercase, they generally do not.

EXAMPLES   *e*nd *o*f *m*onth/e.o.m.
*S*ociety for *T*echnical *C*ommunication/STC
*c*ollect *o*n *d*elivery/c.o.d.
*p*ost *m*eridiem/p.m.

In business and industry, acronyms and initialisms are often used by people working together on particular projects or having the same specialties—as, for example, engineers or chemists. As long as such people are communicating with one another, the abbreviations are easily recognized and understood. If the same acronyms or initialisms were used in **correspondence** to someone outside the group, however, they might be incomprehensible; the reader might have no idea what the expression meant.

Acronyms and initialisms can be convenient—for the reader and the writer alike—if they are used appropriately. Technical writers, however, often overuse them, either as an **affectation** or in a misguided attempt to make their writing concise.

WHEN TO USE ACRONYMS AND INITIALISMS

Here are two sample guidelines to apply in deciding whether or not to use acronyms and initialisms.

1. If you must use a multiword term as much as once each **paragraph,** you should instead use its acronym or initialism. For example, a **phrase** such as "primary software overlay area" can become tiresome if repeated again and again in one piece of writing; it would be better, therefore, to use *PSOA.*

2. If something is better known by its acronym or initialism than by its formal term, you should use the abbreviated form. The initialism *a.m.,* for example, is much more common than the formal *ante meridiem.*

If these conditions do not exist, however, always spell out the full term.

HOW TO USE ACRONYMS AND INITIALISMS

The first time an acronym or initialism appears in a written work, write the complete term, followed by the abbreviated form in **parentheses.**

> EXAMPLE   The transaction processing monitor (TPM) controls all operations in the processor.

Thereafter, you may use the acronym or initialism alone. In a long document, however, you will help your reader greatly by repeating the full term in parentheses after the abbreviation at regular intervals so that he does not have to search back to the first time the acronym or initialism was used to find its meaning.

> EXAMPLE   Remember that the TPM (transaction processing monitor) controls all operations in the processor.

Write acronyms in **capital letters** without **periods.** The only exceptions are those acronyms that have become accepted as **common nouns,** which are written in lowercase letters.

> EXAMPLES   NASA, HUD, laser, scuba

Initialisms may be written either uppercase or lowercase. In general, do not use periods when they are uppercase; always use periods when they are lowercase. Two exceptions are geographic names and academic degrees.

> EXAMPLES   EDP/e.d.p., EOM/e.o.m., OD/o.d.
> U.S.A., U.S.S.R.
> B.S., M.B.A.

Form the plural of an acronym or initialism by adding an *s.* Do not use an *apostrophe* unless confusion would result.

> EXAMPLES   MIRVs, CRTs, seven GI's

activate/actuate

Even linguists disagree on the distinction between these two words, although both mean "make active." *Actuate* is usually applied only to mechanical processes.

> EXAMPLES   The governor *activated* the National Guard.
> The relay *actuates* the trip hammer. (mechanical process)
> The electrolyte *activates* the battery. (chemical process)

actually

Actually is an **adverb** meaning "really" or "in fact." Although it is often used for **emphasis** in speech, such use of the word should be avoided in writing. (See also **intensifiers.**)

> CHANGE Did he *actually* finish the report on time?
> TO Did he finish the report on time?

adapt/adept/adopt

Adapt is a **verb** meaning "adjust to a new situation." *Adept* is an **adjective** meaning "highly skilled." *Adopt* is a verb meaning "take or use as one's own."

> EXAMPLE The company will *adopt* a policy of finding engineers who are *adept* administrators and who can *adapt* to new situations.

ad hoc

Ad hoc is Latin for "for this" or "for this particular occasion." An ad hoc committee is one set up to consider a particular issue, as opposed to a permanent committee. This term has been fully assimilated into English and thus does not have to be italicized (or underlined). (See also **foreign words in English.**)

adjectives

An adjective makes the meaning of a **noun** or **pronoun** more exact by pointing out one of its qualities (descriptive adjective) or by imposing boundaries upon it (limiting adjective.)

> EXAMPLES a *hot* iron (descriptive)
> He is *cold*. (descriptive)
> *ten* automobiles (limiting)
> *his* desk (limiting)

Limiting adjectives include these common and important categories:

Articles (a, an, the)
Demonstrative Adjectives (this, that, these, those)
Possessive Adjectives (my, his, her, your, our, their)
Interrogative and Relative Adjectives (whose, which, what)
Numeral Adjectives (two, first)
Indefinite Adjectives (all, none, some, any)

Of these, the **demonstrative adjectives,** the **possessive adjectives,** and the interrogative and **relative adjectives** are formed from pronouns and are sometimes called pronominal adjectives.

COMPARISON OF ADJECTIVES

Most adjectives add the **suffix** -*er* to show comparison with one other item and the suffix -*est* to show **comparison** with two or more other items. The three degrees of comparison are called the *positive, comparative,* and *superlative.*

> EXAMPLES   The first ingot is *bright.* (positive form)
> The second ingot is *brighter.* (comparative form)
> The third ingot is *brightest.* (superlative form)

However, many two-**syllable** adjectives and most three-syllable adjectives are preceded by the words *more* or *most* to form the comparative or the superlative.

> EXAMPLES   The new facility is *more impressive* than the old one.
> The new facility is the *most impressive* in the city.

A few adjectives have irregular forms of comparison *(much, more, most; little, less, least).*

   Absolute words (round, unique) are not logically subject to comparison (rounder, roundest); however, language is not always logical and these words are sometimes used comparatively.

> EXAMPLE   Phase-locked loop circuits make FM tuner peformance *more exact* by decreasing tuner distortion.

PLACEMENT OF ADJECTIVES

When limiting and descriptive adjectives appear together, the limiting adjectives precede the descriptive adjectives, with the **articles** usually in the first position.

> EXAMPLE   *The ten gray* cars were parked in a row. (article, limiting adjective, descriptive adjective)

Within a **sentence,** adjectives may appear before the nouns they modify (the attributive position) or after the nouns they modify (the predicative position).

> EXAMPLES   The *small* jobs are given priority. (attributive position)
> Test are taken even when exposure is *brief.* (predicative position)

An adjective in an attributive position may shift to a predicative position in a larger, more complex construction.

EXAMPLES   We passed a *big* budget. (attributive position)
We negotiated a contract *bigger by far than theirs*. (predicative position for adjective phrase)

When an adjective follows a **linking verb,** it is called a **predicate adjective.**

EXAMPLES   The warehouse is *full*.
The lens is *convex*.
His department is *efficient*.

A predicate adjective is one kind of **subjective complement.** By completing the meaning of a linking verb, a predicate adjective describes or limits the subject of the **verb.**

EXAMPLES   The job is *easy*.
The manager was very *demanding*.

An adjective that follows a **transitive verb** and modifies its **direct object** is one kind of **objective complement.** (See also **complements.**)

EXAMPLE   The lack of lubricant rendered the bearing *useless*.

USAGE OF ADJECTIVES

Because of the need for precise qualification in technical writing, it is often necessary to use nouns as adjectives.

EXAMPLE   The *test* conclusions led to a redesign of the system.

Technical people frequently string together a series of nouns to form a unit modifier, thereby creating **jammed modifiers.** Be aware of this problem when using nouns as adjectives.

CHANGE   The test control group meeting was held on Wednesday.
TO   The meeting of the test control group was held on Wednesday.

As a rule of thumb, it is better to avoid general *(nice, fine, good)* and trite (a *fond* farewell) adjectives; in fact, it is good practice to question the need for most adjectives in your writing. Often, your writing will not only read as well without an adjective, but it may be even better without it. If an adjective is needed, try to find one that is as exact as possible for your meaning.

adjective clauses

An adjective clause is a subordinate **clause** that functions as an **adjective** by modifying a **noun** or **pronoun** in another clause.

> EXAMPLE   The designer *we commissioned last year* (modifying "designer") has delivered the drawings, *which we have approved* (modifying "drawings").

Adjective clauses are often introduced by **relative pronouns** (*who, that,* and *which*).

> EXAMPLE   Our culture is beginning to produce a breed of people *who are very mobile.*

Adjective clauses are also sometimes introduced by relative **adverbs** (*when, where,* and *why*).

> EXAMPLE   The fourth quarter is the period *when cost control will be crucial.*

Adjective clauses may be **restrictive** or nonrestrictive. If the clause is essential to limiting the meaning of the noun, it is restrictive and not set off by **commas.**

> EXAMPLE   The statistics *that accompany this report* were compiled from a series of questionnaires.

If the clause is not intended to limit the meaning of the noun, but merely provides further information about it, the clause is nonrestrictive and set off by commas.

> EXAMPLE   New statistics, *which will be submitted soon,* suggest a different conclusion.

If adjective clauses are not placed carefully, they may appear to modify the wrong noun or pronoun.

> CHANGE   The factory in the suburbs *that we bought* has increased in value. (This sentence implies that we bought the suburbs rather than the factory.)
> TO   The factory *that we bought* in the suburbs has increased in value.
> OR   The suburban factory *that we bought* has increased in value.

(See also **that/which/who.**)

adjustment letters

An adjustment letter is written in response to a **complaint letter** and tells the customer what your firm intends to do about his complaint.

Although it is sent in response to a problem situation, an adjustment letter actually provides an excellent opportunity to build goodwill for the firm. An effective adjustment letter both repairs the damage that has been done and restores the customer's confidence in your firm.

You should settle claims quickly and courteously, trying always to satisfy the customer at a reasonable cost to your company.

Grant adjustments graciously, for a settlement made grudgingly will do more harm than good. **Tone** is critical. No matter how unpleasant or unreasonable the complaint letter, your response must remain both respectful and positive. Avoid emphasizing the unfortunate situation at hand; put your **emphasis** instead on what you are doing to correct it. Not only must you be gracious, but also you must admit your error in such a way that the buyer will not lose confidence in your firm.

Before granting an adjustment to a claim for which your company is at fault, you must investigate what happened and decide what you can do to satisfy the customer. Treat every claim individually, and lean toward giving the customer the benefit or the doubt. The following guidelines might help you to write adjustment letters.

1. Open with whatever you believe the **reader** will consider good news:
   Grant the adjustment for uncomplicated situations ("Enclosed is a replacement for the damaged part").
   Reveal that you intend to grant the adjustment by admitting that the customer was in the right ("Yes, you were incorrectly billed for the oil delivery"). Then later explain the specific details of the adjustment. This method is good for adjustments that require detailed explanations.
   Apologize for the error ("Please accept our apologies for not acting sooner to correct your account"). This method is effective for adjustments for situations in which the customer's inconvenience is as much an issue as money.
   Use a combination of these techniques. Often, situations requiring an adjustment are unique and do not fit a single pattern.
2. Explain what caused the problem—if such an explanation will help restore your reader's confidence or goodwill.
3. Explain specifically how you intend to make the adjustment—if it is not obvious in your opening.
4. Close with an offer to be of further service.

First you must investigate what happened and do what you can do to satisfy the customer. Treat every claim individually, and lean toward giving the customer the benefit of the doubt.

Express appreciation to the customer for calling your attention to the situation, explaining that this helps your firm to keep the quality of its product or service high. Point out any steps you may be taking to prevent a recurrence of whatever went wrong, giving the customer as much credit as the facts allow.

Close pleasantly—looking forward, not back. Avoid recalling the problem in your closing ("Again, we apologize . . ."). Following is a typical adjustment letter.

Dear Mr. Denlinger:

Thank you for your letter regarding your order for nine TR-5771-3 tuners. Please accept our apologies for not sending the proper tuners and for incorrectly billing you.

Evidently when your package arrived at our loading docks, a dock worker failed to see your letter in the container. He set the box aside with several boxes of parts destined for our Rebuilt Parts Department; therefore, your note did not come to the attention of our parts manager.

We have shipped the correct tuners by United Parcel; you should receive them shortly after you receive this letter. I have also instructed our Billing Department to charge you our preferred-customer rate—normally reserved for orders of more than $2,000.

If I can be of any further help, please let me know.

Sincerely,

You may sometimes need to educate your **reader** about the use of your product or service. Customers sometimes submit claims that are not justified, even though they honestly believe them to be. The customer may actually be at fault—for not following maintenance instructions properly for example. Such a claim is granted only to build goodwill. When you write a letter of adjustment in such a situation, it is wise to give the explanation before granting the claim—otherwise, your reader may never get to the explanation. If your explanation establishes customer responsibility, be sure that it is by implication rather than by outright statement. Look at the following example of such an adjustment letter.

Dear Mr. Ortiz:

Enclosed is your SWELCO Coffee Maker, which you sent to us on August 17.

General Television, Inc.

5521 West 23rd Street
New York, NY 10062

Customer Relations
(212) 574-3894

September 28, 19--

Mr. Fred J. Swesky
7811 Ranchero Drive
Tucson, AZ   85761

Dear Mr. Swesky:

Thank you for your letter regarding the replacement of your KL-71
television set.

You stated in your letter that you used the set on an uncovered patio.
As our local service representative pointed out, this model is not
designed to operate in extreme heat conditions. As the accompanying
instruction manual states and our engineers confirm, such exposure can
produce irreparable damage to this model. Since your set was used in
such extreme heat conditions, therefore, we cannot honor the two-year
replacement warranty.

However, with the enclosed certificate, your local GTI dealer will give
you a trade-in allowance equal to his markup for the set. This means
you can purchase a new set at wholesale, provided you return your
original set.

Sincerely yours,

Susan Siegel

Susan Siegel
Assistant Director

SS/mr
Enclosure

Letter 2

In various parts of the country, tap water may contain a high mineral content. If you fill your SWELCO Coffee Maker with water for breakfast coffee before going to bed, a mineral scale will build up on the inner wall of the water tube—as explained on page two of your SWELCO Instruction Booklet.

We have removed the mineral scale from the water tube of your coffee maker and thoroughly cleaned the entire unit. To ensure the best service from your coffee maker in the future, clean it once a month by operating it with four ounces of white vinegar and eight cups of water. To rinse out the vinegar taste, operate the unit twice with clear water.

With proper care, your SWELCO coffee maker will serve you faithfully and well for many years to come. Should you have any problems with it, however, please let us know.

Sincerely,

Sometimes you may decide to grant a partial adjustment, even though the claim is not really justified, as in Letter 2.

adverbs

An adverb modifies the action or condition expressed by a **verb.**

EXAMPLE   The recording head hit the surface of the disc *hard.* (The adverb tells *how* the recording head hit the disc.)

An adverb may also modify an **adjective,** another adverb, or a **clause.**

EXAMPLES   The graphics department used *extremely* bright colors. (modifying an adjective)
The redesigned brake pad lasted *much* longer. (modifying another adverb)
Surprisingly, the machine failed. (modifying a clause)

An adverb answers one of the following questions:

WHERE? (adverb of place)

EXAMPLE   Move the throttle *forward* slightly.

WHEN? (adverb of time)

EXAMPLE   Replace the thermostat *immediately.*

HOW? (adverb of manner)

EXAMPLE   Add the reagent *cautiously.*

HOW MUCH? (adverb of degree)

EXAMPLE The *nearly* completed report was lost in the move.

An adverb may be a common, a conjunctive, an interrogative, or a numeric **modifier.** Typical common adverbs are *almost, seldom, down, also, now, ever,* and *always.*

EXAMPLE I *rarely* work on the weekend.

Typical **conjunctive adverbs** are *however, therefore, nonetheless, nevertheless, consequently, accordingly,* and *then.*

EXAMPLE I rarely work on the weekend; *however,* this weekend will be an exception.

Typical **interrogative adverbs** are *where, when, why,* and *how.*

EXAMPLE *How* many hours did you work last week?

Typical numeric adverbs are *once* and *twice.*

EXAMPLE I have worked overtime *twice* this week.

COMPARISON OF ADVERBS

Adverbs are normally compared by adding *-er* or *-est* to them or by inserting *more* or *most* in front of them. One-syllable adverbs use the comparative-ending *-er* and the superlative ending *-est.*

EXAMPLES This copier is *faster* than the old one.
This copier is the *fastest* of the three tested.

Most adverbs with two or more **syllables** end in *-ly,* and most adverbs ending in *-ly* are compared by inserting the comparative *more* or the superlative *most* in front of them.

EXAMPLES He moved *more quickly* than the other company's salesman.
Most surprisingly, the engine failed during the test phase.

Less and *least* are **antonyms** of *more* and *most.*

EXAMPLES This filter can be changed *less quickly* than the previous filter.
Least surprisingly, the engine failed during the test phase.

There are a few irregular adverbs that require a change in form to indicate **comparison.**

EXAMPLES The training program functions *well.*
Our training program functions *better* than most others in the industry.
Many consider our training program the *best* in the industry.

PLACEMENT OF ADVERBS

An adverb may appear almost anywhere in a **sentence,** but its position may affect the meaning of the sentence. Avoid placing an adverb between two verb forms where it can be read ambiguously as modifying either.

CHANGE   The operator tried *belatedly* to close the relief valve.
TO   The operator *belatedly* tried to close the relief valve.

The adverb is commonly placed in front of the verb it modifies.

EXAMPLE   The pilot *meticulously* performed the preflight check.

An adverb may, however, follow the verb (or the verb and its **object**) that it modifies.

EXAMPLES   The gauge dipped *suddenly.*
They repaired the computer *quickly.*

The adverb may be placed between a **helping verb** and a main verb.

EXAMPLE   In this temperature range, the pressure will *quickly* drop.

If an adverb modifies only the main verb, and not any accompanying helping verbs, place the adverb immediately before the main verb.

EXAMPLE   The alternative proposal has been *effectively* presented.

An adverb **phrase** should not be inserted between the words of a compound verb.

CHANGE   This suggestion has *time and time again* been rejected.
TO   This suggestion has been rejected *time and time again.*

To place **emphasis** on an adverb, put it before the **subject** of the sentence.

EXAMPLES   *Clearly,* he was ready for the promotion when it came.
Unfortunately, fuel rationing has been unavoidable.

In writing, such adverbs as *nearly, only, almost, just,* and *hardly* are placed immediately before the words they limit. A speaker can place these words earlier and avoid **ambiguity** by stressing the word to be limited; in writing, however, only the appropriate placement of the adverb will ensure **clarity.**

CHANGE   The punch press with the auxilliary equipment *only* costs $47,000.
TO   The punch press with the auxilliary equipment costs *only* $47,000.

The first sentence is ambiguous because it might be understood to mean that *only* the punch press with auxilliary equipment costs $47,000. (See also **misplaced modifiers.**)

adverb clauses

An adverb clause is a subordinate **clause** used to modify a **verb, adjective,** or **adverb** in the main clause.

EXAMPLE   The property is located *where the railroad track crosses the road.*

Like adverbs, adverb clauses normally express such ideas as those of time, place, condition, cause, manner, or **comparison.**

EXAMPLES   It stopped *after we installed the new bearings.* (time)
It stopped *where the previous model stopped.* (place)
If we install the new bearings again, it will stop again. (condition)
It stopped *because we installed the new bearings.* (cause)
It stopped *as though it had run into a wall.* (manner)
The new bearings perform no better *than the old bearings did.* (comparison)

Adverb clauses are often introduced by **subordinating conjunctions** (*because, when, where, since, though,* etc.). If the adverb clause follows the main clause, a **comma** should not normally separate the two clauses.

EXAMPLE   The sonic boom problem caused by the hypersonic transport is greatly alleviated *because the craft makes a steep ascent.*

If the adverb clause precedes the main clause, the adverb clause should be set off by a comma.

EXAMPLE   *Because the craft makes a steep ascent,* the sonic boom problem is greatly alleviated.

Placement of an adverb clause can change the **emphasis** of a **sentence.** Do not allow this to happen inadvertently.

EXAMPLES   Ruins and artifacts are mere objects of curiosity *unless they are used to reconstruct the daily activities of prehistoric peoples.*
Unless they are used to reconstruct the daily activities of prehistoric peoples, ruins and artifacts are mere objects of curiosity.
Ruins and artifacts, *unless they are used to reconstruct the daily activities of prehistoric peoples,* are mere objects of curiosity.

An adverb clause cannot act as the **subject** of a sentence.

CHANGE   *Because medical expenses have increased* is the reason health insurance rates have gone up.

TO   Health insurance rates have gone up *because medical expenses have increased.*

OR   *Because medical expenses have increased,* health insurance rates have gone up.

advice/advise

Advice is a **noun** that means "counsel" or "suggestion."

EXAMPLE   My *advice* is to sign the contract immediately.

Advise is a **verb** that means "give advice."

EXAMPLE   I *advise* you to sign the contract immediately.

affect/effect

Affect is a **verb** that means "influence."

EXAMPLE   The public utility commission's decisions *affect* all state utilities.

Effect can function either as a verb that means "bring about" or "cause," or as a **noun** that means "result." It is best, however, to avoid using *effect* as a verb. A less formal word, such as *made*, is usually preferable.

CHANGE   The new manager *effected* several changes that had a good *effect* on morale.

TO   The new manager *made* several changes that had a good *effect* on morale.

affectation

Affectation is the use of language that is more technical, or showy, than is necessary to communicate information to the **reader.** Writing that is unnecessarily ornate, pompous, or pretentious is usually an attempt to impress the reader by showing off a repertoire of fancy or flashy words. Affected writing forces the reader to work harder to understand the writer's meaning. Affected writing typically contains abstract, highly technical, pseudotechnical, pseudolegal, or foreign words and is often liberally sprinkled with **vogue words. Jargon** can become affectation if it is misused. **Euphemisms** can contribute to affected writing if their purpose is to hide the facts of a situation rather than treat them with dignity or restraint.

The easiest kind of affectation to be lured into is the use of **long variants**: words created by adding **prefixes** and **suffixes** to simpler words (*analyzation* for analysis, *telephonic communication* for telephone call). The practice of **elegant variation**—attempting to avoid repeating the same word in the same **paragraph** by substituting pretentious **synonyms**—is also a form of affectation. Another contributor to affectation is **gobbledygook,** which is wordy, roundabout writing that has many pseudolegal and psuedoscientific terms sprinkled throughout.

Attempts to make the trivial seem important are also causes of affectation. This is apparent in the first version of the following example, taken from a specification soliciting bids from local merchants for the operation of a television repair shop in an Air Force Post Exchange.

CHANGE   In addition to performing interior housekeeping services, the concessionaire shall perform custodial maintenance on the exterior of the facility and grounds. Where a concessionaire shares a facility with one or more other concessionaires, exterior custodial maintenance responsibilities will be assigned by Post Exchange management on a fair and equitable basis. In those instances where the concessionaire's activity is located in a Post Exchange complex wherein predominant tenancy is by Post Exchange-operated activities, then Post Exchange management shall be responsible for exterior custodial maintenance except for those described in 1, 2, 3, and 4 below. The necessary equipment and labor to perform exterior custodial maintenance, when such a responsibility has been assigned to the concessionaire, shall be furnished by the concessionaire. Exterior custodial maintenance shall include the following tasks:

1. Clean entrance door and exterior of storefront windows daily.
2. Sweep and clean the entrance and customer walks daily.
3. Empty and clean waste and smoking receptacles daily.
4. Check exterior lighting and report failures to the contracting officer's representative daily.

TO   The merchant will, with his own equipment, perform the following duties daily:

1. Maintain a clean and neat appearance inside the store.
2. Clean the entrance door and the outsides of the store windows.

3. Sweep the entrance and the sidewalk.
4. Empty and clean wastebaskets and ashtrays.
5. Check the exterior lighting, and report failures to the representative of the contracting officer.

The merchant will also maintain the grounds surrounding his store. Where two or more merchants share the same building, Post Exchange management will assign responsibility for the grounds. Where the merchant's store is in a building that is occupied predominantly by Post Exchange operations, Post Exchange management will be responsible for the grounds.

Affectation is a serious problem in technical writing because many people apparently feel that affectation lends a degree of formality, and hence authority, to their writing. Nothing could be further from the truth. Affectation puts up a smoke screen that the reader must penetrate to get to the writer's meaning. (See also **conciseness/wordiness.**)

affinity

Affinity refers to the attraction of two persons or things to each other.

EXAMPLE   The *affinity* between these two elements can be explained in terms of their valence electrons.

Affinity should never be used to mean "ability" or "aptitude."

CHANGE   She has an *affinity* for chemistry.
TO   She has an *aptitude* for chemistry.
OR   She has a *talent* for chemistry.

aforesaid

Aforesaid means "stated previously," but it is legal **jargon** and should be avoided in writing.

CHANGE   The *aforesaid* problems can be solved in six months.
TO   The three problems explained in the preceding paragraphs can be solved in six months.

agree to/agree with

When you *agree to* something, you are "giving consent."

EXAMPLE   I *agree to* a road test of the new model by August 1.

When you *agree with* something, you are "in accord" with it.

EXAMPLE   I *agree with* the recommendations of the advisory board.

agreement

Agreement, grammatically, means the correspondence in form between different elements of a **sentence** to indicate **person, number, gender,** and **case.** A **pronoun** must agree with its antecedent, and a **verb** must agree with its **subject.**

A subject and its verb must agree in number and in person.

EXAMPLES   The *design is* an acceptable one. (The first person singular subject, *design,* requires the first person singular form of the verb, *is*)
The new *products are* going into production soon. (The third person plural subject, *products,* requires the third person plural form of the verb, *are.*)

A pronoun and its antecedent must agree in person, number, and gender.

EXAMPLES   The *employees* report that *they* become less efficient as the humidity rises. (The third person plural subject, *employees,* requires the third person plural pronoun, *they.*)
Mr. Joiner offered *his* services during negotiations. (The third person singular, masculine subject, *Mr. Joiner,* requires the third person singular, masculine form of the pronoun, *his.*)

See also **agreement of pronouns and antecedents** and **agreement of subjects and verbs.**

agreement of pronouns and antecedents

Every **pronoun** must have an antecedent, or a **noun** to which it refers.

CHANGE   Electronics technicians must constantly struggle to keep up with the professional literature because *it* is a dynamic science.
TO   Electronics technicians must constantly struggle to keep up with the professional literature because *electronics* is a dynamic science.
OR   Electronics technicians must continue to study *electronics* because *it* is a dynamic science.

Using the **relative pronoun** *which* to refer to an idea instead of a specific noun can be confusing.

CHANGE   He acted independently on the advice of his consultant, *which* the others thought unjust. (Was it the fact that he acted independently or was it the advice that the others thought unjust?)

TO   The others thought it unjust for him to act independently.

OR   The others thought it unjust for him to act on the advice of his consultant.

Using the pronouns *it, they,* and *this* can lead to vague or uncertain references.

CHANGE   Studs and thick treads make snow tires effective. *They* are implanted with an air gun.

TO   Studs and thick treads make snow tires effective. *The studs* are implanted with an air gun.

CHANGE   The inadequate quality-control procedure has resulted in an equipment failure. *This* is our most serious problem at present.

TO   The inadequate quality-control procedure has resulted in an equipment failure. Quality control is our most serious problem at present.

OR   The inadequate quality-control procedure has resulted in an equipment failure. The equipment failure is our most serious problem at present.

GENDER

A pronoun must agree in **gender** with its antecedent.

EXAMPLE   Mr. Swivet in the Accounting Department acknowledges *his* share of the responsibility for the misunderstanding, just as Mrs. Barkley in the Research Division must acknowledge *hers.*

Traditionally, a masculine, singular pronoun is used to agree with such indefinite antecedents as *anyone* and *person.*

EXAMPLE   *Each* may stay or go as *he* chooses.

Many people are now sensitive to an implied sexual bias in such usage. When alternatives are available, use them. One solution is to use the plural. Another is to use both feminine and masculine pronouns, although this combination is clumsy when used too often.

CHANGE   Every *employee* must sign *his* time card.

TO   All *employees* must sign *their* time cards.

OR   Every *employee* must sign *his or her* time card.

Do not, however, attempt to avoid expressing gender by resorting to a plural pronoun when the antecedent is singular.

CHANGE A *technician* can expect to advance on *their* merit.
    TO A *technician* can expect to advance on *his* merit.
    OR *Technicians* can expect to advance on *their* merit.
    OR A *technician* can expect to advance on *his or her* merit.

NUMBER

A pronoun must agree with its antecedent in **number.** Many problems of agreement are caused by expressions that are not clear in number.

CHANGE Although the typical *engine* runs well in moderate temperatures, *they* often stall in extreme cold.
    TO Although the typical *engine* runs well in moderate temperatures, *it* often stalls in extreme cold.

Use singular pronouns with the antecedents *everybody* and *everyone* unless to do so would be illogical because the meaning is obviously plural. (See also **everybody/everyone.**)

EXAMPLES *Everyone* pulled *his* share of the load.
             Everyone laughed at my sales slogan, and I really couldn't blame *them.*

A compound antecedent joined by *or* or *nor* is singular if both elements are singular, and plural if both elements are plural.

EXAMPLES Neither the *engineer* nor the *draftsman* could do *his* job until *he* understood the new concept.
             Neither the *executives* nor the *directors* were pleased at the performance of *their* company.

When one of the antecedents connected by *or* or *nor* is singular and the other plural, the pronoun agrees with the nearer antecedent.

EXAMPLES Either the *supervisor* or the *operators* will have *their* licenses suspended.
             Either the *operators* or the *supervisor* will have *his* license suspended.

A compound antecedent with its elements joined by *and* requires a plural pronoun.

EXAMPLE Martha and Joan took *their* layout drawings with *them.*

If both elements refer to the same person, however, use the singular pronoun.

EXAMPLE The respected *scientist and author* departed from *his* prepared speech.

Collective nouns may be singular or plural, depending on meaning.

EXAMPLES   The *committee* arrived at the recommended solutions only after *it* had deliberated for days.

The *committee* quit for the day and went to *their* respective homes.

agreement of subjects and verbs

A **verb** must agree with its **subject.** Do not let intervening **phrases** and **clauses** mislead you.

CHANGE   The use of insecticides, fertilizers, and weed killers, although they offer unquestionable benefits, often result in unfortunate side effects.

TO   The *use* of insecticides, fertilizers, and weed killers, although they offer unquestionable benefits, often *results* in unfortunate side effects. (The verb *results* must agree with the subject of the sentence, *use,* rather than with the subject of the preceding clause, *they.*)

Be careful to avoid making the verb agree with the **noun** immediately in front of it if that noun is not its subject. This problem is especially likely to occur when a plural noun falls between a singular subject and its verb.

EXAMPLES   Only *one* of the emergency lights *was* functioning when the accident occurred. (The subject of the verb is *one,* not *lights.*)

Each of the switches *controls* a separate circuit. (The subject of the verb is *each,* not *switches.*)

Each of the managers *supervises* a very large region. (The subject of the verb is *each,* not *managers.*)

The proper *use* of the tools *requires* practice. (The subject of the verb is *use,* not *tools.*)

Be careful not to let modifying phrases obscure a simple subject.

EXAMPLE   The *advice* of two engineers, one lawyer, and three executives *was* obtained prior to making a commitment. (The subject of the verb is *advice,* not *executives.*)

Sentences with inverted word order can cause problems with agreement between subject and verb.

EXAMPLE   From this work *have come* several important *improvements.* (The subject of the verb is *improvements,* not *work.*)

Such words as *type, part, series,* and *portion* take singular verbs even when they precede a phrase containing a plural noun.

EXAMPLES A *series* of meetings *was* held about the best way to market the new product.

A large *portion* of most industrial annual reports *is* devoted to promoting the corporate image.

Subjects expressing measurement, weight, mass, or total often take singular verbs even though the subject word is plural in form. Such subjects are treated as a unit.

EXAMPLES *Four years is* the normal duration of the training program.

Twenty dollars is the wholesale price of each unit.

Indefinite pronouns such as *some, none, all, more,* and *most* may be singular or plural depending upon whether they are used with a **mass noun** (*oil* in the following examples) or with a **count noun** (*drivers* in the following examples).

EXAMPLES *None* of the oil *is* to be used.

None of the truck drivers *are* to go.

Most of the oil *has* been used.

Most of the drivers *know* why they are here.

Some of the oil *has* leaked.

Some of the drivers *have* gone.

One and *each* are normally singular.

EXAMPLES *One* of the brake drums *is* still scored.

Each of the original founders *is* scheduled to speak at the dedication ceremony.

A verb following the **relative pronouns** *who* and *that* agrees in number with the noun to which the **pronoun** refers (its antecedent).

EXAMPLES Steel is one of those *industries* that *are* hardest hit by high energy costs. (*That* refers to *industries.*)

This is one of those engineering *problems* that *require* careful analysis. (*That* refers to *problems.*)

She is one of those *employees* who *are* rarely absent. (*Who* refers to *employees.*)

The word *number* sometimes causes confusion. When used to mean a specific number, it is singular.

EXAMPLE *The number* of committee members *was* six.

When used to mean an approximate number, it is plural.

EXAMPLE *A number* of people *were* waiting for the announcement.

Relative pronouns (*who, which, that*) may take either singular or plural verbs, depending upon whether the antecedent is singular or plural. (See also **who/whom.**)

EXAMPLES He is a chemist *who takes* work home at night.
He is one of those chemists *who take* work home at night.

Some **abstract nouns** are singular in meaning though plural in form: *mathematics, news, physics,* and *economics.*

EXAMPLES *News* of the merger *is* on page four of the *Chronicle.*
Textiles is an industry in need of import quotas.

Some words are always plural, such as *trousers* and *scissors.*

EXAMPLE His *trousers were* torn by the machine.
BUT A *pair* of trousers *is* on order.

Collective subjects take singular verbs when the group is thought of as a unit, plural verbs when the individuals are thought of spearately. (See also **collective nouns.**)

A singular subject that is followed by a phrase or clause containing a plural noun still requires a singular verb.

CHANGE One in twenty transistors we receive from our suppliers are faulty.
TO *One* in twenty transistors we receive from our suppliers *is* faulty. (The subject is *one,* not *transistors.*)

The number of a **subjective complement** does not affect the number of the verb—the verb must always agree with the subject.

EXAMPLE The *topic* of his report *was* rivers. (The subject of the sentence is *topic,* not *rivers.*)

A book with a plural title requires a single verb.

EXAMPLE Romig's *Binomial Tables is* a useful source.

COMPOUND SUBJECTS

A compound subject is one that is composed of two or more elements joined by a **conjunction** such as *and, or, nor, either . . . or,* or *neither . . . nor.* Usually, when the elements are connected by *and,* the subject is plural and requires a plural verb.

EXAMPLE *Chemistry and accounting are* both prerequisites for this position.

There is one exception to the *and* rule. Sometimes the elements connected by *and* form a unit or refer to the same thing. In this case, the subject is regarded as singular and takes a singular verb.

EXAMPLES   *Bacon and eggs is* a high-cholesterol meal.
The *red, white, and blue flutters* from the top of the capitol.
Our greatest *technical challenge and business opportunity is* the Model MX Calculator.

A compound subject with a singular and a plural element joined by *or* or *nor* requires that the verb agree with the element nearest to it.

EXAMPLES   Neither the office manager nor the *secretaries were* there.
Neither the secretaries nor the office *manager was* there.
Either they or *I am* going to write the report.
Either I or *they are* going to write the report.

If *each* or *every* modifies the elements of a compound subject, use the singular verb.

EXAMPLES   *Each* manager and supervisor *has* a production goal to meet.
Every manager and supervisor *has* a production goal to meet.

all around/all-around/all-round

All-round and *all-around* both mean "comprehensive" or "versatile."

EXAMPLE   The company started an *all-round* training program.

Do not confuse these words with the two-word **phrase** *all around,* as in "The fence was installed *all around* the building."

all right/all-right/alright

All right means "all correct," as in "The answers were all right." In **formal writing** it should not be used to mean "good" or "acceptable." It is always written as two words, with no **hyphen;** *all-right* and *alright* are both incorrect.

CHANGE   The decision that the committee reached was *all right.*
TO   The decision that the committee reached was acceptable.

all together/altogether

All together means "all acting together," or "all in one place."

EXAMPLE   The necessary instruments were *all together* on the tray.

Altogether means "entirely" or "completely."

> EXAMPLE   The trip was *altogether* unnecessary.

allude/elude

Allude means to make an indirect reference to something not specifically mentioned.

> EXAMPLE   The report simply *alluded* to the problem rather than stating it clearly.

Elude means to escape notice or detection.

> EXAMPLES   The discrepancy in the account *eluded* the auditor.
> The leak *eluded* the inspectors.

allude/refer

Allude means to make an indirect reference to something not specifically mentioned.

> EXAMPLES   In his speech he *alluded* to the rumors of widespread safety infractions.
> The memo *alluded* to past equipment failures.

Refer is used to indicate a direct reference to something.

> EXAMPLE   He *referred* to the chart three times during his presentation.

allusion

The use of allusion (implied or indirect reference) promotes economical writing because it is a shorthand way of referring to a body of material in a few words, or of helping to explain a new and unfamiliary process in terms of one that is familiar. Be sure, however, that your **reader** is familiar with the material to which you allude. In the following **paragraph,** the writers sums up the argument he has been developing with an allusion to the Bible. The biblical story is well known, and the allusion, with its implicit reference to "right standing up to might," concisely emphasizes the writer's point.

> EXAMPLE   As it presently exists, the review process involves the consumer's attorney sitting alone, usually without adequate technical assistance, faced by two or three government attorneys, two or three

attorneys from the XYZ Corporation, and large teams of experts who support the government and corporation's attorneys. The entire proceeding is reminiscent of David versus Goliath.

Allusions should be used with restraint. If overdone, they can lead to **affectation.**

allusion/illusion

An *allusion* is an indirect reference to something not specifically mentioned.

> EXAMPLE   The report made an *allusion* to metal fatigue in support structures.

An *illusion* is a mistaken perception or a false image.

> EXAMPLE   County officials are under the *illusion* that the landfill will last indefinitely.

almost/most

Do not use *most* as a colloquial substitute for *almost* in your writing.

> CHANGE   New shipments arrive *most* every day.
> TO   New shipments arrive *almost* every day.

If you can substitute *almost* for *most* in a **sentence,** *almost* is the word you need.

already/all ready

Already is an **adverb** expressing time.

> EXAMPLE   The shipment had *already* been made when the stop order arrived.

All ready is a two-word **phrase** meaning "completely prepared."

> EXAMPLE   He was *all ready* to start work on the project when it was suddenly canceled.

also

Also is an **adverb** that means "additionally."

> EXAMPLE   Two 500,000-gallon tanks have recently been constructed on site. Several 10,000-gallon tanks are *also* available, if needed.

It should not be used as a **connective** in the sense of ''and.''

CHANGE   He brought the reports, letters, *also* the design group supervisor's recommendations.

TO   He brought the reports, letters, *and* the design group supervisor's recommendations.

ambiguity

A word or passage is ambiguous when it is susceptible to two or more interpretations, yet provides the **reader** with no certain basis for choosing among the alternatives.

EXAMPLE   Mathematics is more valuable to an engineer than a computer. (Does this mean that an engineer is more in need of mathematics than a computer is? Or does it mean that mathematics is more valuable to an engineer than a computer is?)

Ambiguity can take many forms: ambiguous **pronoun reference, misplaced modifiers, dangling modifiers,** ambiguous coordination, ambiguous juxtaposition, incomplete **comparison,** incomplete **idiom,** ambiguous **word choice,** and so on.

CHANGE   Inadequate quality-control procedures have resulted in more equipment failures. *These* are our most serious problem at present. (ambiguous pronoun reference; does *these* refer to ''quality-control procedures'' or ''equipment failures''?)

TO   Inadequate quality-control procedures have resulted in more equipment failures. These failures are our most serious problem at present.

OR   Inadequate quality-control procedures have resulted in more equipment failures. Quality control is our most serious problem at present.

CHANGE   Ms. Jones values rigid quality-control standards more than Mr. Rosenblum. (incomplete comparison)

TO   Ms. Jones values rigid quality-control standards more than Mr. Rosenblum *does.*

CHANGE   His hobby was cooking. He was especially fond of cocker spaniels. (missing modifier)

TO   His hobby was cooking. He was *also* especially fond of cocker spaniels.

CHANGE All navigators are *not* talented in mathematics. (misplaced modifier; the implication is that *no* navigator is talented in mathematics.)

TO *Not* all navigators are talented in mathematics.

Ambuiguity is also often caused by thoughtless word choice.

CHANGE The general manager has denied reports that the plant's recent fuel allocation cut will be *restored.* (inappropriate word choice)

TO The general manager has denied reports that the plant's recent fuel allocation cut will be *rescinded.*

amount/number

Amount is used with things thought of in bulk **(mass nouns).**

EXAMPLES The *amount* of electricity available for industrial use is limited.
The *amount* of oxygen was insufficient for combustion.

Number is used with things that can be counted as individual items **(count nouns).**

EXAMPLES A large *number* of stockholders attended the meeting.
The *number* of employees who are qualified for early retirement has increased in recent years.

ampersands

The ampersand (&) is a **symbol** sometimes used to represent the word *and,* especially in the names of organizations.

EXAMPLES Chicago & Northwestern Railway
Watkins & Watkins, Inc.

You may use the ampersand in **footnotes, bibliographies, lists,** and **references.** When writing the name of an organization in **sentences** or in an address, however, spell out the word *and* unless the ampersand appears in the official name of the company. (See also **abbreviations.**)

An ampersand must always be set off by normal word spacing but should never be preceded by a **comma.**

CHANGE Carlton, Dillon, & Manchester, Inc.
TO Carlton, Dillon & Manchester, Inc.

Do not use the ampersand in the titles of articles, journals, books, or other publications.

> CHANGE   Does the bibliography include Knoll's *Radaition Detection & Measurement?*
>
> TO   Does the bibliography include Knoll's *Radiation Detection and Measurement?*

analogy

Analogy is a **comparison** between two objects or concepts to show ways in which they are similar. In effect, analogies say, "A is to B as C is to D."

> EXAMPLE   Pollution is to the environment as cancer is to the body.

The resemblance between the concepts represented in an analogy must be close enough to illuminate the relationship the writer wants to establish. Analogies may be brief or extended, depending on the writer's purpose.

Analogy can be a particularly useful tool to the technical person writing for an intelligent and educated but nontechnical audience, such as top management, because of its effectiveness in **defining terms** and **explaining processes.** Like all figurative language, analogy can provide a shortcut means of communication if it is used with care and restraint.

The following example explains a computer search technique for finding information in a data file by comparing it to the method used by most people to find a word in a **dictionary.**

> The search technique used in this kind of file processing is similar to the search technique used to look up a word in a dictionary. To locate a specific word, you scan the key words located at the top of each dictionary page (these key words identify the first and last words on the page) until you find the key words that confine your word to a specific page. Assume that all the key words that reference the last word on each page of a dictionary were placed in a file, along with their corresponding page numbers, and that all the words in the dictionary were placed in another file. To locate any word, you would simply scan the first file until you found a key word greater in alphabetical sequence than the desired word and go to the place in the second file indicated by the key word to find the desired word. This search technique is called indexed sequential processing.
>
> —*The NCR File Management Manual,* (Dayton: NCR Corporation, 1974), p. 76.

and/or

And/or means that either both circumstances are possible or only one of two circumstances is possible; however, it is clumsy and awkward because it makes the **reader** stop to puzzle over your distinction. Also, writers more often than not use the term inexactly. As Wilson Follett said, "English speakers and writers have managed to express this simple relationship without *and/or* for over six centuries."* Avoid it.

CHANGE   Use A and/or B.
TO   Use A or B or both.

anecdotes

An anecdote is an interesting or humorous incident or story indirectly related to the writer's subject. It is used to develop or clarify a point. For example, Isaac Asimov, discussing prescientific thinking, tells the following anecdote about the tendency of people to explain the outcome of their actions on the basis of astronomical phenomena.

> Thus in 1066, the comet we now call Halley's Comet appeared in the sky just as William of Normandy was making ready to invade England. It predicted catastrophe and that is exactly what came, for the Saxons lost the Battle of Hastings and passed under the permanent rule of the Normans. The Saxons couldn't have asked for a better catastrophe than that.
>
> On the other hand, if the Saxons had won and had hurled William's expeditionary force into the Channel, that would have been catastrophe enough for the Normans.
>
> Whichever side lost, the comet was sure to win.

—Isaac Asimov, *The Stars in Their Courses* (New York: Ace Books, 1971), p. 20.

annual reports

The corporate annual report is a legally required document that is, in effect, a "state of the company" message. It is written primarily for stockholders, although many other audiences must be kept in mind when the report is prepared (bankers, financial press, labor unions, employees, etc.). An annual report usually covers the high points of the previous year's operations and attempts to forecast the coming year's

*Modern American Usage (New York: Hill & Wang, 1966), p. 65.

operations. It may also explain the company's present directions and highlight its present strengths. If weaknesses have developed or failures have occurred, the annual report may analyze them and explain the efforts being made to overcome them.

A wide variety of annual reports are published every year. Some are lavish in presentation and paint the company and its operations in glowing terms; others are spartan financial recitations that merely meet the legal requirements for annual financial reporting. Most annual reports are combinations of legally required financial reporting and narrative articles presenting the company with its best foot forward.

Although they vary greatly, annual reports can be loosely divided into five major divisions: (1) financial highlights, (2) a statement to stockholders or letter from the president, (3) a narrative section containing articles on the company's operations, (4) a financial statement, and (5) a listing of the company's board of directors and officers.

Financial Highlights. The financial highlights section is a quick review of the company's sales and earnings that usually precedes the president's message to the stockholders, sometimes even appearing on the inside of the front cover. The most-read part of the annual report, this section often compares sales and earnings for three years, and it often includes the percentage of change from year to year.

Statement to Stockholders. This section of the annual report is a direct statement to stockholders from the company's president or chairman of the board of directors. This statement (which is sometimes presented in the form of a letter) is second in readership only to the financial highlights. It is the place to set the stage for the rest of the report. It should avoid repeating financial facts already cited in the financial highlights; instead, it can be used to (1) interpret the entire year's performance, (2) touch upon plans and future directions, and (3) give the company's explanations for any failures.

The statement to stockholders may be an in-depth review of the company's operations during the past year, or it may be a brief summary of the entire report. A brief summary of the annual report is sometimes followed by an article in "interview" (question and answer) **format** that reviews the past year's operations as though an actual interview were taking place between the writer and the president of the company.

Avoid stereotyped language and **clichés** when you write this statement. Also avoid technical terms. Use simple **sentences** and short **paragraphs,** and use a straightforward and **informal style.**

Narrative Feature Articles. This section of the annual report is normally used to present company operations and new products or developments in a positive light. Select the **topic** for these articles very carefully, making certain that they are timely and meaningful and that they contribute to the primary objective of the annual report. The following list includes topics that are often dealt with in annual reports.

Major profit factors in last year's performance
Sales trends
The company's performance compared to its competition
International operations
World problems that affect the company
The dependability of the company's overseas markets
Research and development efforts
Significant organizational changes
The company's labor relations outlook
Productivity
Acquisitions and their significance
Prospects for increasing stock dividends
Outlook for next year (for the company as a whole or by divisions)
Significant new products or services
Current market performance of existing products or services
Service and support operations
Social responsibilities—on environmental protection, energy conservation, safety, hiring minorities and the handicapped, and so on. (Be certain to document the company's efforts in this area because overstated claims by a few companies in the past have generated suspicion of all such claims.)

Most **readers** merely scan an annual report, but they stop to look at **photographs** and **illustrations** and to read their captions. Therefore, use photographs and illustrations liberally in this section of the annual report, and make your captions informative and well-written.

Financial Statement. The financial statement should not be forbidding in appearance; it should be uncluttered, inviting, and consistent with the other sections of the annual report. The financial statement may appear at the beginning, middle, or end of the annual report; however, it most often appears at the end. It is frequently prepared as a separate booklet, printed on colored paper, and then stitched into the report.

Minimum normal contents of the financial statement include (1) balance sheet, (2) statement of income, (3) changes in financial position, (4) the auditor's statement, and (5) **footnotes.** Most annual reports also include a fairly comprehensive **comparison** of financial results of the past ten to fifteen years.

Footnotes in the financial section should be written in simple, direct language rather than in accounting terms. The auditor's statement should be limited to no more than one-third of a page.

Board of Directors and Company Officers. The final part of the annual report lists the company's board of directors, along with their corporate affiliations. Many annual reports also include a portrait photograph of each director. Also listed by name and title (sometimes with photographs) are the company's officers. These generally include the chairman of the board of directors, president, vice-presidents, secretary, treasurer, and legal counsel. To avoid hard feelings about some being shown in photographs and others not, establish a logical cut-off point. For example, if there are too many officers to include photographs of them all, show only the executive committee. Do not, however, underestimate the importance of these photographs; all your reading audiences are interested in seeing what your company's top executives look like.

PREPARING THE REPORT

Assuming you are to prepare the annual report, the first thing you must do is interview the president of the company or the chairman of the board of directors to determine the general directions the report is to take. Next, you must interview the various vice-presidents, or division heads, to learn the highlights of each division's operations during the year. Determine from these executives the proper **emphasis** to place on their divisions' performances—but stay within the general direction established by the president or chairman of the board.

The primary objective of the annual report often depends upon which of several audiences (shareholders, bankers, business and financial analysts, the financial press, employees, labor unions, the communities in which your company operates) you are most concerned about. Since you cannot satisfy all these groups with one report, you must establish priorities among them. Having done that, you can compose a list of "must" topics, then a list of secondary topics (such as diversification, expansion of markets, product development, etc.). You can also select the main topics for brief coverage in the statement to stockholders, decide whether to show sales and earnings by division, determine the amount of space to allot to various divisions and subsidiaries, determine the contents of charts and graphs, decide whether to include photographs of the executives of branches, divisions, and subsidiaries, decide whether to include a frank discussion of company problems and solutions, and decide how lavish or spartan the report should be in appearance and costs.

Tone is a critical factor in writing and designing an annual report. Companies are as different as people; some are conservative, and some are aggressively youthful. The annual report should convey the image your company has or wants to establish. Static, formal page layout with trditional typography will suggest a conservative, dignified company, for example; and bold graphics, dynamic design, and unusual typography will suggest an aggressive company. Don't mix the two, however; photographs of executives in dark business suits would not be compatible with bold graphics.

Be selective in your choice of photographs and illustrations, choosing only those that will make the maximum contribution to the report's theme. Remember that most readers merely scan an annual report, looking carefully only at photographs and captions. So both photographs and their captions are critical to the success of your annual report. A photograph of a machine can be dull. But put an operator in it, and the caption can tell what he is doing, and why it is meaningful.

Study your company's annual reports of the past several years for content, **style,** and **format.** Within reason, try to emulate these, since they presumably were successful. Review the writing process outlined in the Checklist of the Writing Process at the beginning of this book, and use all the steps listed there to the best of your ability—your company's annual report could well be the most important writing you will ever be asked to do.

The following information is normally printed on the inside front and back covers of the annual report: (1) notice of the annual meeting, (2) corporate address, (3) names of transfer agents, (4) registrar, (5) stock exchange, and so on. Some annual reports use the inside front cover for the announcement of the annual meeting alone.

Use of Charts and Graphs. **Graphs** and charts enable readers to grasp statistical material quickly and easily—provided the charts and graphs are not so complex that they defeat that purpose. Some companies scatter graphs throughout the annual report at strategic locations near the financial highlights, in the statement to stockholders, throughout the feature articles, and so forth. Others group them in the financial statement.

The subjects that most easily lend themselves to graphs and charts—usually shown in a five-year comparison arrangement—are assets, capital expenditures, dividends, earnings (by product groups or by divisions), industry growth, inventories, liabilities, net worth, prices (trend of), reserves, sales (by product groups or by divisions), and source and disposition of funds (taxes, wages, working capital).

ante-/anti-

Ante- means "before" or "in front of."

EXAMPLES   *Ante*room, *ante*date, *ante*diluvian (before the flood)

Anti- means "against" or "opposed to."

EXAMPLES   *anti*body, *anti*clerical, *anti*neutrino

Anti- is hyphenated when joined to proper nouns or to words beginning with the letter *i*.

EXAMPLES   *anti*-American, *anti*-intellectual

When in doubt, consult your **dictionary.** (See also **prefixes.**)

antonyms

An antonym is a word that is nearly the opposite, in meaning, of another word.

EXAMPLES   good / bad, well / ill, fresh / stale

Many pairs of words that look as if they are antonyms are not. Be careful not to use these words incorrectly.

EXAMPLES   famous / infamous, flammable / inflammable, limit / delimit

When in doubt about the pronunciation, correct use, or exact meaning of a word, use your **dictionary.**

apostrophes

The apostrophe (') is used to show possession, to mark the omission of letters, and to indicate the plural of Arabic numbers and letters.

TO SHOW POSSESSION

An apostrophe is used with an *s* to form the possessive **case** of some **nouns.**

EXAMPLE   A recent scientific analysis of *New York City's* atmosphere concluded that a New Yorker on the street took into his lungs the equivalent in toxic materials of 38 cigarettes a day.

With coordinate nouns, the last noun takes the possessive form to show joint possession.

EXAMPLE   Michelson and *Morley's* famous experiment on the velocity of light was made in 1887.

To show individual possession with coordinate nouns, each noun should take the possessive form.

EXAMPLE   The difference between *Tom's* and *Mary's* test results is statistically insignificant.

Singular nouns ending in *s* may form the possessive either by an apostrophe alone or by *'s*.

EXAMPLES   a waitress' uniform, an actress' career
a waitress's uniform, an actress's career

Singular nouns of one syllable form the possessive by adding *'s*.

EXAMPLE   The boss's desk was cluttered.

Use only an apostrophe with plural nouns ending in *s*.

EXAMPLES   a managers' meeting, the technicians' handbook

When a noun ends in multiple consecutive *s* sounds, form the possessive by adding only an apostrophe.

EXAMPLES   Jesus' disciples, Moses' sojourn.

The apostrophe is not used with possessive **pronouns.**

EXAMPLES   yours, its, his, ours, whose, theirs

It's is a **contraction** of *it is,* not the possessive form of *it.*
In names of places and institutions, the apostrophe is usually omitted.

EXAMPLES   Harpers Ferry, Writers Book Club

TO SHOW OMISSION

An apostrophe is used to mark the omission of letters in a word or date.

EXAMPLES   can't, I'm, I'll
the class of '61

TO FORM PLURALS

An apostrophe and an *s* may be added to show the plural of a word as a word. (The word itself is underlined, or italicized.)

EXAMPLE   There were five *and's* in his first sentence.

If the term is in all **capital letters** or ends with a capital letter, however, the apostrophe is not required to form the plural. (See also **acronyms and initialisms.**)

EXAMPLES   The university awarded seven *Ph.D.s* in engineering last year.
He had included 43 *ADDs* in his computer program.

Do not use apostrophes to indicate the plural of letters and numbers unless confusion would result without one.

EXAMPLES   5s, 30s, two 100s, seven I's

appendix/appendixes

An appendix is a collection of supplementary material at the end of a **formal report** or book. The plural form of the word may be either *appendixes* or *appendices*.

Although not a mandatory part of a report, an appendix can be quite useful for explanations that are too long for **footnotes** but could be helpful to the **reader** seeking further assistance or clarification, passages from documents and laws that reinforce or illustrate the text, long lists of charts and **tables,** letters and other supporting documents, calculations (in full), raw data, case histories, and so forth.

Do not use the appendix as a wastebasket for miscellaneous bits and pieces of information you were unable to work into the text. If you could not fit them into the text, they very likely do not belong in the report.

application letters

The letter of application is essentially a sales letter. You are marketing your skills, abilities, and knowledge. Remember that you may be competing with many other applicants. The immediate **objective** of an application letter is to get the attention of the person who screens and hires job applicants. Your ultimate goal is to obtain a **job interview.**

The successful application letter does three things: catch the **reader's** favorable attention, convince the reader that you are qualified for consideration, and request an interview. It should be concisely written.

A letter of application should provide the following information.

1. Identify an employment area or state a specific job title.
2. List your source of information about the job.
3. Describe your qualifications for the job in summary form—cover specifically education, work experience, and activities showing leadership skills.
4. Refer the reader to your **résumé.**
5. Ask for an interview, stating where you can be reached and when you will be available for an interview.

If you are applying for a specific job, include information pertinent to the position—details not included on the more general résumé.

Personnel directors review many letters each week. To save them time, you should state your job objective directly at the beginning of the letter.

EXAMPLE   Please consider me for a position in an engineering department where I can use my computer science training to solve engineering problems.

If you have been referred to a prospective employer by one of its employees, a placement counselor, a professor, or someone else, however, you might say so before stating your job objective.

EXAMPLE   During the recent NOMAD convention in Washington, D.C., a member of your sales staff, Mr. Dale Jarrett, informed me of a possible opening for a manager in your Dealer Sales Division. My extensive background in the office machine industry, I believe, should qualify me for the position.

In succeeding **paragraphs** expand upon the qualifications you mentioned in your **opening.** Add any appropriate details, highlighting the experience listed on your résumé that is especially pertinent to the job you are seeking.

Type your letter, proofread it carefully for errors, and *keep a copy* for future reference. (See also **proofreading.**)

Three sample letters of application follow. The first one is written by a recent college graduate, the second by a college student about to graduate, and the third by a writer with many years of work experience.

Letter 3 is in response to a local newspaper article about a company's plan to build a new plant. The writer is not applying for a specific job opening, but describes the positions he is looking for. In Letter 4 the writer does not specify where she learned of the opening because she does

4 Washington Boulevard
Detroit, MI 48214
June 14, 19--

Personnel Manager
Loudons, Inc.
4619 Drove Lane
Gary, IN 46409

Dear Sir or Madam:

The <u>Detroit Free Press</u> recently reported that Loudons, Inc.,
is constructing a new assembly plant near Gary. I would like
to apply for a foreman position in the new plant.

I recently graduated from Midlands Community College in
Detroit with an Associate Degree in Industrial Engineering.
In addition, I have worked for the past three years as a
special-assignment worker at Michigan Industries, Inc., on a
part-time basis while at Midlands and now on a full-time
schedule. My duties have required close work with plant
foremen, general foremen, and plant supervisors and have given
me extensive exposure to quality-control techniques and
manufacturing processes. Details of my education and work
experience are included on the enclosed résumé.

I will be happy to meet with you at your convenience. You may
reach me either at my home address or at (313) 233-6312
during regular business hours.

Sincerely,

James L. Nardinski

Enclosure: Résumé

Letter 3

2701 Wyoming Street
Atlanta, GA 30307
May 29, 19--

Ms. Laura Goldman
Chief Engineer
Acton, Inc.
80 Roseville Road
St. Louis, MO 63130

Dear Ms. Goldman:

I am looking for a position in an accounting department where
I can use my degree in computer science. I am writing to
explore the possibility of obtaining such a position with your
firm.

I expect to receive a Bachelor of Science degree in Computer
Science from Georgia Institute of Technology in June. Since
September 19--, I have been participating in the Professional
Training Program at Computer Systems International, in Atlanta,
where I have served in apprentice positions on a rotating
basis in several staff sections. I am currently working as
Programmer Trainee in the Accounting Department and have thus
gained valuable experience in computer applications. Details
of my academic record and of my other work experience may be
found on the enclosed resume.

I look forward to an interview with you soon. You may reach
me at my office phone (404) 866-7000, Extension 312, between
8:30 a.m. and 5:30 p.m., or at my home (404) 256-6320 at other
times.

Sincerely yours,

Victoria T. Fromme

Victoria T. Fromme

Enclosure: Résumé

Letter 4

522 Beethoven Drive
Roanoke, VA 24017
November 15, 19--

Miss Cecilia Smathers
Vice-President, Dealer Sales
Hamilton Office Machines, Inc.
Hampton, VA 23661

Dear Miss Smathers:

During the recent NOMAD convention in Washington, D.C.,
Mr. Dale Jarrett, a member of your sales staff, informed me of
a possible opening for a manager in your Dealer Sales
Division. I believe that my extensive background in the
office-machine industry should qualify me for the position.

I was with the Dealer Division at Technology, Inc., from its
formation in 19-- to its phase-out last year. During this
period I was involved in all areas of dealer sales, both
within Technology, Inc., and through personal contact with a
number of independent dealers. Between 19-- and 19--, I
served as Special Assistant to the Dealer Sales Manager.
Details of my education and work experience are on the enclosed
résumé.

I would be glad to meet with you to discuss my qualifications
in detail. Please write to me at my home or telephone me at
(703) 449-6743 any weekday.

Sincerely,

George Mindukakis

George Mindukakis

Enclosure: Résumé

Letter 5

not know whether a position is actually available. Letter 5 opens with an indication of where the writer learned of the job vacancy. (See also **job search** and **reference letters.**)

appositives

An appositive is a **noun** or noun **phrase** that follows and amplifies another noun or noun phrase. It has the same grammatical function as the noun it complements.

EXAMPLES   Dennis Gabor, *a British scientist,* experimented with coherent light in the 1940s.
The British scientist *Dennis Gabor* experimented with coherent light in the 1940s.
George Thomas, *head of the Economic and Planning Branch of PRC,* summarized the president's speech in a confidential memo to the advertising staff.

For detailed information on the use of **commas** with appositives, see **restrictive and nonrestrictive phrases.**

When in doubt about the **case** of an appositive, you can check it by substituting the appositive for the noun it modifies.

EXAMPLE   My boss gave the two of us, Jim and *me,* the day off. (You wouldn't say, "My boss gave *I* the day off.")

argumentation

See **persuasion.**

articles

As a **part of speech,** articles are considered to be **adjectives** because they modify the items they designate by either limiting them or making them more exact. There are two kinds of articles, indefinite and definite.

Indefinite: *a* and *an* (denotes an unspecified item)

EXAMPLE   *A* program was run on our new computer. (Not a specific program, but an unspecified program. Therefore, the article is indefinite.)

Definite: *the* (denotes a partiular item)

> EXAMPLE *The* program was run on the computer. (Not just any program, but *the* specific program. Therefore, the article is definite.)

The choice between *a* and *an* depends on the sound rather than the letter following the article. Use *a* before words beginning with a consonant sound (*a* person, *a* happy person, *a* historical event). Use *an* before words beginning with a vowel sound (*an* Uncle, *an* hour). With **abbreviations,** use *a* before initial letters having a consonant sound: *a* TWA flight. Use *an* before initial letters having a vowel sound: *an* SLN report (note that the first sound is "ess").

Do not omit all articles from your writing. This is, unfortunately, an easy habit to develop. To include the articles costs nothing; to eliminate them makes the reading more difficult (see **telegraphic style**).

> CHANGE Pass card through punch area for debris.
> TO Pass *a* card through *the* punch area *to clear away* any debris.

> CHANGE There has been decline in domestic output of crude oil.
> TO There has been *a* decline in *the* domestic output of crude oil.

On the other hand, don't overdo it. An article can be superfluous.

> EXAMPLES There is no more difficult *a* subject than calculus. (Eliminate the article).
>
> I'll meet you in *a* half *an* hour. (Choose one and eliminate the other.)
>
> Fill with *a* half *a* pint of fluid. (Choose one and eliminate the other.)

Do not capitalize articles when they appear in titles except as the first word of the title. (See also **capital letters.**)

> EXAMPLE *Time* magazine reviewed *The Nuclear Option in Europe for the Eighties.*

as

Since *as* can mean so many things (*since, because, for, that, at that time, when, while,* etc.) and can be at least four **parts of speech (conjunction, preposition, adverb, pronoun),** it is often overused and misused, especially in speech. In writing, *as* is often weak or even ambiguous. Notice how the following sentence has two possible meanings.

> EXAMPLE *As* we were together, he revealed his plans.
> MEANS *Because* we were together, he revealed his plans.
> OR *While* we were together, he revealed his plans.

The word *as* can also contribute to **awkwardness** by appearing too many times in a **sentence.**

CHANGE   *As* we realized *as* soon *as* we began the project, the problem needed a solution.

TO   We realized the moment we began the project that the problem needed a solution.

(See also **because.**)

as much as/more than

These two **phrases** are sometimes illogically run together, especially when intervening phrases delay the completion of the phrase.

CHANGE   The engineers had *as much,* if not *more,* influence in planning the program *than* the accountants.

TO   The engineers had *as much* influence in planning the program *as* the accountants, if not *more.*

OR   The engineers had *as much* influence *as* the accountants, if not *more,* in planning the program.

as regards/with regard to/in regard to/regarding

With regards to and *in regards to* are incorrect **idioms** for *with regard to* and *in regard to.* *As regards* and *regarding* are both acceptable variants.

CHANGE   *With regards to* the building contract, this question is pertinent.

TO   *With regard to* the building contract, this question is pertinent.

OR   *In regard to* the building contract, this question is pertinent.

OR   *As regards* the building contract, this question is pertinent.

OR   *Regarding* the building contract, this question is pertinent.

Avoid using "with regard to" where "about" would be more specific.

CHANGE   I am writing you *with regard to* your design for a new product.

TO   I am writing you *about* your design for a new product.

as such

The **phrase** *as such* is seldom useful and should be omitted.

CHANGE   The drafting department, *as such,* worked overtime to meet the deadline.

TO   The drafting department worked overtime to meet the deadline.

CHANGE   This program is poor. *As such,* it should be eliminated.
TO   This program is poor and should be eliminated.

as well as

Do not use *as well as* together with *both.* The two expressions have similar meanings; use one or the other.

CHANGE   *Both* General Motors, *as well as* Ford, are developing electric cars.
TO   *Both* General Motors *and* Ford are developing electric cars.
OR   General Motors, *as well as* Ford, is developing an electric car.

as to whether

The **phrase** *as to whether (as to when* or *as to where)* is clumsy and redundant. Either omit it altogether or use only *whether.*

CHANGE   *As to whether* we will redesign the fuel-injection system, we are still undecided.
TO   *Whether* we will redesign the fuel-injection system is still undecided.

Be wary of all phrases starting with *as to;* they are often redundant, vague, or indirect.

CHANGE   *As to* his policy, I am in full agreement.
TO   I am in full agreement with his policy.

attribute/contribute

Attribute (with the accent on the second syllable) is a **verb** that means "point to a cause or source."

EXAMPLE   He *attributes* the plant's improved safety record to the new training program.

Attribute (with the first syllable accented) is a **noun** meaning a quality or characteristic belonging to someone or something.

EXAMPLE   His mathematical skill is his most valuable *attribute.*

Contribute means "give."

EXAMPLE   His mathematical skills will *contribute* much to the project.

audience

See **reader.**

augment/supplement

Augment means to increase or magnify in size, degree, or effect.

EXAMPLE   Many employees *augment* their incomes by working overtime.

Supplement means to add something to make up for a deficiency.

EXAMPLE   The physician told him to *supplement* his diet with vitamins.

average/mean/median

Average, or arithmetical *mean,* is determined by dividing a sum of two or more quantities by the number of quantities. For example, if one report is 10 pages, another is 30 pages, and a third is 20 pages, their *average* (or *mean*) length is 20 pages. It is incorrect, therefore, to say that *"each* averages 20 pages" because *each* report is a specific length.

CHANGE   Each report *averages* 20 pages.
TO   The three reports *average* 20 pages.

A *median* is the middle number in a sequence of numbers.

EXAMPLE   The *median* of the series, 1, 3, 4, 7, 8 is 4.

awhile/a while

Awhile is an **adverb** meaning "for a short time." It is not preceded by *for* because the meaning of *for* is inherent in the meaning of *awhile. A while* is a **noun** phrase that means "a period of time."

CHANGE   Wait for *awhile* before investing more heavily.
TO   Wait for *a while* before investing more heavily.
OR   Wait *awhile* before investing more heavily.

awkwardness

Any writing that strikes the **reader** as awkward—that is, as forced or unnatural—impedes the reader's understanding. Awkwardness has many causes, including overloaded **sentences,** overlapping **subordina-**

tion, grammatical errors, ambiguous statements, overuse of **expletives** or of the passive **voice,** faulty **logic,** unintentional **repetition, garbled sentences,** and **jammed modifiers.**

To avoid awkwardness, make your writing as direct and concise as possible. The following three guidelines will help you to smooth out most awkward passages. In general, you should keep your sentences uncomplicated. Use the active voice unless you have a particular reason to use the passive voice. "Tighten up" your writing by eliminating excess words.

B

bad/badly

Bad is the **adjective** form that follows such **linking verbs** as *feel* and *look*.

EXAMPLES With the flu, you will feel *bad* for three days.
We don't want our department to look *bad* at the meeting.

Badly is an **adverb.**

EXAMPLE The test model performed *badly* during the trial run.

To say "I feel badly" would mean, literally, that your sense of touch was impaired.

balance/remainder

One meaning of *balance* is "a state of equilibrium"; another meaning is "the amount remaining in a bank account after balancing deposits and withdrawals." *Remainder,* in all applications, is "what is left over." *Remainder* is the more accurate word, therefore, to mean "that which is left over."

EXAMPLES The accounting department must attempt to maintain a *balance* between looking after the company's best financial interests and being sensitive to the company's research and development work.

The *balance* in the corporate account after the payroll has been met is a matter for concern.

Round the fraction off to its nearest whole number and drop the *remainder.*

When using these words figuratively (outside of banking and mathematical contexts), use *remainder* to mean "what is left over."

CHANGE Four of the speakers at the conference were from West Germany, and the *balance* were from the United States.

TO Four of the speakers at the conference were from West Germany, and the *remainder* were from the United States.

be sure and/be sure to

The **phrase** *be sure and* is colloquial and unidiomatic when used for *be sure to*. (See also **idioms.**)

CHANGE   When the claxon sounds, *be sure and* phone the guard desk immediately.

TO   When the claxon sounds, *be sure to* phone the guard desk immediately.

because

To express cause, *because* is the strongest and most specific **connective** (others are *for, since, as, inasmuch as, insofar as*). *Because* is unequivocal in stating causal relationship.

EXAMPLE   We didn't complete the project *because* the raw materials became too costly. (The use of *because* emphasizes the cause/effect relationship.)

For can express causal relationships, but it is weaker than *because,* and it allows the **clause** that follows to be a separate **independent clause.**

EXAMPLE   We didn't complete the project, *for* raw materials became too costly. (The use of *for* expresses but does not emphasize the cause/effect relationship.)

As a connective to express cause, *since* is also a weak substitute for *because.*

EXAMPLE   *Since* the computer is broken, paychecks will be delayed.

However, *since* is an appropriate connective when the **emphasis** is on circumstances or conditions rather than on cause and effect.

EXAMPLE   *Since* I was in town anyway, I decided to visit the test site.

As is the least definite connective to indicate cause; its use for this purpose is better avoided.

CHANGE   I left the office early, *as* I had finished my work.

TO   I left the office early *because* I had finished my work.

OR   *Since* I had finished my work, I left the office early.

Inasmuch as implies concession; that is, it suggests that a statement is true "in view of the circumstances."

EXAMPLE   You may as well complete the project, *inasmuch as* you have already begun it.

being as/being that

These **phrases** are **nonstandard English** and should not be used in writing. Use *because* or *since*.

beside/besides

Besides, meaning "in addition to" or "other than," should be carefully distinguished from *beside,* meaning "next to" or "apart from."

> EXAMPLE   *Besides* the two of us from the Systems Department, three people from Production were standing *beside* the president when he presented the award.

between/among

Between is normally used to relate two items or persons.

> EXAMPLES   The roll pin is located *between* the grommet and the knob.
> Preferred stock offers a buyer a middle ground *between* bonds and common stock.

Among is used to relate more than two.

> EXAMPLE   The subcontracting was distributed *among* the three firms.

between you and me

People sometimes use the incorrect expression *between you and I.* Because the **pronouns** are **objects** of the **preposition** *between,* the objective form of the **personal pronoun** *(me)* must be used. (See also **case, grammar.**)

> CHANGE   *Between you and I,* John should be taken off the job.
> TO   *Between you and me,* John should be taken off the job.

bi/semi

When used with periods of time, *bi* means "two" or "every two." *Bimonthly* means "once in two months"; *biweekly* means "once in two weeks."

When used with periods of time, *semi* means "half of" or "occurring twice within a period of time." Semimonthly means "twice a month"; semiweekly means "twice a week."

Both *bi* and *semi* are normally joined with the following element without space or **hyphen.**

biannual/biennial

By conventional usage, *biannual* means "twice during the year," and *biennial* means "every other year." (See also **bi/semi.**)

bibliography

A bibliography is an alphabetically arranged list of books, articles, and other source materials consulted in the preparation of a **report** or research paper. Its purpose is to present the sources of your information in a convenient, standardized form helpful to someone seeking additional information on the subject. Bibliographies are usually placed at the end of the report.

A bibliography that lists only those works referred to directly in the text or notes of a research paper is sometimes labeled "works cited." An annotated bibliography is one that includes a brief description of some or all of the works cited. In scientific and technical writing, works cited are called **references.**

When typing a bibliography, single-space within an entry and double-space between entries. Indent the second and subsequent lines of an entry three spaces so that the author's name stands out.

Details vary from one field to another, but all bibliographic entries should contain (1) the author's name, (2) the full title, and (3) the publication facts. Bibliographic forms in this section are based on the University of Chicago Press' *A Manual of Style* (12th ed.).

Note, however, that the bibliography should be consistent in format with the references if the written work includes both. For a sample bibliography that is consistent with the references format recommended in this book, see "A Selected List of Style Manuals and Guides" at the end of the references entry.

AUTHOR'S NAME

List the author's name, last name first. If there is more than one author, give the name of the first author listed on the title page, last name first *(Doe, John Q.)* in alphabetical sequence with your other bibliography entries, and then list the other authors' names in normal order *(Jane R. Roe).* Separate the names from each other by **commas,** and separate the authors' names from the work's title with a **period.**

Alphabetize all works by the same author according to their titles.

When listing multiple works by the same author, replace the author's name in all entries after the first with a line followed by a period.

TITLE OF WORK

Capitalize the key words in the title; italicize (or underline) the complete title or enclose in **quotation marks** the complete title of an article; and conclude the title with a period.

PUBLICATION FACTS

Include all pertinent facts about the publication of the work, including the publisher as well as the date and city of publication.

> EXAMPLES   Graham, Frank, Jr. *Disaster by Default: Politics and Water Pollution.* New York: M. Evans, 1966.
> _____. *Since Silent Spring.* Boston: Houghton Mifflin, 1970.

BOOKS

For a book include the pertinent facts in this order: (1) the name of the author (or authors), editor, or institution responsible for the work cited, followed by a period; (2) the title of the work, underlined and followed by a period; (3) the title of the series, if applicable, followed by a comma; (4) the volume number, if applicable, followed by a period; (5) the edition, if not the first edition, followed by a period; (6) the number of volumes, followed by a period; (7) the city where the book was published, followed by a **colon;** (8) the publisher's name, followed by a comma; and (9) the year of publication, followed by a period to end the entry.

JOURNAL ARTICLES

For a **journal article,** include the following facts in this order: (1) the name of the author (or authors), followed by a period; (2) the title of the article, in quotation marks and with a period; (3) the name of the periodical in which the article was published (underlined but with no other **punctuation**); (4) the volume number of the periodical; (5) the date of publication (set within **parentheses** and followed by a colon); and (6) the page numbers of the article within the periodical, followed by a period.

EXAMPLE BIBLIOGRAPHIC ENTRIES

BOOK, ONE AUTHOR

Hanley, Wayne. <u>Natural History in America</u>. New York: Quadrangle
    Press/New York Times Books, 1977.

64 **bibliography**

BOOK, TWO OR MORE AUTHORS

Jelley, Herbert M., and Robert O. Herrmann. The American
Consumer: Issues and Decisions. New York: McGraw-Hill, 1973.

Young, Hugo, Bryon Silcock, and Peter Dunn. Journey to
Tranquility. Garden City, N.Y.: Doubleday, 1970.

BOOK EDITION, IF NOT THE FIRST EDITION

Silverman, S.R., and Hollowell Davis. Hearing and Deafness,
2nd ed. New York: Holt, Rinehart and Winston, 1978.

MULTIVOLUME BOOK

Bartholemew, John, ed. Times Atlas of the World. Midcentury
Edition. 5 vols. London: Times Publishing Co., Ltd.,
1955-1959.

Give the full number of volumes in the set following either
the book title or the book's edition.

BOOK IN A SERIES

Cooper, E.L., ed. Invertebrate Immunology. Contemporary Topics
in Immunology, vol. 4. New York: Plenum Publishing
Corporation, 1974.

Give the series title and the volume number following the
title of the book.

EDITOR OF A COLLECTION

Fowles, Jib, ed. Handbook of Futures Research. Westport, Conn.:
Greenwood Press, 1978.

CORPORATE AUTHOR

U.S. Comptroller General. Waste Disposal Practices -- A Threat
to Health and the Nation's Water Supply, Report to Congress.
Washington, D.C.: U.S. General Accounting Office, 1978.

JOURNAL ARTICLE

Thomas, Willard. "Interrelated Video Factors and a Basic Model for System Selection." Technical Communication 27 (1980): 16-18.

SYMPOSIUM OR CONFERENCE PAPER IN A PROCEEDING

Ballard, T. M. "Role of Soil Reconnaissance Surveys in Predicting Use-Impact Susceptibility of Wildlands." In Conference Proceedings on the Recreational Impact on Wildlands ed. R. Ittner, D. R. Potter, J. K. Agee, and A. Anschell, pp. 45-49. Seattle, Washington: U.S. Forest Service, Pacific Northwest Region, 1979.

MAGAZINE ARTICLE

"The Assault on the Federal Trade Commission." Consumer Reports, March 1980, pp. 149-152.

NEWSPAPER ARTICLE

Allan, John H. "Fixed-Income Securities Ease." The New York Times, Late City Ed., 15 June 1979, Sec. D., p. 70.

ENCYCLOPEDIA ARTICLE

"Electricity." Encyclopaedia Britannica. 14th ed.
    Because encyclopedia articles are arranged alphabetically, volume and page numbers are unnecessary. If the article is signed, begin the entry with the author's last name.

REPORT

Office of Noise Abatement and Control. Potential Effectiveness of Barriers Toward Reducing Highway Noise Exposure on a National Scale. Washington, D.C.: U.S. Environmental Protection Agency, 1978.

PAMPHLET OR BOOKLET

Hursh, Laurence M. <u>Coronary Heart Disease: Risk Factors and
the Diet Debate</u>. Rosemont, Ill.: National Dairy Council,
1980.

THESIS OR DISSERTATION

Johnson, M.G. "Infiltration Capacities and Surface Erodibility
Associated with Forest Harvesting Activities in the Oregon
Cascades," Master's Thesis, Oregon State University, 1978.

PERSONAL CORRESPONDENCE

Brady, Robert T. Salvo Corp., July 12, 1980. Letter to author.

INTERVIEW

Denlinger, Virgil, Assistant Chief of Police. Milwaukee,
Wisconsin: March 15, 1980. Interview.

blend words

A blend word is formed by combining part of one word with part of another.

> EXAMPLES motor + hotel = motel
> breakfast + lunch = brunch
> smoke + fog = smog
> electric + execute = electrocute
> chuckle + snort = chortle

Although blend words (sometimes called "portmanteau words") may occasionally be created by a specialist to meet a specific need—such as *stagflation* (a stagnant economy coupled with inflation)—resist creating blend words in your writing unless an obvious need arises. If you must create a blend word, be sure to define it clearly for your **reader**. Otherwise, creating blend words is at best merely "cute" and could be confusing. (See also **new words**.)

both . . . and

Statements using the *both . . . and* construction should always be balanced both grammatically and logically.

EXAMPLE   A successful photograph must be *both* clearly focused *and* adequately lighted.

Notice that *both* and *and* are followed logically by ideas of equal weight and grammatically by identical constructions.

CHANGE   For success in engineering, it is necessary both *to develop* writing skills and *mastering* calculus.

TO   For success in engineering, it is necessary both *to develop* writing skills and *to master* calculus.

Do not substitute *as well as* for *and* in this construction.

CHANGE   For success in engineering, it is necessary *both* to master calculus *as well as* to develop writing skills.

TO   For success in engineering, it is necessary *both* to master calculus *and* to develop writing skills.

(See also **parallel structure** and **correlative conjunctions.**)

brackets

The primary use of brackets is to enclose a word or words inserted by an editor or writer into a **quotation** from another source.

EXAMPLE   The text stated, "Fissile and fertile nuclei spontaneously emit characteristic nuclear radiations [such as neutrons and gamma rays] that are sufficiently energetic to penetrate the container or cladding."

Brackets are also used to set off a parenthetical item within **parentheses.**

EXAMPLE   We should be sure to give Emanuel Foose (and his brother Emilio [1812-1882] as well) credit for his role in founding the institute.

Brackets are also used in academic writing to insert the Latin word *sic*, which indicates that the writer has quoted material exactly as it appears in the original, even though it contains an obvious error.

EXAMPLE   Dr. Smith pointed out that "the earth does not revolve around the son [sic] at a constant rate."

bunch

Bunch refers to like things that grow or are fastened together. Do not use the word *bunch* to refer to people.

CHANGE   A *bunch* of trainees toured the site.
TO   A *group* of trainees toured the site.

C

can/may

In writing, *can* refers to capability, and *may* refers to possibility or permission.

EXAMPLES   I *can* have the project finished by January 1. (capability)
I *may* be in Boston on Thursday. (possibility)
I *can* be in Boston on Thursday. (capability)
May I have an extra week to finish the project? (permission)

cannot/can not

Cannot is one word.

CHANGE   We *can not* meet the deadline specified in the contract.
TO   We *cannot* meet the deadline specified in the contract.

cannot help but

Avoid the **phrase** *cannot help but* in writing. (See also **double negatives.**)

CHANGE   We *cannot help but* cut our staff.
TO   We cannot avoid cutting our staff.

canvas/canvass

Canvas is a **noun** meaning "heavy, coarse, closely woven cotton or hemp fabric." *Canvass* is a **verb** for the act of soliciting votes or opinions.

EXAMPLES   The maintenance crew spread the *canvas* over the equipment.
The executive committee decided to *canvass* the employees.

capital/capitol

Capital may refer either to financial assets or to the city that hosts the government of a state or a nation. *Capitol* refers to the building in which the state or national legislature meets. *Capitol* is often written with a small *c* when it refers to a state building, but it is always capitalized when it refers to the home of the United States Congress in Washington, D.C.

capital letters

The use of capital letters (or upper-case letters) is determined by custom and tradition. Capital letters are used to call attention to certain words, such as **proper nouns** and the first word of a **sentence.** Care must be exercised in using capital letters because they can affect the meaning of words (march/March, china/China, turkey/Turkey). For the same reason, however, capital letters can help eliminate **ambiguity.**

FIRST WORDS

The first letter of the first word in a sentence is always capitalized.

EXAMPLE   Of all the plans you mentioned, the first one seems the best.

The first word after a **colon** may be capitalized if the statement following is a complete sentence or if it introduces a formal resolution or question.

EXAMPLE   Today's meeting will deal with only one issue: What is the firm's role in environmental protection?

If a subordinate element follows the colon, however, or if the thought is closely related, use a lower-case letter following the colon.

EXAMPLE   We had to keep working for one reason: our deadline was upon us.

The first word of a complete sentence in **quotation marks** is capitalized.

EXAMPLE   Dr. Vesely stated, "It is possible to postulate an imaginary world in which no decisions are made until all the relevant information is assembled."

Complete sentences contained as numbered items within a sentence may also be capitalized.

EXAMPLE   To make correct decisions, you must do three things: (1) Identify the information that would be pertinent to the decision anticipated, (2) Establish a systematic program for acquiring this pertinent information, and (3) Rationally assess the information so acquired.

The first word in the salutation and complimentary close of a letter is capitalized. (See also **correpondence.**)

EXAMPLES   Dear Mr. Smith:
Sincerely yours,
Best regards,

SPECIFIC PEOPLE AND GROUPS

Capitalize all personal names.

EXAMPLES   Walter Bunch, Mary Fortunato, Bill Krebs

Capitalize names of ethnic groups and nationalities.

EXAMPLES   American Indian, Italian, Jew, Chicano
Thus Italian immigrants contributed much to the industrialization of the United States.

Do not capitalize names of social and economic groups.

EXAMPLES   middle class, working class, ghetto dwellers

SPECIFIC PLACES

Capitalize the names of all political divisions.

EXAMPLES   Chicago, Cook County, Illinois, Ontario, Iran, Ward Six.

Capitalize the names of geographical divisions.

EXAMPLES   Europe, Asia, North America, the Middle East, the Orient.

Do not capitalize geographic features unless they are part of a proper name.

EXAMPLE   The mountains in this area make television transmission difficult, which is also true in the Great Smoky Mountains.

The words *north, south, east,* and *west* are capitalized when they refer to sections of the country. They are not capitalized when they refer to directions.

EXAMPLES   I may travel south when I relocate to Delaware.
We may build a new plant in the South next year.
State Street runs east and west.

Capitalize the names of stars, constellations, and planets.

EXAMPLES   Saturn, Andromeda, Jupiter, Milky Way

Do not capitalize *earth, sun,* and *moon,* however, except when they are used with the names of other planets.

EXAMPLES   Although the sun rises in the east and sets in the west, the moon may appear in any part of the evening sky when darkness settles over the earth.
Mars, Pluto, and Earth were discussed at the symposium.

SPECIFIC INSTITUTIONS, EVENTS, AND CONCEPTS

Capitalize the names of institutions, organizations, and associations.

> EXAMPLE   The American Society of Mechanical Engineers and the Department of Housing and Urban Development are cooperating in the project.

An organization usually capitalizes the names of its internal divisions and departments.

> EXAMPLES   Faculty, Board of Directors, Engineering Department

Types of organizations are not capitalized unless they are part of an official name.

> EXAMPLES   Our group decided to form a writers' association; we called it the American Association of Writers.
> I attended Post High School. What high school did you attend?

Capitalize historical events.

> EXAMPLE   Dr. Jellison discussed the Boston Tea Party at the last class.

Capitalize words that designate specific periods of time.

> EXAMPLES   Labor Day, the Renaissance, the Enlightenment, January, Monday, the Great Depression, Lent

Do not, however, capitalize seasons of the year.

> EXAMPLES   spring, autumn, winter, summer

Capitalize scientific names of classes, families, and orders, but do not capitalize species or English derivatives of scientific names.

> EXAMPLES   Mammalia, Carnivora/mammal, carnivorous

TITLES OF BOOKS, ARTICLES, PLAYS, AND FILMS

Capitalize the initial letters of the first and last words of a title of a book, article, play, or film, as well as all major words in the title. Do not capitalize **articles** *(a, an, the)*, **conjunctions** *(and, but, if)*, or short **prepositions** *(at, in, on, of)* unless they begin the title. Capitalize prepositions that contain more than four letters *(between, because, until, after)*.

> EXAMPLES   The microbiologist greatly admired the book *The Lives of a Cell*.
> Her favorite article is still "On the Universe Around Us."
> The book *Year After Year* recounts the life story of a great scientist.

Some reference systems for scientific and technical publications do not follow these guidelines. See **references.**

PERSONAL TITLES

Titles preceding proper names are capitalized.

EXAMPLES   Miss March, Professor Galbraith, Senator Church

Appositives following proper names are not normally capitalized. (The word *President* is usually capitalized when it refers to the chief executive of a national government.)

EXAMPLE   Frank Jones, senator from New Mexico (but Senator Jones)

The only exception is an epithet, which actually renames the person.

EXAMPLES   Alexander the Great, Solomon the Wise

Use capital letters to designate family relationships only when they occur before a name or substitute for a name.

EXAMPLES   One of my favorite people is Uncle Fred.
Jim and Mother went along.
Jim and my mother went along.

ABBREVIATIONS

Capitalize **abbreviations** if the words they stand for would be capitalized.

EXAMPLES   OSU (Ohio State University)
p. (page)
Ph.D. (Doctor of Philosophy)

LETTERS

Certain single letters are always capitalized. Capitalize the **pronoun** *I* and the **interjection** *O* (but do not capitalize *oh* unless it is the first word in a sentence).

EXAMPLES   When I say writing, O believe me, I mean rewriting.
When I say writing, oh believe me, I mean rewriting.

Capitalize letters that serve as names or indicate shapes.

EXAMPLES   X-ray, vitamin B, T-square, U-turn, I-beam

MISCELLANEOUS CAPITALIZATIONS

The word *Bible* is capitalized when it refers to the Christian Scriptures; otherwise, it is not capitalized.

EXAMPLE He quoted a verse from the Bible, then read from Blackstone, the lawyer's bible.

All references to deities (Allah, God, Jehovah, Yahweh) are capitalized.

EXAMPLE God is the One who sustains us.

A complete sentence enclosed in **dashes, brackets,** or **parentheses** is not capitalized when it appears as part of another sentence.

EXAMPLES We must make an extra effort in safety this year (accidents last year were up 10%).
Extra effort in safety should be made this year. (Accidents were up 10%.)

Certain units, such as parts and chapters of books, rooms in buildings, etc., when specifically identified by number, are normally capitalized.

EXAMPLES Chapter 5, Ch. 5, Room 72, Rm. 72

Minor divisions within such units are not capitalized unless they begin a sentence.

EXAMPLES page 11, verse 14, seat 12

When in doubt about whether or not to capitalize, check a **dictionary.**

case, grammar

Grammatically, *case* indicates the functional relationship of a **noun** or a **pronoun** to the other words in a **sentence.** Nouns change form only in the possessive case; pronouns may show change for the subjective, the objective, or the possessive case. The case of a noun or pronoun is always determined by its function in its **phrase, clause,** or sentence. If it is the **subject** of its phrase, clause, or sentence, it is in the subjective case; if it is an **object** within its phrase, clause, or sentence, it is in the objective case; if it reflects possession or ownership and modifies a noun, it is in the possessive case. The subjective case indicates the person or thing acting (*He* sued the vendor); the objective case indicates the thing acted upon (The vendor sued *him*); and the possessive case indicates the person or thing owning or possessing something (It was *his* company).

The different forms of a noun or pronoun indicate whether it is functioning as a subject (subjective case), as a **complement** (usually objective case), or as a **modifier** (possessive case).

Subjective Case	Objective Case	Possessive Case
I	me	my, mine
we	us	our, ours
he	him	his, his
she	her	her, hers
they	them	their, theirs
you	you	your, yours
who	whom	whose

SUBJECTIVE CASE

A pronoun is in the subjective case (also called nominative case) when it represents the person or thing acting.

EXAMPLE   *I* wrote a letter to that company before I graduated.

A **linking verb** links a pronoun to its antecedent to show that they identify the same thing. Because they represent the same thing, the pronoun is in the subjective case even when it follows the **verb,** which makes it a **subjective complement.**

EXAMPLES   *He* is the head of the Quality Control Group. (subject)
The head of the Quality Control Group is *he*. (subjective complement)

Whether a pronoun is a subject or a subjective complement, it is in the subjective case. This situation causes much confusion with one sentence in particular—*It is I* versus *It is me*. The sentence consists of a subject *(it)*, a linking verb *(is)*, and a **personal pronoun** (*I* or *me*). Although "It is me" appears to fit the normal subject-verb-object pattern of English sentences, it really does not. The pronoun actually stands for the subject and is therefore a subjective complement—hence "it is I" is the grammatically correct way to express the thought. However, we are so accustomed to the subject-verb-object pattern of English sentences that we instinctively express the objective form of the pronoun *(me)*. For this reason, "It is me" is more natural-sounding in speech and is therefore also appropriate to the type of writing in which the expression is likely to appear. (See **complements.**)

The subjective case is used after the words *than* and *as* because of the understood (although unstated) portion of the clauses in which these words appear.

EXAMPLES   George is as good a designer as *I* [am].
Our subsidiary can do the job better than *we* [can].

OBJECTIVE CASE

A pronoun is in the objective case when it indicates the person or thing receiving the action expressed by the verb. (Objective case is also called accusative case.)

EXAMPLE   They informed *me* by letter that they had received my résumé.

A pronoun is in the objective case when it is the object of a verb, **gerund,** or **preposition,** and when it is the subject of an **infinitive.** Pronouns that follow prepositions must be in the objective case.

EXAMPLES   Between you and *me*, his facts are questionable.
Many of *us* attended the conference.

Pronouns that follow action verbs (which excludes all forms of the verb *be*) must be in the objective case. Don't be confused by an additional name.

EXAMPLES   The company promoted *me* in June.
The company promoted John and *me* in June.

Pronouns that follow gerunds must be in the objective case.

EXAMPLE   Training *him* was the best thing I could have done.

Subjects of infinitives must be in the objective case.

EXAMPLE   We asked *them* to recalibrate the instruments.

For determining the case of an object, English does not differentiate between **direct objects** and **indirect objects;** both require the objective form of the pronoun. (See also **complements.**)

EXAMPLES   The interviewer seemed to like *me*. (direct object)
They wrote *me* a letter. (indirect object)

POSSESSIVE CASE

A noun or a pronoun is in the possessive case when it represents a person or thing owning or possessing something.

EXAMPLE   Dr. Peterson's risk calculations appear in Appendix A of *his* report.

Nouns. Although exceptions are relatively common, it is a good rule of thumb to use the *'s* form of the possessive case with nouns referring to persons and living things, and to use an *of* phrase for the possessive case of nouns referring to inanimate objects.

EXAMPLES   The *chairman's* address was well received.
The leaves *of the tree* look healthy.

If this rule leads to awkwardness or wordiness, however, be flexible.

EXAMPLES   The *company's* pilot plants are doing well.
The *plane's* landing gear failed.

Established **idiom** calls for the possessive case in many stock phrases.

EXAMPLES   A day's journey, a day's work, a moment's notice, at his wit's end, the law's delay

In a few cases, idiom even calls for a double possessive employing both the *of* and *'s* forms.

EXAMPLE   That colleague *of* George*'s* was at the conference.

Plural words ending in *s* need only add an **apostrophe** to form the possessive case.

EXAMPLE   the *laborers'* union

When several words compose a single term, add the *'s* to the last word only.

EXAMPLES   The *Chairman of the Board's* statement was brief.
The *Department of Energy's* fiscal 19— budget shows increased revenues of $192 million for uranium enrichment.

To show individual possession with coordinate nouns, make both nouns possessive.

EXAMPLE   The *Senate's and House's* chambers were packed.

To show joint possession with coordinate nouns, make only the last possessive.

EXAMPLE   The *Senate and House's* joint declaration was read to the press.

Pronouns. The use of possessive pronouns does not normally cause problems except with gerunds and **indefinite pronouns.** Several indefinite pronouns (*all, any, each, few, most, none,* and *some*) require *of* phrases to form the possessive case.

> EXAMPLE   Both dies were stored in the warehouse, but rust had ruined the surface *of each.*

Others, however, use the apostrophe.

> EXAMPLE   *Anyone's* contribution is welcome.

Only the possessive form of a pronoun should be used with a gerund.

> EXAMPLES   The safety officer insisted on *my* wearing a respirator.
> *Our* monitoring was not affected by changing weather conditions.

Pronouns in compound constructions should be in the same case.

> EXAMPLES   This is just between *them* and *us.*
> Both *they* and *we* must agree to the arrangement.

APPOSITIVES

An **appositive** should be in the same case as the noun with which it is in apposition.

> EXAMPLES   Two design engineers, Jim Knight and *I,* were asked to review the drawings. (subjective case)
> The group leader selected two members to represent the department—Jim Knight and *me.* (objective case)
> We all came—Jim and Carol and *I.* (subjective case)
> He gave us both, Rod and *me,* a week to make up our minds. (objective case)

TIPS ON DETERMINING THE CASE OF PRONOUNS

One test to determine the proper case of a pronoun is to try it with some **transitive verb** such as *resembled* or *hit.* If the pronoun would logically precede the verb, use the subjective case; if it would logically follow the verb, use the objective case.

> EXAMPLES   *She* (he, they) resembled her father. (subjective case)
> Angela resembled *him* (her, them). (objective case)

In the following type of sentence, try omitting the noun to determine the case of the pronoun.

EXAMPLES   *(We/Us)* pilots fly our own airplanes.

Us [pilots] fly our own airplanes. (This incorrect usage is obviously wrong.)

We [pilots] fly our own airplanes. (This correct usage sounds right.)

To determine the case of a pronoun that follows *as* or *than,* try mentally adding the words that are normally omitted.

EXAMPLES   The other operator is not paid as well *as she* [is paid]. (You would not write, "*Her* is paid.")

His partner was better informed that *he* [was informed]. (You would not write, "*Him* was informed.")

If compound pronouns cause problems, try using them singly to determine the proper case.

EXAMPLES   *(We/Us)* and the Johnsons are going to the Grand Canyon.

We are going to the Grand Canyon. (You would not write, "*Us* are going to the Grand Canyon.")

WHO/WHOM

Who and *whom* cause much trouble in determining case. *Who* is the subjective case form, whereas *whom* is the objective case form. When in doubt about which form to use, try substituting a personal pronoun to see which one fits. If *he* or *they* fits, use *who.*

EXAMPLES   *Who* is the congressman from the 45th district?

He is the congressman from the 45th district.

If *him* or *them* fits, use *whom.*

EXAMPLES   It depended upon *them.*

It depended upon *whom?*

It was they upon *whom* it depended.

It is becoming common to use *who* for the objective case when it begins a clause or sentence, although some writers still object to such an "ungrammatical" construction.

EXAMPLE   *Who* should I call to report a fire?

The best advice is to know your **reader.**

case, usage

The word *case* is often merely filler. Be critical of the word, and eliminate it if it contributes nothing.

> CHANGE　An exception was made in the *case* of those connected with the project.
>
> TO　An exception was made for those connected with the project.

cause-and-effect method of development

When your purpose is to explain why something happened, or why you think something will happen, the cause-and-effect method of development is a useful writing strategy.

The goal of the cause-and-effect method of development is to make the relationship between a situation and either its cause or its effect as plausible as possible. The conclusions you draw about the relationships should be based on the evidence you have gathered. Because not all evidence will be of equal value to you, keep some guidelines in mind for evaluating evidence.

EVALUATING EVIDENCE

The facts and arguments you gather should be pertinent to your **topic.** Be careful not to draw a conclusion from your evidence that it does not lead to or support. You may have researched some statistics, for example, which show that an increasing number of Americans are licensed to fly small planes. But you cannot use this information as evidence that there is a slowdown in interstate highway construction in the United States—the evidence does not lead to that conclusion. Statistics on the increase in small-plane licensing may be relevant to other conclusions, however. You could argue that the upswing has occurred because small planes save travel time, provide easy access to remote areas, and, once they are purchased, are economical to operate.

Your evidence should be adequate. Incomplete evidence can lead to false conclusions.

> EXAMPLE　Driver training classes do not help prevent auto accidents. Two people I know who completed driver training classes were involved in accidents.

Although the evidence cited to support the conclusion may be accurate, there is not enough of it. A thorough investigation of the usefulness of driver training classes in keeping the accident rate down would require many more than two examples. And it would require a comparison of the driving records of those who had completed driver training with drivers who had not.

Your evidence should be representative. If you conduct a survey to obtain your evidence, be sure that you do not solicit responses only from individuals or groups whose views are identical to yours; that is, be sure you obtain a representative sampling.

Your evidence should also be plausible. Two events that occur close to each other in time or place may or may not be causally related. Thunder and black clouds do not always signal rain, but they do so often enough that if we are outdoors and the sky darkens and we hear thunder, we seek shelter. If you sprain your ankle after walking under a ladder, however, you cannot conclude that a ladder brings bad luck. Merely to say that *X* caused *Y* (or will cause *Y*) is inadequate. You must demonstrate the relationship with pertinent facts and arguments.

LINKING CAUSES TO EFFECTS

To show a true relationship between a cause and an effect, you must demonstrate that the existence of the one *requires* the existence of the other. It is often difficult to establish beyond any doubt that one event was *the* cause of another event. More often, a result will have more than one cause. As you research your subject, your task is to determine which cause or causes are most plausible.

When several probable causes are equally valid, report your findings accordingly, as in the following **paragraph** on the use of an energy-saving device called a furnace-vent damper. The damper is a metal plate fitted inside the flue or vent pipe of natural-gas or fuel-oil furnaces. When the furnace is on, the damper opens to allow the gases to escape up the flue. When the furnace shuts off, the damper closes, thus preventing warm air from escaping up the flue stack. The dampers are potentially dangerous, however. If the dampers fail to open at the proper time, they could allow poisonous furnace gases to back into the house and asphyxiate anyone in a matter of minutes. Tests run on several dampers showed a number of probable causes for their malfunctioning.

EXAMPLE   One damper was sold without proper installation instructions, and another was wired incorrectly. Two of the units had slow-opening dampers (15 seconds) that prevented the [furnace] burner from firing. And one damper jammed when exposed to a simulated fuel temperature of more than 700 degrees.

—Don De Bat, "Save Energy But Save Your Life, Too," *Family Safety* (Fall 1978), 27.

The investigator located all the causes of damper malfunctions and reported on them. Without such a thorough account, recommendations to prevent similar malfunctions would be based on incomplete evidence.

By substituting *problem* for "cause" and *solution* for "effect," you can use the same approach to develop a **report** dealing with a solution to a problem.

The following report is developed from effect to cause.

MEMORANDUM

```
To:   James K. Arburg, Safety Officer
From: Lawrence T. Baker, Foreman of Section A-40
Date: November 30, 19--

Subject:  Personal-Injury Accident in Section A-40
          October 10, 19--

On October 10, 19--, at 10:15 P.M., Jim
Hollander, operating punch press #16, accidentally
brushed the knee switch of his punch press with his
right knee as he swung a metal sheet over the
punching surface.  The switch activated the
punching unit, which severed Hollander's left thumb
between the first and second joints as his hand
passed through the punch station.  While an
ambulance was being summoned, Margaret Wilson, R.N.,
administered first aid at the plant dispensary.
There were no witnesses to the accident.
     The ambulance arrived from Mercy Hospital at
10:45 P.M., and Hollander was admitted to the
emergency room at the hospital at 11:00 P.M.  He
was treated and kept overnight for observation, then
released the next morning.
     Hollander returned to work one week later, on
October 17.  He has been given temporary duties in
the tool room until his injury heals.
```

Conclusions About the Cause of the Accident

```
The Maxwell punch press on which Hollander was
working has two switches, a hand switch and a knee
switch, and both must be pressed to activate the
punch mechanism.  The hand switch must be pressed
first, and then the knee switch, to trip the punch
mechanism.  The purpose of the knee switch is to
leave the operator's hands free to hold the panel
being punched.  The hand switch, in contrast, is a
safety feature.  Because the knee switch cannot
activate the press until the hand switch has been
pressed, the operator cannot trip the punching
mechanism by touching the knee switch accidentally.
```

```
          Inspection of the punch press that Hollander
   was operating at the time of the accident made it
   clear that Hollander had taped the hand switch of
   his machine in the ON position, effectively
   eliminating its safety function.  He could then
   pick up a panel, swing it onto the machine's
   punching surface, press the knee switch, stack the
   newly punched panel, and grab the next unpunched
   panel, all in one continuous motion--eliminating
   the need to let go of the panel, after placing it
   on the punching surface, in order to press the hand
   switch.
          To prevent a recurrence of this accident, I
   have conducted a brief safety session with all punch
   press operators, at which I described Hollander's
   experience and cautioned them against tampering with
   the safety features of their machines.
```

center around

Be careful always to substitute *on* or *in* for *around* in this redundant and illogical expression.

CHANGE   The experiments *center around* the new discovery.

TO   The experiments *center on* the new discovery.

Usually the idea intended by *center around* is best expressed by *revolved around*.

EXAMPLE   The subcommittee hearings on mandatory fire sprinklers *revolved around* occupant safety.

chairman/chairwoman/chairperson

Although *chairwoman* is acceptable and *chairperson* (or *chair*) is common, in business and industry *chairman* is still widely used as a title for a presiding officer of either sex. Know your reading audience, however, because many people have become sensitive to any word that may imply sexual bias.

EXAMPLES   Mary Roberts preceded John Stevens as *chairman* of the executive committee.

Mary Roberts was *chairwoman* of the executive committee before John Stevens became *chairman*.

Mary Roberts preceded John Stevens as *chairperson* of the executive committee.

character

Character, used in the sense of "nature" or "quality," is often an unnecessary and inexact filler that should be omitted from your writing. Choose instead a word that conveys your meaning more specifically.

CHANGE   The modifications changed the whole *character* of the engine.
TO   The modifications changed the performance of the engine.

chronological method of development

The chronological method of development arranges the events under discussion in sequential order, beginning with the first event and continuing chronologically to the last event. **Trip reports,** work schedules, some **minutes of meetings,** and certain **trouble reports** are among the types of writing in which information is organized chronologically.

In the following **report** a fire fighter describes a fire that took place at a lumber mill. After providing important background information, the writer presents the events as they occurred chronologically.

Woodworking Plant Fire*

| Exposed Building Destroyed | July 24, 19— |
| Notification Delayed | Burney, California |

Setting   Wood bark, sawdust, and wood chips were stored in three piles about 100 feet south of one building at this lumber mill and 150 feet west of a second building. The second building, called the "panel plant," consisted of one story and a partial attic and was used in part as an electric shop, and in part for the storage of finished lumber. About six pallet loads of Class I flammable liquids in 55-gallon drums were stored in the western section. The building contained a sprinkler system.

Cause of fire   Fire, caused by spontaneous ignition of the piled bark, spread to sawdust and chip piles, then to the chip-loading facilities, and finally to the panel plant. A 40-mph wind was blowing in the direction of the panel plant.

First fire noticed   A watchman first noticed the fire in the bark pile about 6 a.m. He notified the plant superintendent, who arrived more than an hour later, hosed down the smoldering bark pile, and set up several irrigation sprinklers to wet the area.

At about 1:20 p.m., smoke was seen at the farther end of the bark pile. The hose was not long enough to reach this area and

*"Bimonthly Fire Record," *Fire Journal* 72 (March 1977), 24.

Fire department
called

the local fire department was called, nearly eight hours after the fire was originally discovered.

The fire burned up into the hollow joisted roof of the panel plant. The sprinklers were on a dry system and, from accounts of witnesses, it is estimated that the fire pump was not started until after the fire had been burning in the panel plant for 15 to 30 minutes.

Start of pumping
equipment delayed

The plant was a $350,000 loss. . . .

cite/site/sight

Cite means "acknowledge" or "quote an authority"; *site* is the place or plot of land where something is located; *sight* is the ability to see.

EXAMPLES   The speaker *cited* several famous economists to support his prediction about the stock market.

The *site* for the new factory is three miles from the middle of town.

After the accident, his vision was blurred and he feared that he might lose his *sight*.

clarity

No element is more essential to writing than clarity. You should strive to make all of your writing direct, orderly, and precise. Many factors contribute to clarity just as many other elements can defeat it. Logical development, unity, coherence, emphasis, subordination, pace, transition, an established point of view, conciseness, and word choice contribute to clarity. **Ambiguity, awkwardness,** vagueness, poor use of idiom, clichés, and inappropriate level of usage detract from clarity.

It is surely evident that a logical **method of development** and a good outline are essential to clarity. Without a logical method of development and an outline, you may communicate only isolated thoughts to your **reader,** and your **objective** cannot be achieved by a jumble of isolated thoughts. You must use a method of development that puts your thoughts together in a logical, meaningful sequence. Only then will your writing achieve the **unity** and **coherence** so vital to clarity.

Proper **emphasis** and **subordination** are mandatory if you wish to achieve clarity. If you do not use these two complementary techniques wisely, your **clauses** and **sentences** may all appear to be of equal importance. Your reader will be forced to guess which are most important, which are least important, and which fall between the two extremes. At the very least, this will puzzle and annoy your reader; at worst, it will render your writing incoherent.

The **pace** at which you present your ideas is important to clarity because if the pace is not carefully adjusted to both the **topic** and the reader, your writing will appear cluttered and unclear.

Point of view establishes through whose eyes, or from what vantage point, the reader views the subject. A consistent point of view is essential to clarity; if you switch from the first person to the third person in midsentence, you are certain to confuse your reader.

Clear **transition** contributes to clarity by providing the smooth flow that enables the reader to connect your thoughts with one another without conscious effort. This enables the reader to concentrate solely on absorbing your ideas.

That **conciseness** is a requirement of clearly written communication should be evident to anyone who has ever attempted to decipher an insurance policy or legal contract. Although words are our chief means of communication, too many of them can impede communication just as effectively as too many cars on a highway can impede traffic. For the sake of clarity, prune excess verbiage from your writing.

The selection of precise words over **vague words** is **word choice.** Thoughtful choice of the right word advances clarity by defeating ambiguity and awkwardness.

Another important contributor to clarity is a careful and methodical approach to your writing project, including proper application of all the steps of the writing process—**preparation, research, organization, writing the draft,** and **revision.**

clauses

A clause is a syntactical construction, or group of words, that contains a **subject** and a **predicate** and functions as part of a **sentence.** A clause that could stand alone as a **simple sentence** is an **independent clause.**

EXAMPLE   *The scaffolding fell* when the rope broke.

A clause that could not stand alone if the rest of the sentence were deleted is a **dependent clause.**

EXAMPLE   I was at the St. Louis branch *when the decision was made.*

Every subject-predicate word group in a sentence is a clause. Unlike a **phrase,** a clause can make a complete statement because it contains a finite **verb** (as opposed to **verbals**) as well as a subject. Every sentence must contain at least one independent clause, with the obvious exception of minor sentences—**sentence fragments** that are acceptable because the missing part is clearly understood, such as "At last." or "So much for that."

A clause may function as a **noun,** an **adjective,** or an **adverb** in a larger sentence, or it may be modified by one or more other clauses that are subordinate to it.

EXAMPLE   While I was in college, I studied differential equations.

While I was in college is an **adverb clause** modifying the independent clause *I studied differential equations.*

A clause may be connected with the rest of its sentence by a **coordinating conjunction,** a **subordinating conjunction,** a **relative pronoun,** or a **conjunctive adverb.**

EXAMPLES   It was 500 miles to the facility, *so* we made arrangements to fly. (coordinating conjunction)
Mission control will have to be on the alert *because* at launch the space laboratory will contain a highly flammable fuel. (subordinating conjunction)
It was Robert M. Fano *who* designed and developed the earliest "Multiple Access Computer" system at M.I.T. (relative pronoun)
It was dark when we arrived; *nevertheless,* we began the tour of the factory. (conjunctive adverb)

INDEPENDENT CLAUSES

Unlike a dependent clause, which is part of a larger construction, an independent clause is complete in itself; it could stand alone as a separate sentence if taken out of its larger sentence.

EXAMPLE   *We abandoned the project* because the cost was excessive.

DEPENDENT CLAUSES

A dependent, or subordinate, clause is a group of words that has a subject and a verb but must nonetheless depend upon a main clause to complete its meaning. Within the sentence as a whole, a dependent clause can function as a noun, an adjective, or an adverb.

A **noun clause** is a subordinate clause that functions as a noun. It can be a subject, an **object,** or a **complement.**

EXAMPLES   *That the noise level on Eighth Avenue in New York City is as loud as an alarm clock ringing three feet away* is more readily perceived by a sound meter than by a native New Yorker. (subject)
Mr. Yen told me *that you are a key-punch operator.* (direct object)
That is not *what I meant.* (subjective complement)

An **adjective clause** is a subordinate clause that functions as an adjective by modifying a noun or pronoun in the main clause.

EXAMPLES   The laboratory technician *we met yesterday* showed us his equipment, *which included an electron microscope.*
The man *who called earlier* is here.

An **adverb clause** is a subordinate clause used to modify a verb, an adjective, or another adverb in another clause or phrase. As an adverb, a dependent clause may express a relationship of time, cause, result, degree, or contrast.

EXAMPLES   I go fishing *when I need to forget the pressures of my job.* (time)
I left work early *because I felt sick.* (cause)
I worked late *so I could finish the project on time.* (result)
He was hurt so badly *that he limped for a week.* (degree)
The office seemed smaller *than when I last saw it.* (contrast)

ELLIPTICAL CLAUSES

An elliptical clause is a clause that is clearly understood even though one or more words are not stated.

EXAMPLE   *While* [he was] *achieving great success at his research,* Dr. Brandt was backed by an outstanding research team.

This construction works, of course, only if the subject of the clause and the sentence are the same (*he* and *Dr. Brandt*). (See also **dangling modifiers.**)

clichés

A cliché is an expression that has been used for so long that it is no longer fresh (although some clichés were, at one time, fresh **figures of speech**). Because they have been used continuously over a long period of time, clichés come to mind easily. In addition to being stale, clichés are usually wordy and often vague. Each of the following clichés is followed by better, more direct words, or expressions.

EXAMPLES   quick as a flash / quickly, in five minutes
straight from the shoulder / frank
last but not least / last, finally
as plain as day / clear, obvious
abreast of the times / up to date, current

Clichés are often used in an attempt to make writing elegant or impressive (see **affectation**). Because they are wordy and vague, however, they slow communication and can even irritate your reader. So, although clichés come to mind easily while you are **writing the draft,** they normally should be eliminated during the **revision** phase of the writing process.

On rare occasions, certain clichés, because they are so much a part of the language, may provide (much like **jargon**) a time-saving and efficient means of relating an idea.* For example, "guardedly optimistic" or "inextricably interwoven" may, in appropriate contexts, best express your idea—provided you are certain that the expression is meaningful and acceptable to your reader. Be on guard, however, because clichés can too easily become the pattern of your **word choice.** The best advice is to avoid clichés if *any* other choice of words will work. Consider the following **paragraphs,** first with clichés, then rewritten without them.

CHANGE Our new computer system will have a positive impact on the company *as a whole.* It will keep us *abreast of the times* and make our competition *green with envy.* The committee deserves a *pat on the back* for its *herculean efforts* in convincing management that it was *the thing to do.* I'm sure that their *untiring efforts* will *not go unrewarded.*

TO Our new computer system will have a positive impact throughout the company. It will keep our operations up-to-date and make our competition envious. The committee deserves credit for their efforts in convincing management of the need for the computer. I'm sure that the value of their efforts will be recognized.

clipped forms of words

When the beginning or end of a word is cut off to create a shorter word, the result is called a clipped form.

EXAMPLES dorm, lab, demo, phone

The work *specification,* for example, is often shortened to **spec.** Although acceptable in conversation, most clipped forms should not appear in writing unless they are commonly accepted as part of the special vocabulary of an occupational group.

---

*Charles Sukor, "Clichés: A Re-assessment," *College Composition and Communication,* 26 (May 1975): 159-162.

Apostrophes are not normally used with clipped forms of words (not *'phone,* but *phone*). Since they are not strictly **abbreviations,** clipped forms are not followed by **periods** (not *lab.,* but *lab*).

Do not use clipped forms of **spelling** (*thru, nite,* etc.).

coherence

Writing is coherent when its ideas are logically ordered and consistent: Each idea should relate clearly to the others, with one idea flowing smoothly to the next. Many elements contribute to smooth and coherent writing; however, the major components are (1) a logical sequence of ideas and (2) clear **transitions** between ideas.

A logical sequence of presentation is the most important single requirement in achieving coherence, and the key to achieving the most logical sequence of presentation is the use of a good outline. The outline forces you to establish a beginning **(introduction),** a middle (body), and an end **(conclusion),** and this alone contributes greatly to coherence. The outline also enables you to lay out the most direct route to your **objective**—without digressing into interesting but only loosely related side issues, a habit that inevitably defeats coherence. Drawing up an outline permits you to experiment with different sequences and choose the best one.

Thoughtful transition is also essential to coherence, for without it your writing cannot achieve the smooth flow from **sentence** to sentence and from **paragraph** to paragraph that is required for coherence. Notice the difference between the following two paragraphs; the first has no transition, the second has transition added.

CHANGE   The moon had always been an object of interest to human beings. Until the 1960s, getting there was only a dream. Some thought that we were not meant to go to the moon. In 1969 Neil Armstrong stepped onto the lunar surface. Moon landings became routine to the general public.

TO   The moon had always been an object of interest to human beings, *but* until the 1960s, getting there was only a dream. *In fact,* some thought that we were not meant to go to the moon. *However,* in 1969 Neil Armstrong stepped onto the lunar surface. *After that* moon landings became routine to the general public.

The transitional words and expressions of the second paragraph fit the ideas snugly together, making that paragraph read more smoothly than the first. Attention to transition in longer works is essential if your reader is to move smoothly from point to point in your writing.

Check your draft carefully for coherence during **revision;** if your writing is not coherent, you are not really communicating with your **reader.**

collective nouns

A collective noun names a group or collection of persons, places, things, concepts, actions, or qualities.

EXAMPLES   army, committee, crowd, team, public, class, jury, humanity

When a collective noun refers to a group as a whole, it takes a singular **verb** and **pronoun.**

EXAMPLE   The jury *was* deadlocked; *it* had to be disbanded.

When a collective noun refers to individuals within a group, it takes a plural verb and pronoun.

EXAMPLE   The jury *were* allowed to go to *their* homes for the night.

Another way to emphasize the individuals in the jury would be to use the **phrase** *members of the jury.*

EXAMPLE   The members of the jury *were* allowed to go to *their* homes for the night.

Some collective nouns regularly take singular verbs *(crowd);* others do not *(people).*

EXAMPLES   *The crowd was* growing impatient.
Many *people were* able to watch the first space shuttle land safely.

Some collective nouns have regular plural forms *(team, teams);* others do not *(sheep).* For additional information about the **case, number,** and function of collective nouns, see **nouns.**

colons

The colon is a mark of anticipation and introduction that halts the **reader,** then connects the first statement to what follows it.

A colon may be used to connect a list or series to a **clause,** word, or **phrase** with which it is in apposition.

EXAMPLE   Three decontamination methods are under consideration: a zeolite-resin system, an evaporation and resin system, and a filtration and storage system.

Do not, however, place a colon between a **verb** and its **objects.**

CHANGE   The three fluids for cleaning pipettes include: water, alcohol, and acetone.

TO   The three fluids for cleaning pipettes include water, alcohol, and acetone.

One common exception is made when a verb is followed by a stacked list.

EXAMPLE   The corporations that manufacture computers include:

Sperry Rand	CDC
NCR	IBM
Burroughs	Honeywell

Do not use a colon between a **preposition** and its object.

CHANGE   I would like to be transferred to: Tucson, Boston, or Miami.

TO   I would like to be transferred to Tucson, Boston, or Miami.

A colon may be used to link one statement to another that develops, explains, amplifies, or illustrates the first. A colon may be used in this way to link two **independent clauses.**

EXAMPLE   Any large organization is confronted with two separate, though related, information problems: it must maintain an effective internal communication system, and it must see that an effective overall communication system is maintained.

A colon may be used to link an **appositive** phrase to its related statement if greater **emphasis** is needed.

EXAMPLE   There is only one thing that will satisfy Mr. Sturgess: our finished report.

Colons are used to link numbers signifying different identifying **nouns.**

EXAMPLES   Genesis 10:16 (chapter 10, verse 16)
9:30 a.m. (9 hours, 30 minutes)

In proportions, the colon indicates the ratio of one amount to another.

EXAMPLE   The cement is mixed with the water and sand at 7:5:14. (In this case, the colon replaces *to*.)

Colons are often used in mathematical ratios.

EXAMPLE $7:3 = 14:x$

In **bibliography, footnote,** and **reference** citations, colons link the place of publication with the publisher and perform other specialized functions.

EXAMPLE Wine, R. L. *Statistics for Scientists and Engineers.* Englewood Cliffs, N.J.: Prentice-Hall, 1964.

A colon follows the salutation in business letters, even when the salutation refers to a person by name.

EXAMPLES Dear Ms. Jeffers:
Dear Sir:
Dear George:

The initial **capital letter** of a **quotation** is retained following a colon if the quoted material originally began with a capital letter.

EXAMPLE The senator stated: "We are not concerned about the present. We are worried about the future."

A colon always goes outside **quotation marks.**

EXAMPLE This was the real meaning of his "suggestion": the division must show a profit by the end of the year.

When quoting material that ends in a colon, drop the colon and replace it with **ellipses.**

CHANGE "Any large corporation is confronted with two separate, though related, information problems:"

TO "Any large corporation is confronted with two separate, though related, information problems . . ."

The first word after a colon may be capitalized if (1) the statement following is a complete **sentence** or (2) it introduces a formal resolution or question.

EXAMPLE This year's conference attendance was low: We did not advertise widely enough.

If a subordinate element follows the colon, however, use a lower-case letter following the colon.

EXAMPLE There is only one way to stay within our present budget: to reduce expenditures for research and development.

commas

The comma has a wide variety of uses: it can link, enclose, separate, and show omissions. Effective use of the comma depends upon your understanding of how **ideas** fit together. Used with care, the comma can add **clarity** and **emphasis** to your writing.

The comma can prevent **ambiguity** by separating **sentence** elements that might otherwise be misunderstood. Compare the second sentence in each of the following examples.

> CHANGE The sale of 134 units to the Lane Company was due to Bill Hendrick's efforts; so, possibly, were the 550 units sold to Dayco, Inc. *If so, that is 684 sales out of 1,368 that were accounted for by our sales district.* (Ambiguous. Were 684 or 1,368 accounted for by our district?)
>
> TO The sale of 134 units to the Lane Company was due to Bill Hendrick's efforts; so, possibly, were the 550 units sold to Dayco, Inc. *If so, that is 684 sales, out of 1,368, that were accounted for by our sales district.* (Clear. It is obvious that 684 were accounted for by our district.)

TO LINK

Use a comma between **independent clauses** that are linked by a **coordinating conjunction** *(and, but, for, or, nor, so, yet)*. The comma precedes the **conjunction.**

> EXAMPLE Human beings have always prided themselves on their unique capacity to create and manipulate symbols, but today computers are manipulating symbols.

When the **clauses** are short and closely related, many writers omit the comma.

> EXAMPLE The cable snapped and the power failed.

TO ENCLOSE

Commas are used to enclose nonrestrictive clauses and parenthetical elements. (For other means of punctuating parenthetical elements, see **dashes** and **parentheses.**)

> EXAMPLES Our new Detroit factory, *which began operations last month,* should add 25 percent to total output. (nonrestrictive clause)
>
> We can, *of course,* expect their lawyer to call us. (parenthetical element)

Yes and *no* are set off by commas in such uses as the following.

EXAMPLES I agree with you, *yes.*
No, I do not think we can finish as soon as we would like.

A **direct address** should be enclosed in commas.

EXAMPLE You will note, *Mark,* that the surface of the brake shoe complies with the specifications.

Phrases in apposition (which identify another expression) are enclosed in commas.

EXAMPLE Our company, *the Blaylok Precision Company,* did well this year.

Commas enclose nonrestrictive **participial phrases.**

EXAMPLE The lathe operator, *working quickly and efficiently,* finished early.

TO SEPARATE

Commas are used to separate introductory elements from the rest of the sentence, to separate items in a series, to separate subordinate clauses from main clauses, and to separate certain elements for clarity or emphasis.

To Separate Introductory Elements. It is generally a good rule of thumb to put a comma after an introductory clause or phrase unless it is very short. Identifying where the introductory element ends helps indicate where the main part of the sentence begins.

EXAMPLE *Since many rare fossils seem never to occur free from their matrix,* it is wise to scan every slab with a hand lens.

When long modifying phrases precede the main clause, they should always be followed by a comma.

EXAMPLE *During the first series of field-performance tests last year at our Colorado proving ground,* the new motor failed to meet our expectations.

When an introductory phrase is short and closely related to the main clause, the comma may be omitted.

EXAMPLE *In two seconds* a 20°F temperature is created in the test tube.

A comma should always follow an introductory absolute phrase.

EXAMPLE *The tests completed,* we organized the data for the final report.

A **prepositional phrase,** if long, should be followed by a comma.

> EXAMPLE *In spite of much talk about the preeminent importance of the individual,* we seem always to expect the individual to sacrifice personal interests to those of the smooth operation of the administrative machinery.

Certain types of introductory words are followed by a comma. One such is a **noun** used in direct address.

> EXAMPLE *Bill,* enclosed is the article you asked me to review.

An introductory **interjection** (such as *oh, well, why, indeed, yes,* and *no*) is followed by a comma.

> EXAMPLES *Yes,* I will make sure your request is approved.
> *Indeed,* I will be glad to send you further information.

A transitional word or phrase like *moreover* or *furthermore* is usually followed by a comma.

> EXAMPLE *Moreover,* steel can withstand a humidity of 99%, provided that there is no chloride or sulphur dioxide in the atmosphere.

Introductory **adverbs** and adverb phrases are set off with a comma when they serve the dual function of modifying the idea that follows and connecting it to the idea in the previous sentence. (See also **transition.**)

> EXAMPLE We can expect a better balance of payments in the coming year. *In addition,* we should look for a better world market as a result. *However,* we should expect some shortages due to the overall economic climate.

When adverbs closely modify the **verb** or the entire sentence, however, they should not be followed by a comma.

> EXAMPLE *Perhaps* we can still solve the environmental problem. *Certainly* we should try.

Interrupting transitional words or phrases are usually set off with commas.

> EXAMPLE We must wait for the written authorization to arrive, *however,* before we can begin work on the project.

Commas are omitted when the word or phrase does not interrupt the continuity of thought.

> EXAMPLE I *therefore* suggest that we begin construction.

To Separate Items in a Series.   Commas should be used to separate words in a series.

> EXAMPLE   Basically, plants control the wind by *obstruction, guidance, deflection, and filtration.*

Phrases and clauses in coordinate series, like words, are punctuated with commas.

> EXAMPLE   It is well known that plants absorb noxious gases, act as receptors of dirt particles, and cleanse the air of other impurities.

When **adjectives** modifying the same noun can be reversed and make sense, or when they can be separated by *and* or *or,* they should be separated by commas.

> EXAMPLE   The drawing was of a *modern, sleek, swept-wing* airplane.

When an adjective modifies a phrase, no comma is required.

> EXAMPLE   He was investigating his *damaged radar beacon system. (damaged* modifies the phrase *radar beacon system.)*

Never separate a final adjective from its noun.

> CHANGE   He is a conscientious, honest, reliable, worker.
>   TO   He is a conscientious, honest, reliable worker.

Although the comma before the last word in a series is sometimes omitted, it is generally clearer to include it. The confusion that may result from omitting the comma is illustrated in the following sentence.

> EXAMPLE   Random House, Irwin, Doubleday and Dell are publishing companies.

Is "Doubleday and Dell" one company or two? "Random House, Irwin, Doubleday, and Dell" removes the doubt.
   Commas are conventionally used to separate distinct items. Use commas between the elements of an address written on the same line.

> EXAMPLE   Walter James, 4119 Mill Road, Dayton, Ohio 45401

Use a comma to separate the elements of a **date** written on the same line. However, when the day is omitted, the comma is unnecessary.

> EXAMPLES   July 2, 1949
>   July 1949

Use commas to separate the elements of Arabic **numbers.**

EXAMPLE   1,528,200

Use a space rather than a comma in metric values, because many countries use the comma as the decimal marker.

EXAMPLE   1 528 200

A comma may be substituted for the **colon** in a personal letter. Do not use a comma in a business letter, however, even if you use the person's first name.

EXAMPLES   Dear John, (personal letter)
Dear John: (business letter)

Use commas to separate the elements of geographical names.

EXAMPLE   Toronto, Ontario, Canada

Use a comma to separate names that are reversed.

EXAMPLE   Smith, Alvin

Use commas to separate certain elements of **footnote, reference,** and **bibliography** entries.

EXAMPLES   Bibliography—Fowles, Jib, ed. *Handbook of Futures Research.* Westport, Conn.: Greenwood Press, 1978.
Footnote—[1]Jib Fowles, ed. *Handbook of Futures Research* (Westport, Conn.: Greenwood Press, 1978), p. 30.
Reference—1. Fowles, Jib, ed. Handbook of futures research, Westport, Conn.: Greenwood Press; 1978.

To Separate Subordinate Clauses.   Use a comma between the main clause and a subordinate clause when the subordinate clause comes first.

EXAMPLE   When they artificially stimulated the electrochemical action of the brain, scientists learned more about the brain.

Use a comma following an independent clause that is only loosely related to the **dependent clause** that follows it.

EXAMPLE   The plan should be finished by July, even though I lost time because of illness.

In all cases, use a comma following a long introductory dependent clause.

EXAMPLE   Because the vertical wing extensions produce more lift, the engines can be throttled back and fuel can be saved.

To Separate Elements for Clarity or Emphasis. **Conjunctive adverbs** *(however, nevertheless, consequently, for example, on the other hand)* joining independent clauses are preceded by a **semicolon** and followed by a comma. Such adverbs function as both **modifiers** and **connectives.**

EXAMPLES   Your idea is good; *however,* your format is poor.
He has held the project together; *moreover,* he has helped everyone's morale.

Commas are often used before coordinating conjunctions like *but, for,* and *yet*—even though the conjunctions are not between independent clauses—because they separate contrasting ideas.

EXAMPLE   I generally agreed with him, but not completely.

Use a comma to separate two contrasting thoughts or ideas.

EXAMPLES   The project was finished on time, but not within the budget.
The specifications call for 100-ohm resistors, not 1000-ohm resistors.
The project was finished on time, wasn't it?
It was Bill, not Matt, who decided to change the design.

Use a comma to separate a direct quotation from its introduction.

EXAMPLE   Morton and Lucia White said, ''Men live in cities but dream of the countryside.''

Do not use a comma, however, when giving an indirect quotation.

EXAMPLE   Morton and Lucia White said that men dream of the countryside even though they live in cities.

Sometimes commas are necessary to make something clear that might otherwise be confusing.

CHANGE   When you see an airport fly directly over it at an altitude of 1,500 feet.
TO   When you see an airport, fly directly over it at an altitude of 1,500 feet.

Use a semicolon to separate phrases or clauses in a series when one or more of the phrases or clauses contain commas.

EXAMPLE   Among those present were John Howard, president of the Howard Paper Company; Thomas Martin, president of Copco Corp.; and Larry Stanley, president of Stanley Papers.

If you find you need a comma to separate the consecutive use of the same word to prevent misreading, rewrite the sentence.

CHANGE   The assets we had, had surprised us.

TO   We were surprised at the assets we had when the company was founded.

TO SHOW OMISSIONS

A comma sometimes replaces a verb in certain elliptical constructions.

EXAMPLE   Some were punctual; *others, late.* (replaces *were*)

CONVENTIONAL USE WITH OTHER PUNCTUATION

A comma always goes inside **quotation marks.**

EXAMPLE   The operator placed the discharge bypass switch at "normal," which triggered a second discharge.

When an introductory phrase or clause ends with a parenthesis, the comma separating the introductory phrase or clause from the rest of the sentence always appears outside the parenthesis.

EXAMPLE   Although we left late (at 7:30 p.m.), we arrived in time for the keynote address.

Except with **abbreviations,** a comma should not be used with a **period, question mark, exclamation mark,** or **dash.**

CHANGE   "I have finished the project.," he said.

TO   "I have finished the project," he said. (omit the period)

CHANGE   "Have you finished the project?," I asked.

TO   "Have you finished the project?" I asked. (omit the comma)

SUPERFLUOUS COMMAS

A number of common writing errors involve placing commas where they do not belong. These errors often occur because writers assume that a pause in a sentence should be indicated by a comma. It is true that commas usually signal pauses, but it is not true that pauses *necessarily* call for commas.

Be careful not to place a comma between a **subject** and verb, or between a verb and its **object.**

CHANGE   The cold conditions at the test site in the Arctic, made accurate readings difficult.

TO   The cold conditions at the test site in the Arctic made accurate readings difficult.

CHANGE He has often said, that one company's failure is another's
opportunity.
TO He has often said that one company's failure is another's opportunity.

Do not use a comma between the elements of a compound subject or a
compound **predicate** consisting of only two elements.

CHANGE The director of the engineering department, and the supervisor
of the quality-control section were both opposed to the new
schedules.
TO The director of the engineering department and the supervisor of
the quality-control section were both opposed to the new
schedules.

CHANGE The director of the engineering department listed five major
objections, and asked that the new schedule be reconsidered.
TO The director of the engineering department listed five major
objections and asked that the new schedule be reconsidered.

An especially common error is the placing of a comma after a coordinating conjunction such as *and* or *but* (especially *but*).

CHANGE The chairman formally adjourned the meeting, but, the members of the committee continued to argue.
TO The chairman formally adjourned the meeting, but the members
of the committee continued to argue.

CHANGE I argued against the proposal. And, I gave good reasons for my
position.
TO I argued against the proposal. And I gave good reasons for my
position.

Do not place a comma before the first item or after the last item of a
series.

CHANGE We are considering a number of new products, such as, calculators, typewriters, and cameras.
TO We are considering a number of new products, such as calculators, typewriters, and cameras.

CHANGE It was a fast, simple, inexpensive, process.
TO It was a fast, simple, inexpensive process.

The following chart summarizes most of the "do's and don'ts" of
comma usage.

"DO'S AND DON'TS" OF COMMA USAGE

Use commas to LINK two independent clauses joined by *and, but, for, or, so, nor,* or *yet.* (page 94)	DO NOT separate compound subjects, compound predicates, compound objects, a subject and its verb, or a verb and its object.
He needed the report, so he called the district office and asked for a copy.	The advertising campaign, and the national promotions failed to increase sales. (compound subject)
The office was open on Monday, but it was closed on Tuesday because the air conditioner failed.	The public relations staff organize special programs, and film sales presentations. (compound predicate)
The new engineer works very hard, yet we are still understaffed.	The secretary typed the minutes, and the special proposal. (compound object)
Our Brazilian iron ore project promises greater success, and we are optimistic about next year's profits.	Clerks who have taken courses in accounting, are eligible for promotion. (separating a subject and its verb)
	The manager reported, that production had increased. (separating a verb and its object)

Rule	Examples	
Use commas to SEPARATE introductory phrases, clauses, and words from the rest of the sentence. (page 95)	Having made the decision, the manager concentrated on other matters. (introductory phrase) When the tests are made public, the researchers expect national recognition. (introductory clause) Therefore, all travel reimbursement must be approved by the comptroller. (introductory word)	DO NOT separate a phrase or clause that is the subject of the sentence from the rest of the sentence. To be able to type 80 words a minute, is the goal of the new stenographer. Working on the weekend, will enable us to finish the illustrations.
Use commas to SEPARATE items in a series and adjectives modifying the same noun. (page 97)	We plan to fill requests for sleds, toboggans, skates, and ski poles. (items in a series) The president spoke of the challenging, exciting, untapped economic prospects in Paraguay. (adjectives modifying the same noun)	DO NOT separate an adjective from the phrase or word it is modifying. The company purchased two, large jets. (phrase) It was a fast, simple, inexpensive, process. (word)

"DO'S AND DON'TS" OF COMMA USAGE, continued

Use commas to SET OFF interrupting words or phrases. (page 96)	Tomorrow, for example, we'll ship three products. I'm afraid, however, that your letters are not effective.	DO NOT set off (with a comma) an interrupting transitional element that joins two independent clauses. USE a semicolon instead.	*Change:* In your trip report, you can explain the problem in production, however, you should not try to assign blame. *To:* In your trip report, you can explain the problem in production; however, you should not try to assign blame.
Use commas to SET OFF phrases and clauses that are not essential to the basic meaning of the sentence (nonrestrictive). (page 94)	Your second report, on the other hand, was lengthy and quite complete. (phrase) Our new product, which is to be released next month, will be sold in several foreign countries. (clause)	DO NOT set off phrases or clauses that are essential to the meaning of the sentence (restrictive).	The salesman, who writes orders for one million dollars, will win the trip to Madrid. We will ship the parts, that have been inspected, to Houston.
Use commas to SET OFF names, titles, addresses, dates, and quotations.	The Advertising Manager, Marjorie Howard, made the presentation in Dayton, Ohio, on May 16, 19—. The chairman said to his secretary, "I am the only company officer speaking at the stockholders' meeting."	DO NOT separate indirect quotations.	The chairman said, that he was the only company officer speaking at the stockholders' meeting.

comma splice

Do not attempt to join two **independent clauses** with only a **comma; this is called a "comma splice."**

EXAMPLE   It was five hundred miles to the facility, we made arrangements to fly.

Such a comma splice could be corrected in several ways.

1. Substitute a **semicolon,** or a semicolon and a **conjunctive adverb.**

   CHANGE   It was five hundred miles to the facility, we made arrangements to fly.
   TO   It was five hundred miles to the facility; we made arrangements to fly.
   OR   It was five hundred miles to the facility; *therefore,* we made arrangements to fly.

2. Add a **conjunction** following the comma.

   EXAMPLE   It was five hundreds miles to the facility, *so* we made arrangements to fly.

3. Create two **sentences.** (Be aware, however, that putting a **period** between two closely related and brief statements may result in two weak sentences.)

   EXAMPLE   It was five hundred miles to the facility. We made arrangements to fly.

4. Subordinate one **clause** to the other.

   EXAMPLE   *Because* it was five hundred miles to the facility, we made arrangements to fly.

When a conjunctive adverb connects two independent clauses, the conjunctive adverb *must* be preceded by a semicolon and followed by a comma.

   CHANGE   It was five hundred miles to the facility, therefore, we made arrangements to fly.
   TO   It was five hundred miles to the facility; therefore, we made arrangements to fly.

committee

Committee is a **collective noun** that takes a singular **verb.**

   EXAMPLE   The *committee is* to meet at 3:30 p.m.

If you wish to emphasize the individuals on the committee, use *the members of the committee* with the plural verb form.

EXAMPLE   The *members of the committee were* all in agreement.

common nouns

A common noun names general classes or categories of persons, places, things, concepts, actions, or qualities. Common nouns can be **abstract** *(justice),* **concrete** *(jail),* or **collective** *(jury).*

Common Nouns	Proper Nouns
boy	Toby Wilson
city	Chicago
company	General Electric
day	Tuesday
document	Declaration of Independence

Common nouns are not capitalized unless they begin a **sentence.** (For information about the **case, number,** and function of common nouns, see **nouns.**)

comparative degree

Most **adjectives** and **adverbs** can be compared. Their common forms of **comparison** are as follow.

EXAMPLES   The new copier is *fast.* (positive)
The new copier is *faster* than the old one. (comparative)
The new copier is the *fastest* I have ever used. (superlative)

Most one-syllable words use the comparative ending *-er* and the superlative ending *-est.*

EXAMPLES   Ours was the *faster* of the two cars tested.
Ours was the *fastest* car on the track.

Most words with more than one syllable form the comparative degree by using the word *more* and the superlative degree by using the word *most.*

EXAMPLES   She is *more* talented than the rest of the writers on the staff.
She is the *most* talented writer on the staff.

Some words may use either means of expressing degree.

EXAMPLES   He was the *most* able technician at the meeting.
He was the *ablest* technician at the meeting.

A few words have irregular forms of comparison

EXAMPLES   much, more, most
good, better, best

Writers often use the comparative degree when comparing two persons or things and the superlative degree when comparing three or more.

EXAMPLES   This is the *newer* of the two designs.
This is the *newest* of the three designs.

The superlative form is sometimes also used in expressions that do not really express comparison (*best* wishes, *deepest* sympathy, *highest* praise, and *most* sincerely).

compare/contrast

When you *compare* things, you point out similarities or both similarities and differences. When you *contrast* things you point out only the differences. In either case, you *compare* or *contrast* only things that are part of a common category.

EXAMPLES   He *compared* all the features of the two brands before buying.
Their styles of selling *contrasted* sharply.

When *compare* is used to establish a general similarity, it is followed by *to*.

EXAMPLE   *Compared to* the computer, the abacus is a primitive device.

When *compare* is used to indicate a close examination of similarities or differences, it is followed by *with* in formal **usage.**

EXAMPLE   We *compared* the features of the new capacitor very carefully *with* those of the old one.

Contrast is normally followed by *with*.

EXAMPLE   The new policy *contrasts* sharply *with* the earlier one in requiring that sealed bids be submitted.

When the **noun** form of *contrast* is used, one speaks of the *contrast between* two things or of one thing being *in contrast to* the other.

EXAMPLES   There is a sharp *contrast between* the old and new policies.
The new policy is *in* sharp *contrast to* the earlier one.

comparison

When making a comparison, make certain that both or all the elements being compared are clearly evident to your **reader.**

CHANGE   The third-generation computer is *better.*

TO   The third-generation computer is *better than the second-generation computer.*

The things being compared must be of the same kind.

CHANGE   *Imitation alligator hide* is almost as tough as a *real alligator.*

TO   *Imitation alligator hide* is almost as tough as *real alligator hide.*

Be sure to point out the parallels or differences between the things being compared. Don't assume your reader will know what you mean.

CHANGE   Washington is farther from Boston *than Philadelphia.*

TO   Washington is farther from Boston *than it is from* Philadelphia.

OR   Washington is farther from Boston *than Philadelphia is.*

A double comparison in the same **sentence** requires that the first be completed before the second is stated.

CHANGE   The discovery of electricity was *one of the great if not the greatest* scientific discoveries in history.

TO   The discovery of electricity was *one of the great* scientific discoveries in history, *if not the greatest.*

Do not attempt to compare things that are not comparable.

CHANGE   Farmers say that storage space is reduced by 40% compared with baled hay. (*Storage space* is not comparable to *baled hay.*)

TO   Farmers say that baled hay requires 40% less storage space than loose hay requires.

comparison method of development

As a method of development, comparison points out similarities and differences between the elements of your subject. The comparison **method of development** can be especially effective because it can be used to explain a difficult or unfamiliar subject by relating it to a simpler or more familiar one.

You must first determine the basis for the **comparison.** For example, if you were responsible for the purchase of chain saws for a logging company, you would have a number of factors to take into account in order to establish your bases for comparison. Because loggers use the equipment daily, you would have to select durable saws with the appropriate size engines, chain thicknesses, and bar lengths for the type of wood most frequently cut. Since chain saws produce noise and vibration, you would want to compare the quality and cost of the various silencers on the market. You would not include in your comparison such irrelevant factors as color or place of manufacture. Taking all of the important elements into account, however, you would establish a number of bases for choosing from among the available chain saws—engine size, chain thickness, bar length, and noise mufflers.

Once you have determined the basis (or bases) for comparison, you must decide how to present it. In the whole-by-whole method, all the relevant characteristics of one item are discussed before those of the next item are considered. In the part-by-part method, the relevant features of each item are compared one by one. The following discussion of typical woodworking glues, organized according to the whole-by-whole method, describes each type of glue and its characteristics before going on to the next type.

> *White glue* is the most useful all-purpose adhesive for light construction, but it cannot be used on projects that will be exposed to moisture, high temperature, or great stress. Wood that is being joined with white glue must remain in a clamp until the glue dries, which will take about 30 minutes.
>
> *Aliphatic resin glue* has a stronger and more moisture-resistant bond than white glue. It must be used at temperatures above 50°F. The wood should be clamped for about 30 minutes. . . .
>
> *Plastic resin glue* is the strongest of the common wood adhesives. It is highly moisture resistant—though not completely waterproof. Sold in powdered form, this glue must be mixed with water and used at temperatures above 70°F. It is slow setting and the joint should be clamped for four to six hours. . . .
>
> *Contact cement* is a very strong adhesive that bonds so quickly it must be used with great care. It is ideal for mounting sheets of plastic laminate on wood. It is also useful for attaching strips of veneer to the edges of plywood. Since this adhesive bonds immediately when two pieces are pressed together, clamping is not necessary, but the parts to be joined must be very carefully aligned before being placed together. Most brands are quite flammable and the fumes can be harmful if inhaled. To meet current safety standards, this type of glue must be used in a well-ventilated area, away from flames or heat.
>
> —*Space and Storage* (Alexandria, Va.: Time-Life Books, 1977), p. 61.

As is often the case when the whole-by-whole method is used, the purpose of this comparison is to weigh advantages and disadvantages of each glue for certain kinds of woodworking. The comparison could be expanded, of course, by the addition of other types of glue. If, on the other hand, your purpose was to consider, one at a time, the various characteristics of all the glues, the information might be arranged according to the part-by-part method, as in the following example.

Woodworking adhesives are rated primarily according to their bonding strength, moisture resistance, and setting times.

Bonding strengths are categorized as very strong, moderately strong, or adequate for use with little stress. Contact cement and plastic resin glue bond very strongly, while aliphatic resin glue bonds moderately strongly. White glue provides a bond least resistant to stress.

Moisture resistance of woodworking glues is rated as high, moderate, and low. Plastic resin glues are highly moisture-resistant. Aliphatic resin glues are moderately moisture-resistant, white glue is least moisture-resistant.

Setting times for these glues vary from an immediate bond to a four-to-six hour bond. Contact cement bonds immediately and requires no clamping. Because the bond is immediate, surfaces being joined must be carefully aligned before being placed together. White glue and aliphatic resin glue set in thirty minutes; both require clamping to secure the bond. Plastic resin, the strongest wood glue, sets in four to six hours and also requires clamping.

The part-by-part method could accommodate further comparison. Comparisons might be made according to temperature ranges, special warnings, common uses, and so on.

complaint letters

Businesses sometimes err in providing goods and services to customers, and so customers write complaint letters asking that such situations be corrected. The **tone** of such a letter is important; the most effective complaint letters do not sound angry. Do not use a complaint letter to vent your anger; remember that the **reader** of your letter probably had nothing to do with whatever went wrong, and berating that person is not likely to achieve anything positive. In most cases, you need only state

BAKER MEMORIAL HOSPITAL

Television Services
501 Main Street
Springfield, OH 45321
(513) 683-8100

September 23, 19--

Manager, Customer Relations
General Television, Inc.
5521 West 23rd Street
New York, NY 10062

On July 9th I ordered nine TV tuners for your model MX-15 color
receiver. The tuner part number is TR-5771-3.

On August 2nd I received from your Newark, New Jersey, parts warehouse
seven tuners, labeled TR-413-7. I immediately returned these tuners with
a note indicating the mistake that had been made. However, not only have
I failed to receive the tuners I ordered, but I have also been billed
repeatedly.

Would you please either send me the tuners I ordered or cancel my order.
I have enclosed a copy of my original order letter and the most recent
bill.

Sincerely,

Paul Denlinger

Paul Denlinger
Manager

PD:sj
Enclosures

Letter 6

your claim, support it with all the pertinent facts, and then ask for the desired adjustment. Most companies are very willing to correct whatever went wrong.

The **opening** of your complaint letter should include all identifying data concerning the transaction: item, date of purchase, place of purchase if pertinent, cost, invoice number, and so on.

The body of your letter should explain logically and clearly what happened. You should present any facts that prove the validity of your claim. Be sure of your facts, and present them concisely and objectively, carefully avoiding any overtones of accusation or threat. You may, however, wish to state any inconvenience or loss created by the problem, such as a broken machine stopping an entire assembly line.

Your **conclusion** should be friendly, and it should request action. State what you would like your reader to do to solve your problem.

Large organizations often have special departments to handle complaints. If you address your letter to one of these departments—for example, to Customer Relations or Consumer Affairs—it should reach someone who can respond to your claim. In smaller organizations you might write to a vice-president in charge of sales or service. For a very small business, write directly to the owner.

Letter 6 is a typical complaint letter.

complements

A complement is a word, **phrase,** or **clause** used in the **predicate** of a **sentence** to complete the meaning of the sentence.

EXAMPLES   Pilots fly *airplanes.* (word)
To live is *to risk death.* (phrase)
John knew *that he would be late.* (clause)

Four kinds of complements are generally recognized: **direct object** (which completes the sense of a **transitive verb**); **indirect object** (which completes the meaning of a transitive verb and the verb's direct object); **objective complement** (which completes the meaning of a verb's object); and **subjective complement** (which completes the meaning of the subject).

A direct object is a **noun** or noun equivalent that receives the action of a transitive verb; it answers the question *what* or *whom* after the verb.

EXAMPLES   John built *an antenna.* (noun)
I like *to work.* (verbal)
I like *it.* (pronoun)
I like *what I saw.* (noun clause)

An indirect object is a noun or noun equivalent that occurs with a direct object after certain kinds of transitive verbs such as *give, wish, cause,* and *tell.* It answers the question *to whom* or *for whom* (or *to what* or *for what*).

EXAMPLES   Give *John* a wrench. ("wrench" is the direct object)
We should buy *the Milwaukee office* a computer. ("computer" is the direct object)

An objective complement completes the meaning of a sentence by revealing something about the object of its transitive verb. An objective complement may be either a noun or an **adjective.**

EXAMPLES   They call him *a genius.* (noun)
We painted the building *white.* (adjective)

A subjective complement, which follows a **linking verb** rather than a transitive verb, describes the **subject.** A subjective complement may be either a noun or an adjective.

EXAMPLES   His sister is *an engineer.* (noun)
His brother is *ill.* (adjective)

complement/compliment

Complement means "anything that completes a whole." It is used as either a **noun** or a **verb.**

EXAMPLES   A *complement* of four employees would bring our staff up to its normal strength. (noun)
The two programs *complement* one another perfectly. (verb)

Compliment means "praise." It too is used as either a noun or a verb.

EXAMPLES   The manager *complimented* the staff on its efficient job (verb)
The manager's *compliment* boosted staff morale. (noun)

complex sentences

The complex sentence provides a means of subordinating one thought to another (or, put another way, of emphasizing one thought over another) because it contains one **independent clause** and at least one **dependent clause** that expresses a subordinate idea.

EXAMPLE   We lost some of our efficiency (independent clause) when we moved (dependent clause).

Normally, the independent clause carries a main point, and the dependent clause carries a related subordinate point. A dependent clause may occur before, after, or within the independent clause. The dependent clause can serve within the **sentence** as its **subject,** as an **object,** or as a **modifier.**

EXAMPLES   *What he proposed* is irrelevant. (subject)
We know *where it is supposed to be.* (object)
The computer, *which can make more calculations in one hour than thousands of scientists could make in a lifetime,* is perhaps one of the greatest challenges people have ever had to face. (modifier)

Complex sentences offer more variety than **simple sentences.** And frequently the meaning of a **compound sentence** can be made more exact by subordinating one of the two independent clauses to the other to create a complex sentence (thereby establishing the relationship of the two parts more clearly). (See also **subordination.**)

EXAMPLES   We moved *and* we lost some of our efficiency. (compound sentence with **coordinating conjunction**)
When we moved, we lost some of our efficiency. (complex sentence with **subordinating conjunction**)

Learn to handle the complex sentence well; it can be a useful tool with which to express your thoughts clearly and exactly. However, be on guard against placing the main idea of a complex sentence in a dependent clause; this is one of the most common errors made with the complex sentence. For example, do not write the first sentence below if what you mean is expressed by the second sentence.

EXAMPLES   Production was not as great as we expected, although we met our quota. (The main idea expressed here is that production fell short of expectations.)
Although production was not as great as we expected, we met our quota. (The main idea expressed here is that the production quota was met.)

compose/comprise

Compose means "create" or "make up the whole." The parts *compose* the whole.

EXAMPLE   The 13 moving parts *compose* (or make up) the machine.

Comprise means "include." The whole *comprises* the parts.

EXAMPLE   The machine (the whole) *comprises* 13 moving parts.

compound sentences

A compound sentence combines two or more related **independent clauses** that are of equal importance.

EXAMPLE   Drilling a well is the only sure way to determine the presence of oil, *but* it is a very costly endeavor.

The independent clauses of a compound **sentence** may be joined by a **comma** and a **coordinating conjunction,** by a **semicolon,** or by a **conjunctive adverb** preceded by a semicolon and followed by a comma.

EXAMPLES   People deplore violence, *but* they have an insatiable appetite for it on television and in films. (coordinating conjunction)
There is little similarity between the chemical composition of sea water and river water; the various elements are present in entirely different proportions. (semicolon)
People deplore violence; *however,* they have an insatiable appetite for it on television and in films. (conjuctive adverb)

A compound sentence is balanced when its clauses are of similar length and construction.

EXAMPLES   The plan was sound, and the staff was eager to begin.
Fingerprints were used for personal identification in 200 B.C., but they were not used for criminal identification until about A.D. 1880.

compound-complex sentences

A compound-complex sentence consists of two or more **independent clauses** and at least one **dependent clause.**

EXAMPLE   At the same time *that it cools and lubricates the bit and brings the cuttings to the surface,* the "drilling mud" deposits a sheath of mud cake on the wall of the hole to prevent cave-ins; it reduces the friction *which is created by the drill string's rubbing against the wall of the hole;* and, *since the weight of the mud column bears against the wall of the hole,* it helps to contain formation pressures and prevent a blowout.

Here, the three parallel independent clauses are joined in a series linked by **semicolons** and the **coordinating conjunction** *and;* each independent clause contains a dependent clause (in **italics**). The first two dependent clauses are **adjective clauses** modifying *time* and *friction,* and the third is an **adverb clause** that modifies its independent clause as a whole.

The compound-complex sentence offers a vehicle for a more elaborate grouping of ideas than other **sentence** types permit. It is not really recommended for inexperienced writers because it is difficult to handle skillfully and often leads them into sentences that become so complicated that they lose all **logic** and **coherence.** The more technically complex the subject, the more difficult the compound-complex sentence is to control effectively.

compound words

A compound word is made from two or more words that are either hyphenated or written as one word. (If you are not certain whether a compound word should be hyphenated, check a **dictionary.**)

EXAMPLES   nevertheless, mother-in-law, courthouse, run-of-the-mill, low-level, high-energy

Be careful to distinguish between compound words and words that frequently appear together but do not constitute compound words, such as *high school* and *post office.* Also be careful to distinguish between compound words and word pairs that mean different things, such as *greenhouse* and *green house.*

Plurals of compound words are usually formed by adding an *s* to the last letter.

EXAMPLES   bedrooms, masterminds, overcoats, cupfuls

When the first word of the compound is more important to its meaning than the last, however, the first word takes the *s* (when in doubt, check your dictionary).

EXAMPLES   editors-in-chief, fathers-in-law

Possessives are formed by adding 's to the end of the compound word.

EXAMPLES the *vice-president's* speech, his *brother-in-law's* car, the *pipeline's* diameter, the *antibody's* action

concept/conception

A *concept* is a thought or an idea. A *conception* is the sum of a person's ideas, or concepts, on a subject.

EXAMPLES This final *conception* of the whole process evolved from many smaller *concepts*.
From the *concept* of combustion evolved the *conception* of the internal combustion engine.

conciseness/wordiness

Good writing is concise. Make all words, **sentences,** and **paragraphs** count by eliminating unnecessary words and **phrases.** Wordiness results from needless **repetition** of the same idea in different words and in different sentences within your writing.

CHANGE Modern students *of today* are more sophisticated than their parents. (The phrase *of today* repeats the thought already expressed by the adjective *modern.*)
TO Modern students are more sophisticated than their parents.

CHANGE The walls were sky-blue *in color.* (The phrase *in color* is redundant.)
TO The walls were sky-blue.

Good writers go through their writing removing every word, phrase, **clause,** or sentence they can omit. In doing so, they are striving to be as concise as **clarity** permits—but note that conciseness is not a **synonym** for brevity. Brevity may or may not be desirable in a given passage (depending upon the writer's **objective**), but it is always good to be concise. The writer must distinguish between language that is used for effect and mere wordiness that stems from lack of care or judgment. The following **anecdote** was related by Benjamin Franklin to Thomas Jefferson as members of the Continental Congress undertook to trim excess words and phrases from the Declaration of Independence.

> The wise old Pennsylvanian told Jefferson, in an aside, about a hatter who was opening a shop and wanted a signboard for it. What the new proprietor had in mind was a message reading,

"John Thompson, Hatter, makes and sells hats for ready money," but one of his friends suggested that the word "hatter" was superfluous. Another told him that no buyer would care who made the hats, so the word "makes" was omitted. Someone else advised him to leave out "ready money,"since nobody expected to buy on credit, and yet another man said that since Thompson did not propose to give the hats away the word "sells" should go. When the sign was finally erected, Franklin smiled, all that remained was the name "John Thompson" and the picture of a hat.

—Richard M. Ketchum, *The Winter Soldiers* (New York: Doubleday, 1973), p. 19.

A concise sentence is not guaranteed to be effective, but a wordy sentence always loses some of its readability and **coherence** because of the extra load it must carry. Wordiness is understandable in a first draft, but it should never survive **revision.** "I would have written a shorter letter if I'd had more time" is a truism.

CAUSES OF WORDINESS

Modifiers that repeat an idea already implicit or present in the word being modified contribute to wordiness.

EXAMPLES

round circles	isolated *by himself*
basic essentials	*tall* skyscrapers
hot boiling water	descended *down*
advance planning	circle *around*
small *in size*	square *in shape*
balance *of equilibrium*	worthy *of merit*
a dream *come true*	the reason *is because*
visible *to the eye*	cooperate *together*

Another cause of wordiness is the "it . . . that . . ." construction. Not only is this construction wordy, but often it forces you to use the passive **voice** (a major cause of wordiness in its own right).

CHANGE   *It* is agreed *that* our new design will strive for simplicity.

TO   We agree that our new design will strive for simplicity

The **gobbledygook** addict is fond of such words and phrases as *factor, case, basis, elements, field, phenomenon, in terms of, in the nature of, with reference to.* Key words that seem to invite redundancy are *situation, angle, line, factor, aspect, element, consideration, considering.* Although all these words have legitimate uses, technical people seem to have a particular weakness for their overuse and inexact use, under the mistaken impression that they add formality to writing.

Coordinating synonyms that merely repeat one another contribute to wordiness.

EXAMPLE   We must decide *finally* and *for good* on this year's budget.

Wordiness can be caused by unnecessary or redundant phrases and clauses.

EXAMPLE   Our control room simulators, *which are manufactured in this country*, are more sophisticated than Russian simulators. (unnecessary clause)

The use of **expletives, relative pronouns,** and **relative adjectives,** although they have legitimate purposes, often results in wordiness.

CHANGE   *There are* (expletive) many supervisors in the area *who* (relative pronoun) are planning to attend the workshop *which* (relative adjective) is scheduled for Friday.

TO   Many supervisors in the area plan to attend the workshop scheduled for Friday.

Circumlocution (a long, indirect, roundabout way of expressing things) is a leading cause of wordiness.

CHANGE   We urge you to submit any suggestions you may have on the subject, and you may be certain that your suggestions will be given our most careful attention.

TO   Please give us your suggestions. We will consider them carefully.

HOW TO ACHIEVE CONCISENESS

Conciseness can be achieved by effective use of **subordination.** This is, in fact, the best means of tightening wordy writing.

CHANGE   The chemist's report was carefully illustrated and it covered five pages.

TO   The chemist's five-page report was carefully illustrated.

Conciseness can be achieved by using simple words and phrases.

CHANGE   It is the policy of the company to provide the proper equipment to enable each employee to conduct the telephonic communication necessary to discharge his responsibilities; such should not be utilized for personal communications.

TO   Your telephone is provided for company business; do not use it for personal calls.

Conciseness can be achieved by eliminating undesirable **repetition.**

CHANGE   Postinstallation testing, which is offered to all our customers at no further cost to them whatsoever, is available with each Line Scan System One purchased from this company.

TO   Free postinstallation testing is offered with each Line Scan System One.

Conciseness can be achieved by making sentences positive. (See also **positive writing.**)

CHANGE   If the error does not involve data transmission, the special function is not used.

TO   The special function is used only when the error involves data transmission.

Conciseness can sometimes be achieved by changing a sentence from the passive to the active voice, or by changing a sentence from the indicative to the imperative **mood.** The following example does both.

CHANGE   Card codes are normally used when it is known that the cards are to be processed by a computer, and control punches are normally used when it is known that the cards are designed to be processed at a tab installation.

TO   Use card codes when the cards are to be processed by a computer, and use control punches when they are to be processed at a tab installation.

Eliminate wordy introductory phrases or pretentious words and phrases of any kind.

CHANGE   *In connection with* your recent accident, we must *discontinue* your insurance *due to the fact that* this is your third claim within the past year. *Inasmuch* as you were insured with us *at the time of the accident, we are prepared* to honor your claim *along the lines* of your policy.

TO   Concerning your recent accident, we are canceling your insurance because you have filed three claims in the past year. Since you were insured with us at the time of the accident, however, we will honor your claim.

Overuse of **intensifiers** (such as *very, more, most, best, quite*) contributes to wordiness; conciseness can be achieved by eliminating them. The same is true of excessive use of **adjectives** and **adverbs.**

conclusions

The conclusion to a piece of writing not only ties together all the main ideas but can do so emphatically by making a final significant point. This final point may be to recommend a course of action, to make a prediction, to offer a judgment, to speculate on the implications of your ideas, or merely to summarize your main points.

The way you conclude depends both on the purpose of your writing and the needs of your **reader.** For example, a committee **report** about possible locations for a new manufacturing plant could end with a recommendation. A report on a company's annual sales figures might conclude with a judgment about why sales are up or down. A letter about consumer trends could end by speculating on the implications of these trends. A document that is particularly lengthy will often end with a summary of its main points. Study the following examples.

<table>
<tr><td>recommendation</td><td>These results indicate that you may need to alter your testing procedure to eliminate the impurities we found in specimens A through E.</td></tr>
<tr><td>prediction</td><td>Although my original estimate on equipment ($20,000) has been exceeded by $2,300, my original labor estimate ($60,000) has been reduced by $3,500; therefore, I will easily stay within the limits of my original bid. In addition, I see no difficulty in having the arena finished for the December 23 Christmas program.</td></tr>
<tr><td>judgment</td><td>Although our estimate calls for a substantially higher budget than in the three previous years, we believe that it is justified by our planned expansion.</td></tr>
<tr><td>summary</td><td>As this letter has indicated, we could attract more recent graduates by (1) increasing our advertising in local student newspapers, (2) resuming our co-op program, (3) sending a representative to career day programs at local colleges, (4) inviting local college instructors to teach in-house courses here at the plant, and (5) encouraging our employees to attend evening classes at local colleges.</td></tr>
</table>

The closing may merely present ideas for consideration, but also it may call for action or deliberately provoke thought.

<table>
<tr><td>ideas for consideration</td><td>The new prices become effective the first of the year. Price adjustments are routine for the company, but some of your customers will not consider them so. Please consider the needs of</td></tr>
</table>

both your customers and the company as you implement these new prices.

call for action

Send us a check for $250 now if you wish to keep your account active. If you have not responded to our previous letters because of some special hardship, I will be glad to personally work out a solution with you.

thought-provoking statement

Can we continue to accept the losses incurred by careless workmanship? Must we accept it as inevitable? Or should we consider steps to control it firmly now?

Because your conclusion is in a position of great **emphasis,** be careful to avoid closing with a cliché. Some such expressions do say what you want to say so precisely that they are considered acceptable closings, however.

EXAMPLES   It has been a pleasure to be of assistance.
May we thank you again for your cooperation.

Be especially careful not to introduce a new **topic** in your conclusion. A conclusion should always relate to and reinforce the ideas presented in your writing.

concrete nouns

Concrete nouns, as opposed to **abstract nouns,** identify those things that can be perceived by the five senses (such as *wrench, book, house, scissors, gold, water, electricity*). The difference between abstract and concrete nouns is the difference between *durability* (abstract noun) and *stone* (concrete noun).

There are two kinds of concrete nouns: **count nouns** and **mass nouns.** Count nouns identify things that can be counted (one *wrench,* two *books,* three *dogs,* four *houses,* five *pairs of scissors*); mass nouns identify concrete materials that cannot be counted (*coffee, air, cement, corn, steel, gold*). For information about the **case, number,** and function of concrete nouns, see **nouns.** (See also **abstract words/concrete words.**)

conjunctions

A conjunction connects words, **phrases,** or **clauses.** A conjunction can also indicate the relationship between the two elements it connects (*and* joins together, *or* selects and separates).

TYPES OF CONJUNCTIONS

A **coordinating conjunction** is a word that joins two **sentence** elements that have identical functions. The coordinating conjunctions are *and, but, or, for, nor, yet,* and *so.*

EXAMPLES Nature *and* technology are only two conditions that affect petroleum operations around the world. (joining two **nouns**)
To hear *and* to obey are two different things. (joining two phrases)
He would like to include the test results, *but* that would make the report too long. (joining two clauses)

Correlative conjunctions are coordinating conjunctions that are used in pairs. The correlative conjunctions are *either . . . or, neither . . . nor, not only . . . but also, both . . . and,* and *whether . . . or.*

EXAMPLE The inspector will arrive *either* on Wednesday *or* on Thursday.

A **subordinating conjunction** connects sentence elements of different weights, normally **independent clauses** and **dependent clauses.** The most frequently used are *so, although, after, because, if, where, than, since, as, unless, before, that, though, when,* and *whereas.*

EXAMPLE I left the office *after* I had finished writing the report.

A **conjunctive adverb** is an **adverb** that has the force of a conjunction because it is used to join two independent clauses. The most common conjunctive adverbs are *however, moreover, therefore, further, then, consequently, besides, accordingly, also, thus.*

EXAMPLE The engine performed well in the laboratory; *however,* it failed under road conditions.

PUNCTUATING CONJUNCTIONS

Two independent clauses separated by a coordinating conjunction should have a **comma** immediately preceding the coordinating conjunction if the clauses are relatively long.

EXAMPLE The Alpha project was her third assignment, *and* it was the project that made her reputation as a project leader.

If two independent clauses that are joined by a coordinating conjunction have commas within them, a **semicolon** may precede the conjunction.

EXAMPLE Even though the schedule is tight, we must meet our deadline; *and* all the staff, including programmers, will have to work overtime this week.

Two main clauses that are joined by a conjunctive adverb require a semicolon before and a comma after the conjunctive adverb. The conjunctive adverb both connects and modifies. As a **modifier,** it is part of one of the two clauses that it connects. The use of the semicolon makes it a part of the clause it modifies.

> EXAMPLE   The building was not finished on the scheduled date; *nevertheless,* Rogers moved in and began to conduct the business of the department from the unfinished offices.

CONJUNCTIONS IN TITLES
Conjunctions in the titles of books, articles, plays, movies, and so on should not be capitalized unless they are the first or last word in the title. (See also **capital letters.**)

> EXAMPLE   The book *Technical and Professional Writing* was edited by Herman Estrin.

conjunctive adverbs

A conjunctive adverb is an **adverb** because it modifies the **clause** that it introduces; it operates as a **conjunction** because it joins two **independent clauses.** The most common conjunctive adverbs are *however, moreover, therefore, further, then, consequently, besides, accordingly, also,* and *thus.*

> EXAMPLE   The engine performed well in the laboratory; *however,* it failed under road conditions.

The two independent clauses that are joined by a conjunctive adverb require a **semicolon** before and a **comma** after the conjunctive adverb. The conjunctive adverb both connects and modifies. As a *modifier,* it is part of one of the two clauses that it connects, and the use of the semicolon makes it a part of the clause that it modifies. (See also **transition.**)

> EXAMPLE   The new project is almost completed and is already over cost estimates; *however,* we must emphasize the importance of adequate drainage.

connected with/in connection with

Connected with and *in connection with* are wordy **phrases** that can usually be replaced by *in* or *with.*

CHANGE   He is *connected with* the TFT Corporation.
  TO   He is with the TFT Corporation.
  OR   He is employed by the TFT Corporation.

CHANGE   The fringe benefits *in connection with* (or *connected with*) the job are quite good.
  TO   The fringe benefits with the job are quite good.

connectives

The word *connective* is a general term for any word or **phrase** that is used to tie the parts of a **sentence** together or to indicate **subordination** or coordination between the parts of a sentence. Connectives include **conjunctions, prepositions,** and **conjunctive adverbs.**

TYPES OF CONNECTIVES

A **coordinating conjunction** joins two sentence elements that have identical functions and equal importance within the sentence. The most common coordinating conjunctions are *and, but, or, for, nor, yet,* and *so.*

    Correlative conjunctions are coordinating conjunctions that are used in pairs to connect parallel constructions. The most common correlative conjunctions are *either . . . or, neither . . . nor, not only . . . but also, both . . . and,* and *whether . . . or.*

    A **subordinating conjunction** connects sentence elements of different importance, often **independent clauses** and **dependent clauses.** The most common subordinating conjunctions are *so, although, after, because, if, where, than, since, as, unless, before, that, though, when,* and *whereas.*

    A preposition is a connective that is used to show the relationship of a **noun** or noun equivalent, the **object** of the preposition, to some other part of the sentence. The most common prepositions are *after, about, because, before, beside, but, for, since, until, across, among, against, into, at, by, between, with, in, on, to, during, under, over, through, throughout,* and *without.* (See also **prepositional phrases.**)

    A conjunctive adverb is an **adverb** that has the force of a conjunction because it is used to join two independent clauses. The most common conjunctive adverbs are *however, moreover, therefore, further, then, consequently, besides, also, accordingly,* and *thus.*

PUNCTUATING CONNECTIVES

Two independent clauses separated by a coordinating conjunction should have a **comma** immediately preceding the coordinating conjunction if the clauses are relatively long.

EXAMPLE I am glad I accepted the job at IBM, *and* I hope to advance rapidly there.

If two independent clauses that are joined by a coordinating conjunction have commas within them, a **semicolon** may precede the conjunction.

EXAMPLE Even though the schedule is tight, we must meet our deadline; *and* all the staff, including programmers, will have to work overtime this week.

Two main clauses that are joined by a conjunctive adverb require a semicolon before and a comma after the conjunctive adverb. The conjunctive adverb both connects and modifies. As a **modifier,** it is part of one of the two clauses that it connects. Use of the semicolon makes it a part of the clause it modifies.

EXAMPLE The new office building was not finished on the scheduled date; *nevertheless,* the department moved in and began to conduct business from the unfinished offices.

Subordinating conjunctions, prepositions, and correlative conjunctions do not normally require **punctuation.**

EXAMPLES The tests were postponed *because* the equipment had not arrived. (subordinating conjunction)
He did not receive the contract because the competition was too much *for* him. (preposition)
It seemed that every piece tested was *either* too soft *or* too brittle. (correlative conjunction)

connotation/denotation

The terms *connotation* and *denotation* refer to two different ways of interpreting a word's meaning. The *denotation* of a word is its literal and objective meaning. Defined this way, *communism* is a socioeconomic system that aims at a classless society, and a *bureaucrat* is an unelected public official who administers government policy. Of course both words bring to mind a host of secondary or implied meanings that arise from our emotional reactions to them. Many Americans react negatively to the word *communism,* associating it with a totalitarian form of government that is antireligious and committed to world conquest. Likewise, the word *bureaucrat* frequently conjures up visions of someone who insists on rigid adherence to arbitrary rules and regulations. These secondary meanings are a word's connotations.

In your writing you must be sure to consider the connotations of your words. You know what you mean to say, but you must be careful to say it in words that will not be misinterpreted by your **reader**. For example, you might not want to refer to an item as *cheap,* meaning inexpensive, for the word *cheap* also implies shoddy workmanship or poor materials. The **usage** notes in a reputable dictionary can help you to make these distinctions. Technical writing in particular requires an objective and impartial presentation of information, and so you should always be as careful as possible to use words without misleading or unwanted connotations.

On the other hand, a word may have several equally valid denotative meanings. Context will make the meaning clear, as in the following varied uses of the word *check.*

EXAMPLES   The bank verified the *check.*
The driver was instructed to *check* the oil before every trip.
The chess champion was in *check* three times, but he eventually won the match.
The hockey player executed a bruising *check.*

(See also **defining terms.**)

consensus of opinion

Since *consensus* normally means "harmony of opinion," the **phrase** *consensus of opinion* is redundant. The word *consensus* can only be used of a group, never one or two people.

CHANGE   The *consensus of opinion* of the members was that the committee should change its name.
TO   The *consensus* of the members was that the committee should change its name.

continual/continuous

Continual means "happening over and over" or "frequently repeated."

EXAMPLE   Writing well requires *continual* practice.

Continuous means "occurring without interruption" or "unbroken."

EXAMPLE   The *continuous* roar of the machines was deafening.

contractions

A contraction is a shortened **spelling** of a word or **phrase** with an **apostrophe** substituting for the missing letters.

> EXAMPLES   cannot / can't
> will not / won't
> have not / haven't
> it is / it's

Contractions are often used in speech but should be used discriminatingly in **reports,** formal letters, and most technical writing. (See also **formal writing style** and **informal writing style.**)

coordinating conjunctions

A coordinating conjunction is a word that joins two **sentence** elements that have identical functions. The coordinating conjunctions include *and, but, or, for, nor, yet,* and *so.*

> EXAMPLE   The first impact of the cutback was on the purchasing department, *but* it is now making itself felt in our department as well.

Coordinating conjunctions can join words, **phrases,** or **clauses** of equal rank.

> EXAMPLES   Only crabs *and* sponges managed to escape the red tide. (words)
> Our twin objectives are to increase power *and* to decrease noise levels. (phrases)
> The average city dweller in the United States now has 0.17 parts per million of lead in his blood, *and* the amount is increasing yearly. (clauses)

Under normal circumstances, use coordinating conjunctions only to connect **parallel structures.**

copyright

Copyright is the right of exclusive ownership by an author of the benefits resulting from the creation of his or her work. This right gives authors, or others to whom they transfer ownership of copyright, control over the reproduction and dissemination of their works. Once copyrighted, a work cannot be indiscriminately reproduced unless the copyright owner is compensated by royalties or other arrangements.

This right is protected for the life of the author plus 50 years by a federal law that became effective on January 1, 1978. This law protects any work from the moment of its creation, regardless of whether it is ever published and regardless of whether it contains a notice of copyright. At the end of the copyright period, the work becomes "public domain" and may be published or reproduced by anyone without permission or compensation to the copyright owner's estate.

The copyright law provides a "fair use" provision that allows teachers, librarians, reviewers, and others to reproduce copyrighted materials for educational and illustrative purposes without compensation to the copyright owners. The law refers to such purposes as "criticism, comment, news reporting, teaching (including multiple copies for classroom use), scholarship, or research." Although the law does not exactly define "fair use," it sets out four criteria for determining whether a given use is fair: (1) the purpose and nature of the use, (2) the amount of material used in relation to the copyrighted work as a whole, (3) the nature of the copyrighted work, and (4) the effect of the use on the potential market for or value of the copyrighted work. Publications that explain the copyright law in detail are available from the Copyright Office, Library of Congress, Washington, D.C. 20559.

If you use copyrighted materials in your written work, be aware of the following guidelines.

1. A small amount of material from a copyrighted source may be used in your written work without permission or payment as long as the use is reasonable and not harmful to the rights of the copyright owner.
2. A work created after January 1, 1978, receives copyright protection whether or not it bears a notice of copyright, although all published materials contain such a notice, usually on the back of the title page. (For an example of a copyright notice, turn to the back of the title page of this book.)
3. All publications of U.S. government agencies are in the public domain—that is, they are *not* copyrighted.
4. One way to determine whether or not a work from which you wish to quote is in the public domain is to write to the Copyright Office at the address cited in this entry. There is a search fee for this service.

correlative conjunctions

Correlative conjunctions are **coordinating conjunctions** that are used in pairs. The correlative conjunctions are *either . . . or, neither . . . nor, not only . . . but also, both . . . and,* and *whether . . . or.*

EXAMPLES　The shipment will arrive *either* on Wednesday *or* on Thursday.
The shipment contains the parts *not only* for the seven machines scheduled for delivery this month *but also* for those scheduled for delivery next month.

To add not only symmetry but logic to your writing, follow correlative conjunctions with parallel **sentence** elements (such as *on Wednesday* and *on Thursday* in the previous example).

A problem can develop with correlative conjunctions joining a singular **noun** to a plural noun. The problem is with the **verb.** If you use this construction, make the verb agree with the nearest noun.

EXAMPLE　Ordinarily, neither the pressure switch nor the relay *coils fail* the test. ("Ordinarily, neither the pressure switch nor the relay coils fails the test" would be grammatically incorrect.)

A common solution is to change the construction of the sentence.

EXAMPLE　Ordinarily, the test is failed by neither the pressure switch nor the relay coils.

correspondence

All effective business letters have certain characteristics. The following list summarizes the basic steps to complete when planning and writing a letter.

1. Establish your **objective.**
2. **Research** your **topic,** if necessary.
3. Profile your **reader.**
4. Prepare an **outline,** even if only a list of topics to be covered.
5. Write a first draft.
6. Revise the draft—if possible, after a "cooling" period of a couple hours.

These guidelines will help you write a clear, well-organized letter. (You might also find it valuable to look at the "Five Steps to Successful Writing" on page xiii.) Keep in mind, however, that one very important element in business letters is the impression they leave on the reader. To convey the right impression—of yourself as well as of your company— you must take particular care with both the **tone** and the **style** of your writing.

TONE

Letters are generally written directly to another person who is identified by name. You may or may not know the person, but never forget that you are writing to another human being. Letters are always more personal than **reports** or other forms of technical writing. Carefully composed, they can leave your reader with exactly the impression you want to communicate. Through your letters you can let your reader know that you are informed, appreciative, or perhaps simply looking forward to seeing him at some future meeting.

Strive to be courteous and sincere. Ask yourself, how you would feel to receive your letter. Then tailor your writing accordingly. Suppose, for example, that you were a department store manager responding to a refund request from a customer who forgot to enclose his or her receipt. In your letter, you might say, "The sales receipt must be enclosed with the merchandise before the refund can be processed." Put yourself in your reader's place, however, and you would probably word the request more like this: "Please enclose the sales receipt with the merchandise so that we can send your refund promptly."

How would you react to Letter 7? The letter has a distinctly curt, nasty tone. Even though the letter addresses the reader directly, the writer has apparently not considered how the recipient will feel as she reads the letter. There is no expression of regret that Mrs. Mauer is being rejected for the position, nor any appreciation of her efforts in applying for the job. Such a letter gives the impression that its writer is unpleasant and has bad manners.

How might the letter have appeared if the writer had been courteous and considerate? Look at Letter 8. This letter carries the same disappointing news as does the first, but the writer is careful to thank the reader for her time and effort, to explain why she was not accepted for the job, and to offer her encouragement in finding another position.

Southtown Dental Center
3221 Ryan Road
San Diego, California 92217
(714) 321-1579

November 11, 19--

Mrs. Barbara L. Mauer
157 Beach Drive
San Diego, CA 92113

Dear Mrs. Mauer:

Your application for the position of office manager at Southtown Dental Center has been rejected. We have found someone more qualified than you.

Sincerely,

Mary Hernandez
Personnel Director

MH/bt

Letter 7

Southtown Dental Center
3221 Ryan Road
San Diego, California 92217
(714) 321-1579

November 11, 19--

Mrs. Barbara L. Mauer
157 Beach Drive
San Diego, CA 92113

Dear Mrs. Mauer:

Thank you for your time and effort in applying for the
position of office manager at Southtown Dental Center.

Since we needed someone who can assume the duties here with a
minimum of training, we have selected an applicant with over
ten years of experience.

I am sure that with your excellent college record you will
find a position in another office.

Sincerely yours,

Mary Hernandez
Mary Hernandez
Personnel Director

MH/bt

Letter 8

The following tips should help you achieve a tone that shows consideration for your reader.

1. Be respectful, not demanding.

DEMANDING   Submit your answer in one week.
RESPECTFUL   I would appreciate your answer in one week.

2. Be modest, not arrogant.

ARROGANT   My report is thorough, and I'm sure that you won't be able to continue efficiently without it.
MODEST   I have tried to be as thorough as possible in my report, and I hope you find it useful.

3. Be polite, not sarcastic.

SARCASTIC   I just received the shipment we ordered *six months ago*. I'm sending it back—we can't use it now. Thanks!
POLITE   I am returning the shipment we ordered on March 12, which arrived on August 4. Unfortunately, it arrived too late for us to be able to use it.

4. Be positive and tactful, not negative and condescending.

NEGATIVE   Your complaint about our prices is way off target. Our prices are no higher than those of our competitors.
POSITIVE   Thank you for your suggestion concerning our prices. We believe, however, that our prices are competitive with, and in some cases below, those of our competitors.

WRITING STYLE

Letter-writing style may legitimately vary in formality—you may be very informal in a letter to a close business associate, while remaining more formal, or restrained, in a letter to someone you do not know. (Even if you are writing a business letter to a close associate, of course, you must always follow the rules of standard grammar, spelling, and punctuation.)

INFORMAL   It worked! The new process is even better than we had dreamed.
RESTRAINED   I am pleased to report that the new process is even more effective than we had expected.

You will probably use the **formal style** more frequently than the **informal style,** since an obvious attempt to sound casual may strike the reader as insincere. Do not adopt such a formal style, however, that your letters read like legal contracts. Using legalistic-sounding words in an effort to impress your reader will make your writing seem stuffy and pompous—and may well intimidate or irritate your reader. Consider how stiff Letter 9 sounds. The writing style is full of old-fashioned business jargon—expressions like *I wish to state, be advised that,* and *enclosed herewith*—that are both out-of-date and pretentious. Good business letters today have a more personal, down-to-earth style, as Letter 10 illustrates.

Many writers achieve a more informal style by imagining that the person they are writing to is seated across the desk from them; the writer then "speaks" directly to the reader. The advantage of this approach is that you are instinctively more conscious and considerate of your reader, particularly in your use of the pronoun *you.*

CHANGE   We will deliver the order by February 9.
TO   You will receive your order by February 9.

CHANGE   We can permit you to attend classes on company time only when the course is related to your work.
TO   You may attend classes on company time as long as the course is related to your work.

ACCURACY

Since a letter is a written record, it must be accurate. Facts, figures, dates, and explanations that are incorrect or misleading may cost time, money, and goodwill. Remember that when you sign a letter, you are responsible for its contents. Always allow yourself time to review a letter before mailing it. Whenever possible, ask someone who is familiar with its contents to review an important letter. Listen with an open mind to the criticism of others about what you have said, and make any changes you believe necessary.

A second kind of accuracy to check for is the mechanics of writing—**punctuation,** grammar, and **spelling.**

APPEARANCE

It is important that your business letters always be neat and attractive. Type on unruled white bond paper of standard size, and use envelopes of the same quality. Center the letter on the page vertically and horizontally. Be sure that the space between the top of the page and the first line of typing is about equal to the space between the last line of typing and the

---

Amex Laboratories

327 Wilson Avenue
Birmingham, Alabama 35211
(205) 743-6218

September 7, 19--

Mr. Roland E. Forbes
772 South Wilton Street
Birmingham, AL 35207

Dear Mr. Forbes:

In response to your query, I wish to state that we no longer
have an original copy of the brochure requested. Be advised
that a photographic reproduction is enclosed herewith.

Address further correspondence to this office for assistance
as required.

Sincerely yours,

E.T. Hillman

E. T. Hillman

ETH:knt
Enclosure

Letter 9

Amex Laboratories

327 Wilson Avenue
Birmingham, Alabama 35211.
(205) 743-6218

September 7, 19--

Mr. Roland E. Forbes
772 South Wilton Street
Birmingham, AL 35207

Dear Mr. Forbes:

Because we are currently out of original copies
of our brochure, I am sending a photocopy of
it.

If I can be of further help, please let me know.

Sincerely,

E.T. Hillman

E. T. Hillman

ETH:knt
Enclosure

Letter 10

lower edge. When you use company letterhead, consider the bottom of the letterhead as the top edge of the paper.

Remember that a neat appearance cannot improve a poorly written letter, but a sloppy appearance will detract from a well-written letter.

PARTS OF A BUSINESS LETTER

If your employer recommends or requires a particular **format** and typing style, use it. Otherwise, follow the guidelines provided here.

Heading. The writer's full address (street, city, state, and ZIP code) and the date are given in the heading. Because the writer's name appears at the end of the letter, it need not be included in the heading. Words like *street, avenue, first,* or *west* should be spelled out rather than abbreviated. You may either spell out the name of the state in full or use the U.S. Postal Service abbreviations given in **abbreviations.** The date usually goes directly beneath the last line of the address. Do not abbreviate the name of the month.

EXAMPLE   1638 Parkhill Drive East
Great Falls MT 59407
April 8, 19—

Align the heading on the page just to the right of center. If you are using company letterhead that gives the address, type in only the date, two spaces below the last line of printed copy.

The Inside Address. The recipient's full name and address is given in the inside address. You can begin the inside address two spaces below the date if the letter is long, or four spaces below if the letter is quite short. The inside address should be aligned with the left margin—and the left margin should be at least one inch wide. Include the reader's full name and title (if you know them) and his or her full address, including ZIP code.

EXAMPLE   Ms. Gail Silver
Production Manager
Quicksilver Printing Company
14 President Street
Sarasota FL 33546

The Salutation. Place the salutation, or greeting, two spaces below the inside address, also aligned with the left margin. In most business letters, the salutation contains the recipient's title (*Mr., Ms., Dr.,* etc.) and last name, followed by a **colon.** If you are on a first-name basis with the recipient, you would include his or her title and full name on the inside address, but use only the first name in the salutation. Notice that the titles *Mr., Ms., Mrs.,* and *Dr.* may be abbreviated, but that other titles, such as *Captain* or *Professor,* should always to be spelled out.

EXAMPLES Dear Ms. Silver:
Dear Dr. Lee:
Dear Captain Ortiz:
Dear Professor Murphy:
Dear George:

Address women without a professional title by *Ms.,* whether they are married or unmarried. If a woman has expressed a preference for *Miss* or *Mrs.,* however, honor her preference. In cases where you do not know whether the recipient is a man or a woman, you may use a title appropriate to the context of the letter. The following are examples of the kinds of titles you may find suitable.

EXAMPLES Dear Customer:
Dear Homeowner:
Dear Parts Manager:

In the past, correspondence to large companies or organizations was customarily addressed to "Gentlemen." Today, however, writers who do not know the name or title of the recipient often address the letter to the appropriate department. Notice, in the following examples, that the name of the department may appear either in an attention line or in a subject line.

EXAMPLES National Business Systems
501 West National Avenue
Minneapolis, Minnesota 55407

Attention: Customer Relations Department

I am returning three calculators that failed to operate. . . .

National Business Systems
501 West National Avenue
Minneapolis, Minnesota 55407

Subject: Defective Parts for SL-100

I am returning three calculators that failed to operate. . . .

The Body. The body of the letter should begin two spaces below the salutation (or directly below the heading if there is no salutation). Single-space within paragraphs and double-space between paragraphs. If a letter is very short and you want to suggest a fuller appearance, you may instead double-space throughout and indicate paragraphs by indenting the first line of each paragraph five spaces from the left. The right margin should be approximately as wide as the left margin. (In very short letters, you may increase both margins to about an inch and a half.)

Two very important elements in the body are the **opening** and closing. One effective way to arrange your letter is to open with a short **paragraph,** followed by one or more longer paragraphs for the body itself, and close with another short paragraph for a **conclusion.** This is called a diamond arrangement, and it can help to focus the contents of your letter. Never underestimate the importance of a letter's opening and closing; in fact, their positions of **emphasis** make them particularly significant.

In your opening you should identify your subject so as to focus its relevance for the reader. Remember that your reader may not immediately recognize or see the importance of your topic—indeed, he or she may be preoccupied with some other business completely. Therefore, it is important to focus his or her attention on the subject at hand. Be particularly careful to get directly to the point; leave out any less important details.

EXAMPLE   I am writing to suggest ways that we at Benson Tubular Steel might attract more interest from graduates of local colleges and universities. As you know, this year only twelve local college graduates have applied to work at our firm. Last year over thirty graduates applied, and the year before fifty applied. After talking with several college counselors, I am confident that we can solve the problem of decreasing applications.

First, we should resume our advertisements in local college newspapers . . .

Your closing should tie together all the letter's important points and can do so emphatically. Your most significant points can be restated here. For example, a letter about possible locations for a new plant might end with a recommendation. Or, a letter about a company's annual sales figures could end with a judgment about why sales were up or down. Lengthy letters often close by summarizing the main points. You can use a closing for different purposes: You may merely present ideas for consideration, but you can also call for action or deliberately provoke thought.

Because a closing is in a position of emphasis, be especially careful to avoid **clichés.** Of course, some very commonly used closings are so precise that they are hard to replace.

EXAMPLES It has been a pleasure to be of assistance.
If I can answer any questions about the enclosed report, please don't hesitate to call me.
We thank you again for your cooperation.

Be sure not to introduce a new topic in the closing; remember that a conclusion must relate to and reinforce ideas already presented in the letter.

The Complimentary Closing. Type the complimentary closing two spaces below the body. Use a standard expression like *Yours truly, Sincerely,* or *Sincerely yours.* (If the recipient is a friend as well as a business associate, you can use a less formal closing, such as *Best wishes* or *Best regards.*) Capitalize only the initial letter of the first word, and follow the expression with a **comma.** Type your full name four spaces below, aligned with the closing at the left. On the next line you may type in your business title, if it is appropriate to do so. Write your signature in the space between the complimentary closing and your typed name. If you are writing to someone with whom you are on a first-name basis, it is acceptable to sign only your given name; otherwise, sign you full name.

EXAMPLE Sincerely,

Thomas R. Castle

Thomas R. Castle
Treasurer

Sometimes the complimentary closing is followed by the name of the firm. Then comes the actual signature, between the name of the firm and the typed name.

EXAMPLE Sincerely yours,
VIKING SUPPLY COMPANY, INC.

Laura A. Newland

Laura A. Newland
Controller

```
Ms. M. C. Marks
Page 2
March 12, 19--
```

```
Ms. M. C. Marks                    2              March 12, 19--
```

A Second Page. If a letter requires a second page, always carry at least two lines of the body text over to that page; do not use a continuation page to type only the letter's closing. The second page should be typed on plain bond paper, and should have a heading with the recipient's name, the page number, and the date. The heading may go either in the upper left-hand corner or across the page.

Additional Information. Business letters sometimes require the typist's initials, an enclosure notation, or a notation that a copy of the letter is being sent to one or more people. Place any such information at the left margin, two spaces below the last line of the complimentary closing in a long letter, four spaces below in a short letter.

The *typist's initials* should follow the letter writer's initials, and the two sets of initials should be separated by either a colon or a slash. The letter writer's initials should be in capital letters, and the typist's initials should be in lower-case letters. (When the writer is also the typist, no initials are needed.)

EXAMPLES   CBG:pbg
           APM/sjl

Enclosure notations indicate that the writer is sending material along with the letter (an invoice, an article, and so on). They may take several forms, as illustrated below; choose the form that seems most helpful to your readers.

EXAMPLES   Enclosure: Preliminary report invoice
           Enclosures (2)
           Enc. (or Encs.)

Enclosure notations are included in long, formal letters or in any letters where the enclosed items would not be obvious to the reader. Remember, though, that an enclosure notation cannot stand alone: You must mention the enclosed material in the body of the letter.

Copy notations tell the reader that a copy of the letter is being sent to one or more named individuals. You should add a copy notation for both carbon copies and photocopies.

EXAMPLE   cc: Ms. Marlene Brier
              Mr. David Williams
              Ms. Bonnie Ng
              Ms. Robin Horton

A business letter may, of course, contain all three items of additional information.

EXAMPLE   Sincerely yours,

Jane T. Rogers

Jane T. Rogers

JTR/pst
Enclosure: Preliminary report invoice
cc: Ms. Marlene Brier
    Mr. David Williams
    Ms. Bonnie Ng
    Ms. Robin Horton

Letterhead

EVANS & ASSOCIATES
520 Niagara Street
Lexington, KY 40502

(502) 787-1175
TELEX 5-72118

Date

May 15, 19--

Inside
address

Mr. George W. Nagel
Director of Operations
Boston Transit Authority
57 West City Avenue
Boston, MA 02110

Salutation

Dear Mr. Nagel:

Enclosed is our final report evaluating the safety
measures for the Boston Intercity Transit System.

Body

We believe the report covers the issues you raised and
that it is self-explanatory. However, if you have any
further questions, we would be happy to meet with you at
your convenience.

We would also like to express our appreciation to
Mr. L. K. Sullivan of your committee for his generous
help during our trips to Boston.

Complimentary
close

Sincerely,

Signature

Carolyn Brown

Typed name

Carolyn Brown, Ph.D.
Director of Research

Title

Additional
information

CB/ls
Enclosure: Final Safety Report
cc: ITS Safety Committee Members

Full Block Letter Style, With Letterhead

LETTER FORMATS

The two most common business letter formats are the full block style and
the modified block style. The *full block style,* though easier to type because
every line begins at the left margin, is suitable only with letterhead
stationery. In the *modified block style* the return address, the date, and the

Heading just right of center	3814 Oak Lane Lexington, KY 40514 December 8, 19--
Inside address	Dr. Carolyn Brown Director of Research Evans & Associates 520 Niagara Street Lexington, KY 40502
Salutation	Dear Dr. Brown:
Body	Thank you very much for allowing me to tour your testing facilities. The information I gained from the tour will be of great help to me in preparing the report for my class at Marshall Institute. The tour has also given me some insight into the work I may eventually do as a laboratory technician. I especially appreciated the time and effort Steve Dement spent in showing me your facilities. His comments and advice were most helpful.
Complimentary close aligned with heading	Again, thank you.
Signature	Sincerely, *Leslie Warden*
Typed name	Leslie Warden

Modified Block Letter Style, Without Letterhead

complimentary closing are all placed just to the right of the center of the page, and the other elements are aligned at the left margin. All other letter styles are variations of these two basic styles. Again, if your employer recommends or requires a particular style, follow it carefully. Otherwise, choose the style you are most comfortable with and follow it consistently.

count nouns

Count nouns, as opposed to **mass nouns,** identify persons, places, or things that can be separated into individual units and counted.

EXAMPLE   There were four *calculators* in the office.

For information about the **case, number,** and function of count nouns, see **nouns.** (See also **fewer/less.**)

credible/creditable

Something is *credible* if it is believable.

EXAMPLE   The statistics in this report are *credible.*

Something is *creditable* if it is worthy of praise or credit.

EXAMPLE   The chief engineer did a *creditable* job.

criterion/criteria/criterions

Criterion means "an established standard for judging or testing." *Criteria* and *criterions* are both acceptable plural forms of *criterion,* but *criteria* is generally preferred.

EXAMPLE   In evaluating this job, we must use three *criteria.* The most important *criterion* is quality of workmanship.

critique

A *critique* **(noun)** is a written or oral evaluation of something. Avoid using *critique* as a **verb** meaning "criticize."

CHANGE   Please *critique* his job description.
TO   Please prepare a *critique* of his job description.

D

dangling modifiers

Verbal phrases **(gerund, participle, infinitive)** that do not clearly and logically refer to the proper **noun** or **pronoun** are called *dangling modifiers*.

CHANGE   While eating lunch in the cafeteria, the computer malfunctioned. (The problem, of course, is that the sentence neglects to mention *who* was eating lunch in the cafeteria.)

TO   While *the operator* was eating lunch in the cafeteria, the computer malfunctioned.

DANGLING GERUND PHRASES

A dangling gerund phrase can be corrected by adding a noun or a pronoun for it to modify.

CHANGE   After finishing the research, the job was easy.

TO   After finishing the research, *we* found the job to be easy.

Another way to correct a dangling gerund phrase is to make the **phrase** a **clause.**

CHANGE   After finishing the research, the job was easy.

TO   After we finished the research, the job was easy.

OR   The job was easy after we finished the research.

DANGLING PARTICIPIAL PHRASES

A dangling **participial phrase** (a participial phrase that has no noun or pronoun for the phrase to modify) can be corrected by providing a subject that stands near the phrase.

CHANGE   Keeping busy, the afternoon passed swiftly.

TO   Keeping busy, *I* felt that the afternoon passed swiftly.

A dangling participial phrase may also be corrected by making the phrase a clause.

CHANGE   Keeping busy, the afternoon passed swiftly.

TO   Because I kept busy, the afternoon passed swiftly.

CHANGE   Entering the gate, the administration building is visible.

TO   As you enter the gate, the administration building is visible.

DANGLING INFINITIVE PHRASES

Correct a dangling **infinitive phrase** by supplying the missing noun or pronoun that provides the subject of the infinitive phrase.

CHANGE   To improve typing skills, practice is needed.
TO   To improve typing skills, *you* must practice.

CHANGE   To evaluate the feasibility of the project, the centralized plan will be compared with the present system of dispersing facility sites.
TO   To evaluate the feasibility of the project, the *committee* will compare the centralized plan with the present system of dispersing facility sites.

DANGLING SUBORDINATE CLAUSES

Occasionally a **subject** and **verb** are omitted from a **depending clause,** forming what is known as an *elliptical clause.* If the omitted subject of the elliptical clause is not the same as the subject of the main clause, the construction dangles. Simply adding the subject and verb to the elliptical clause solves the problem.

CHANGE   When ten years old, his father started the company.
TO   When *Bill* was ten years old, his father started the company.
OR   Bill was ten years old when his father started the company.

(See also **misplaced modifiers** and **verbals.**)

dashes

The dash is a versatile, yet limited, mark of **punctuation.** It is versatile because it can perform all the duties of punctuation (to link, to separate, and to enclose). It is limited because it is an especially emphatic mark that is easily overused. Use the dash cautiously, therefore, to indicate more informality or **emphasis** (a dash gives an impression of abruptness) than would be achieved by the conventional punctuation marks. In some situations, a dash is required; in others, a dash is a forceful substitute for other marks.

A dash can emphasize a sharp turn in thought.

EXAMPLE   That was the end of the project—unless the company provided additional funds.

A dash can indicate an emphatic pause.

EXAMPLES   Consider the potential danger of a household item that contains mercury—a very toxic substance.
The job will be done—after we are under contract.

Sometimes, to emphasize contrast, a dash is also used with *but*.

EXAMPLE   We may have produced work more quickly—but the result was not as good.

A dash can be used before a final summarizing statement or before **repetition** that has the effect of an afterthought.

EXAMPLE   It was hot near the ovens—steaming hot.

Such a thought may also complete the meaning.

EXAMPLE   We try to speak as we write—or so we believe.

A dash can be used to set off an explanatory or **appositive** series.

EXAMPLE   Three of the applicants—John Evans, Mary Fontana, and Thomas Lopez—seem well qualified for the job.

Dashes set off parenthetical elements more sharply and emphatically than do **commas**. Unlike dashes, **parentheses** tend to reduce the importance of what they enclose. Compare the following **sentences.**

EXAMPLES   Only one person—the president—can authorize such activity.
Only one person, the president, can authorize such activity.
Only one person (the president) can authorize such activity.

Use dashes for **clarity** when commas appear within a parenthetical element; this avoids the confusion of too many commas.

EXAMPLE   Retinal images are patterns in the eye—made up of light and dark shapes, in addition to areas of color—but we do not see patterns; we see objects.

The first word after a dash is never capitalized unless it is a proper noun. In typing, use two consecutive **hyphens** (--) to indicate a dash.

data

The debate over whether *data* should be treated as a plural **noun** or a collective singular continues. In most technical writing, *data* is considered a collective singular. In strictly formal scientific and scholarly writing, however, *data* is generally used as a plural, with *datum* as the singular form.

EXAMPLES   The *data are* voluminous in support of a link between smoking and lung cancer. (formal)
The *data is* now ready to be evaluated. (less formal)

Base your decision upon whether your **reader** should consider the data as a single collection or as a group of individual facts. Whatever you decide, be sure that your **pronouns** and **verbs** agree in **number** with the selected **usage.**

EXAMPLES   The *data are* voluminous. *They indicate* a link between smoking and lung cancer.
The *data is* now ready for evalution. *It is* in the mail.

dates

In business and industry, dates have traditionally been indicated by the month, day, and year, with a **comma** separating the figures.

EXAMPLE   October 26, 1981

The day-month-year system used by the military does not require commas.

EXAMPLE   26 October 1981

A date can be written with or without a comma following the year, depending on how the date is expressed. If the date is in the month-day-year format, set off the year with commas.

EXAMPLE   October 26, 1981, was the date the project began.

If the date is in the day-month-year format, do not set off the date with commas.

EXAMPLE   26 October 1981 was the date the project began.

The strictly numerical form for dates (10/26/81) should be used sparingly, and never on business letters or formal documents, since it is less immediately clear. When this form is used, the order in American usage is always month/day/year. For example, 5/7/81 is May 7, 1981.

CENTURIES

Confusion often occurs because the spelled-out names of centuries do not correspond to the numbers of the years. (See also **numbers.**)

EXAMPLES   The twentieth century is the 1900s (1900-1999).
The nineteenth century is the 1800s (1800-1899).
The fifth century is the 400s (400-499).

decided/decisive

When something is *decided,* it is "clear-cut," "without doubt," "unmistakable." When something is *decisive,* it is "conclusive" or "has the power to settle a dispute."

EXAMPLE   The executive committee's *decisive* action gave our firm a *decided* advantage over the competition.

decreasing-order-of-importance method of development

Decreasing order of importance is a **method of development** in which the writer's major points are arranged in descending order of importance. It begins with the most important fact or point, then goes to the next most important, and so on, ending with the least important. It is an especially appropriate method of development for a **report** addressed to a busy decision-maker, who may be able to reach a decision after considering only the most important points, or for a report addressed to various **readers,** some of whom may be interested only in the major points and others in all of the points.

The advantages of this method of development are that (1) it gets the reader's attention immediately by presenting the most important point first, (2) it makes a strong initial impression, and (3) it ensures that the hurried reader will not miss the most important point.

The following example uses decreasing order of importance.

MEMORANDUM

```
To:   Mary Vincenti, Chief, Personnel Department
From:  Frank W. Russo, Chief, Claims Department
Date:  November 13, 19--
Subject:  Selection of Chief of the Claims
          Processing Section
```

The most qualified candidate for Chief of the
Claims Processing Section is Mildred Bryand, who
is at present Acting Chief of the Claims
Processing Section. In her twelve years in the
Claims Department, Mrs. Bryand has gained wide
experience in all facets of the department's
operations. She has maintained a consistently
high production record and has demonstrated the
skills and knowledge that are required for the
supervisory duties she is now handling in an
acting capacity. A number of additional qualifications also make her the best candidate: her
valuable contributions to and cooperation during
the automation of the department's claims
records; her performance on several occasions as
acting chief in other sections of the Claims
Department; the continual rating of "outstanding"
she has received in all categories of her
job-performance appraisals.

Michael Bastik, Claims Coordinator, our second
choice, also has strong potential for the
position. An able administrator, he has been
with the company for seven years. For the past
year he has been enrolled in several management
training courses at the university. He is ranked
second, however, because he lacks supervisory
experience and because his most recent work has
been with the department's maintenance and supply
components. He would be the best person to fill
many of Mildred Bryand's responsibilities if she
should be made full-time Chief of the Claims
Processing Section.

Jane Fine, our third-ranking candidate, has
shown herself a skilled administrator in her
three years with the Claims Consideration Section.
Despite her obvious potential, my main objection
to her is that, compared with the other top
candidates, she lacks the breadth of experience
in claims processing that would be required of
someone responsible for managing the Claims
Processing Section. Jane Fine also lacks on-
the-job supervisory experience.

de facto/de jure

De facto means that someting "exists" or is "a fact" and therefore must
be accepted for practical purposes. *De jure* means that something "legally" or "rightfully" should or does exist.

EXAMPLE   The law states that no signs should be erected along Highway
127. Store owners have disregarded this law, and, therefore,
many signs exist along Highway 127. The presence of the
signs along Highway 127 is *de facto* but not *de jure*—their
presence is "a fact," but it is not "lawful."

defective/deficient

If something is *defective* it is faulty.

EXAMPLE   The wiring was *defective*.

If something is *deficient* it lacks a necessary ingredient.

EXAMPLE   The compound was found to be *deficient* in calcium.

defining terms

One of the most basic rules of good writing is to be sure that your **reader** understands and follows what he is reading. You must always keep in mind his familiarity with the topic and be careful to define any terms that he may not understand.

Terms can be defined either formally or informally, depending on your purpose and on your reader. (For techniques on expanding a definition to explain the meaning of a term, see **definition method of development**.)

A formal definition is a form of classification. A term is defined by placing it in a category and then identifying what features distinguish it from other members of the same category. Following are several examples of formal definitions.

Term	*Category*	*Distinguishing Features*
A spoon is	an eating utensil	that consists of a small shallow bowl on the end of a handle.
An auction is	a public sale	in which property passes to the highest bidder through successively increased offers.
An annual is	a plant	that completes its life cycle, from seed to natural death, in one growing season.
A contract is	a binding agreement	between two or more persons or parties.
A lease is	a contract	that conveys real estate for a term of years at a specified rent.

An *informal definition* explains a term by giving a familiar word or **phrase** as a **synonym.**

EXAMPLES   An *invoice* is a bill.
Many states have set up wildlife *habitats* (or living spaces).
Plants have a *symbiotic,* or mutually beneficial, relationship with certain kinds of bacteria.

Definitions should normally be positive; focus on what the term is rather than on what it is not. For example, "In a legal transaction, real property is not personal property," although a true statement, does not tell the reader exactly what real property *is*. The definition could just as easily be stated positively: "Real property is legal terminology for the right or interest a person has in land and the permanent structures on that land." (For a discussion of when negative definitions are appropriate, see **definition method of development.**)

When writing a definition, keep in mind the following pitfalls that may result in confusing, inaccurate, or incomplete definitions.

Avoid circular definitions, which merely restate the term to be defined and therefore fail to clarity it.

CIRCULAR   *Spontaneous combustion* is fire that begins spontaneously.
REVISED   *Spontaneous combustion* is the self-ignition of a flammable material through a chemical reaction like oxidation and temperature buildup.

Avoid "is when" and "is where" definitions. Such definitions are too indirect: They say when or where rather than what.

"IS WHEN"   A *contract* is when two or more people agree to something.
REVISED   A *contract* is a binding agreement between two or more people.

"IS WHERE"   A *day-care center* is where working parents can leave their preschool children during the day.
REVISED   A *day-care center* is a facility at which working parents can leave their preschool children during the day.

Avoid definitions that include terms that may be unfamiliar to your readers. Even informal writing will occasionally require terms used in a special sense that your readers may not understand; such terms should always be defined.

EXAMPLE   In these specifications the term *safety can* refers to an approved container of not more than five-gallon capacity having a spring-closing spout cover designed to relieve internal pressure when the can is exposed to fire.

definite/definitive

Definite and *definitive* both apply to what is "precisely defined," but *definitive* more often refers to what is complete and authoritative.

EXAMPLE   When the committee took a *definite* stand, the president made it the *definitive* company policy.

definition method of development

To define something is basically to identify precisely its fundamental qualities. In technical writing, as in all writing, clear and accurate definitions are critical. Sometimes simple, dictionary-like definitions suffice (see **defining terms**), but other times definitions need to be expanded with additional details, examples, comparisons, or other explanatory devices.

When more than a **phrase** or a **sentence** or two is needed to explain an idea, use an *extended definition,* which explores a number of qualities of the item being defined. How an extended definition is developed depends on your reader's needs and on the complexity of the subject. A **reader** familiar with a **topic** or an area might be able to handle a long, fairly technical definition, whereas one less familiar with a topic would require simpler language and more basic information.

Probably the easiest way to give an extended definition is with specific examples. Listing examples gives your reader easy-to-picture details that help him to see and thus to understand the term being defined.

EXAMPLE   Form, which is the shape of landscape features, can best to represented by both small-scale features, such as *trees and shrubs,* and by large-scale elements, such as *mountains and mountain ranges.*

Another useful way to define a difficult concept, especially when you are writing for nonspecialists, is to use an **analogy** to link the unfamiliar concept with a simpler or more familiar one. An analogy is a resemblance in certain aspects between things otherwise unalike, and it can help the reader to understand an unfamiliar term by showing its similarities with a more familiar one. Defining radio waves in terms of the length (long) and frequency (low), a writer develops an analogy to show why a low frequency is advantageous.

EXAMPLE   The low frequency makes it relatively easy to produce a wave having virtually all its power concentrated at one frequency. Think, for example, of a group of people lost in a forest. If they hear sounds of a search party in the distance, they all will begin to shout for help in different directions. Not a very efficient process, is it? But suppose that all the energy which went into the production of this noise could be concentrated into a single shout or whistle. Clearly the chances that the group will be found would be much greater.

Some terms are best defined by an explanation of their causes. Writing in a professional journal, a nurse describes an apparatus used to monitor blood pressure in severely ill patients. Called an indwelling catheter, the device displays blood-pressure readings on an oscilloscope and on a numbered scale. Users of the device, the writer explains, must understand what a *dampened wave form* is.

EXAMPLE   The *dampened wave form*, the smoothing out or flattening of the pressure wave form on the oscilloscope, is *usually caused by an obstruction* that prevents blood pressure from being freely transmitted to the monitor. The obstruction may be a *small clot or bit of fibrin* at the catheter tip. More likely, *the catheter tip* has become *positioned against the artery wall* and is preventing the blood from flowing freely.

Sometimes a formal definition of a concept can be made simpler by breaking the concept into its component parts. In the following example, the formal definition of *fire* is given in the first paragraph, and the component parts are given in the second.

EXAMPLE   Fire is the visible heat energy released from the rapid oxidation of a fuel. A substance is "on fire" when the release of heat energy from the oxidation process reaches visible light levels.

The classic fire triangle illustrates the elements necessary to create fire: *oxygen, heat,* and *burnable material* or *fuel.* Air provides sufficient oxygen for combustion; the intensity of the heat needed to start a fire depends on the characteristics of the burnable material or fuel. A burnable substance is one that will sustain combustion after an initial application of heat to start it.

Under certain circumstances, the meaning of a term can be clarified and made easier to remember by an exploration of its origin. Scientific and medical terms, because of their sometimes unfamiliar Greek and Latin roots, benefit especially from an explanation of this type. Tracing the derivation of a word can also be useful when you want to explain why a word has favorable or unfavorable associations—particularly if your goal is to influence your reader's attitude toward an idea or an activity.

EXAMPLE   Efforts to influence legislation generally fall under the head of *lobbying,* a term that once referred to people who prowl the lobbies of houses of government, buttonholing lawmakers and trying to get them to take certain positions. Lobbying today is all of this, and much more, too. It is a respected—and necessary—activity. It tells the legislator which way the winds of public opinion are blowing, and it helps inform him of the implications of certain bills, debates, and resolutions he must contend with.

—Bill Vogt, *How to Build a Better Outdoors* (New York: David McKay Company, Inc., 1978), p. 93.

Sometimes it is useful to point out what something is not in order to clarify what it is. A negative definition is effective only when the reader is familiar with the item with which the defined item is contrasted. If you say *x* is not *y,* your readers must understand the meaning of *y* for the explanation to make sense. In a crane operators' manual, for instance, a negative definition is used to show that, for safety reasons, a hydraulic crane cannot be operated in the same manner as a lattice boom crane.

EXAMPLE   A hydraulic crane is *not* like a lattice boom crane in one very important way. In most cases, the safe lifting capacity of a lattice boom crane is based on the *weight needed to tip the machine.* Therefore, operators of friction machines sometimes depend on signs that the machine might tip to warn them of impending danger. This practice is very dangerous with a hydraulic crane. . . .

—*Operator's Manual* (Model W-180), Harnishfeger Corporation.

demonstrative adjectives

The demonstrative adjectives are *this, these, that,* and *those.* They are **adjectives** that "point to" the thing they modify, specifying its position in space or time. *This* and *these* specify closer position; *that* and *those* specify more remote position.

EXAMPLES   *This* proposal is the one we accepted.
That proposal would have been impracticable.
These problems remain to be solved.
Those problems are not insurmountable.

Notice that *this* and *that* are used with singular **nouns** and that *these* and *those* are used with plural nouns. Demonstrative adjectives often cause problems when they modify the nouns *kind, type,* and *sort.* Demonstrative adjectives used with these nouns should agree with them in **number.**

EXAMPLES   this kind / these kinds
            that type / those types
            this sort / these sorts

Confusion often develops when the **preposition** *of* is added ("this kind *of,*" "these kinds *of* ") and the **object** of the preposition is not made to conform in number to the demonstrative adjective and its noun.

CHANGE   *This kind* of hydraulic *cranes* is best.
    TO   *This kind* of hydraulic *crane* is best.

CHANGE   *These kinds* of hydraulic *crane* are best.
    TO   *These kinds* of hydraulic *cranes* are best.

The error can be avoided by remembering to make the demonstrative adjective, the noun, and the object of the preposition—all three—agree in number. This **agreement** not only makes the sentence correct but more precise.

Using demonstrative adjectives with words like *kind, type,* and *sort* can easily lead to vagueness. It is better to be more specific.

(For recommendations on the placement and usage of demonstrative adjectives, see **adjectives.**)

demonstrative pronouns

Demonstrative pronouns are actually the same words as **demonstrative adjectives** but they differ in **usage** (a demonstrative adjective modifies a **noun,** a demonstrative pronoun substitutes for a noun). The antecedent of a demonstrative pronoun must be the last noun of the preceding sentence—not the idea of the sentence—or it must be clearly identified within the same sentence.

EXAMPLES   *This* is the *specification* as we wrote it.
            These are the *statistics* we needed to begin the project.
            That was a *proposal* for drilling a water well.
            Those were *salesmen.*

Notice that, as with demonstrative adjectives, *this* and *that* are used for singular nouns and *these* and *those* for plural nouns.

dependent clauses

A dependent (or subordinate) clause is a group of words that has a **subject** and a **predicate** but nonetheless depends upon a main **clause** to complete its meaning. A dependent clause can function in a **sentence** as a **noun**, an **adjective**, or an **adverb**. As nouns, dependent clauses may function in the sentence as subjects, **objects**, or **complements**.

EXAMPLES   *That human beings can learn to control their glands and internal organs by direct or indirect means* is an established fact. (subject)
I learned *that drugs ordered by brand name can cost several times as much as drugs ordered by generic name.* (direct object)
The trouble is *that we cannot finish the project by May.* (subjective complement)

As adjectives, dependent clauses can modify nouns or **pronouns**. Dependent clauses are often introduced by **relative pronouns** and **relative adjectives** *(who, which, that).*

EXAMPLE   The man *who called earlier* is here. (modifying "man")

As adverbs, dependent clauses may express relationships of time, cause, result, or degree.

EXAMPLES   You are making an investment *when you buy a house.* (time)
A title search was necessary *because the bank would not issue a loan without one.* (cause)
The cost of financing a home will be higher *if discount points are charged.* (result)
Monthly mortgage payments should not be much more *than a person earns in one week.* (degree)

Dependent clauses are useful in making the relationship between thoughts more precise and succinct than if the ideas were presented in a series of **simple sentences** or **compound sentences**.

CHANGE   The sewage plant is located between Millville and Darrtown. Both villages use it. (two thoughts of approximately equal importance)
TO   The sewage plant, *which is located between Millville and Darrtown,* is used by both villages. (one thought subordinated to the other)

CHANGE   Title insurance is a policy issued by a title insurance company *and* it protects against any title defects, such as outside claimants. (two thoughts of approximately equal importance)

TO   Title insurance, *which* protects against any title defects, such as outside claimants, is a policy issued by a title insurance company. (one thought subordinated to the other)

Subordinate clauses are especially effective, therefore, for expressing thoughts that describe or explain another statement. Too much **subordination,** or a string of dependent clauses, however, may be worse than none at all.

CHANGE   Since interest rates on conventional loans tend to follow general conditions in the money market, *if* money is plentiful for lending and the demand for loans is low, interest rates will be forced down, *whereas* if money is scarce and the demand for loans is high, interest rates will be forced up.

TO   Interest rates on conventional loans tend to follow general conditions in the money market: if money is plentiful for lending and the demand for loans is low, interest rates will be forced down; if money is scarce and the demand for loans is high, interest rates will be forced up.

CHANGE   He had selected teachers *who* taught classes *that* had a slant *that* was specifically directed toward students *who* intended to go into business.

TO   He had selected teachers *who* taught classes *that* were specifically directed to business students.

description

Effective description uses words to transfer a mental image from the writer's mind to the reader's. The key to effective description is the accurate presentation of details. To select appropriate details, determine what use your **reader** will make of the description. Will he use it to identify an object? Will he use it to repair or assemble the device you are describing?

The following **paragraph** describes the assembly of an electronic typewriter that feeds coded tape or punch cards into the machine. Intended for the typewriter mechanic, this description includes an **illustration** of the assembly mechanism.

The Die Block Assembly consists of two machined block sections, eight Code Punch Pins, and a Feed Punch Pin. The larger section, called the Die Block, is fashioned of a hard, noncorrosive beryllium-copper alloy. It houses the eight Code Punch Pins and the smaller Feed Punch Pin in nine finely machined guide holes. The guide holes at the upper part of the Die Block are made smaller to conform to the thinner tips of the Punch Pins. Extending over the top of the Die Block and secured to it at one end is a smaller, arm-like block called the Stripper Block. The Stripper Block is made from hardened tool steel, and it also has been drilled through with nine finely machined guide holes. It is carefully fitted to the Die Block at the factory so that its holes will be precisely above those in the Die Block and so that the space left between the blocks will measure .015″ (plus or minus .003″). This space should be barely wide enough to allow the passage of a tape or edge-card. It is here, as the Punch Pins are driven up through the media and into the Stripper Block, that the actual cutting, or punching, of the code and feed holes takes place. The residue (chad) from the hole punching operation is pushed out through the top of the Stripper Block and guided out of the assembly by means of a plastic Chad Collector and Chad Collector Extender.

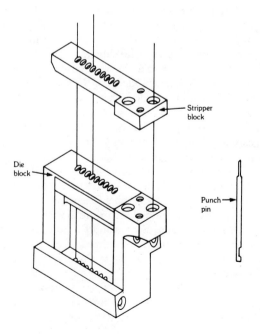

In technical writing you may need to write a description of a process as well as of a mechanism. In describing a mechanical device, for example, you might want to provide your reader with a general description of the whole device. Such a description should include the primary functions and the main parts of the device. This provides a frame of reference for understanding the more specific details that follow—the physical description of the various parts and each part's location and function in relation to the whole. The description would conclude with an explanation of the way the parts work together to get their particular jobs done, as in the following example.

General description of the whole device

An internal combustion engine is one in which the fuel is burned directly inside the machine which converts the heat energy into work. Practically all motor cars use this kind of engine, the fuel being petrol (gasoline). The engine consists of several cylinders in which pistons are fitted. The top of each cylinder is covered over with a head which contains a hollowed-out . . . combustion chamber. A rod connects each piston with a . . . crankshaft. The burning of fuel and air in the combustion chamber produces heat energy which causes expansion of the gases formed by the process of combustion. The gases push the pistons down the cylinder and thus cause the crankshaft to rotate. A flywheel on the end of the crankshaft keeps the latter rotating smoothly between the power impulses from each cylinder; it also transmits the power developed by the engine to the clutch.

Detailed description of the principles of the engine's operation

The easiest way to understand how an engine works is to keep in mind only one cylinder. For the engine to run and do work, certain things must take place over and over again in a regular order. The fuel must be sucked in, it must be compressed and burned; the resulting gases expand, and thus do work, and after expansion they must be removed. These steps make up what is known as a cycle; in most motor car engines it takes four strokes of the piston to make up one cycle. During this time the crankshaft will have made two complete revolutions. This type of engine is said to operate on a four-stroke cycle. The four strokes of the piston are named according to what they accomplish. The first is the suction, the second the compression, the third the power, and the fourth the exhaust stroke.

At the start of the suction stroke, the piston is at the top of the cylinder. It moves downward and the inlet valve opens. This sucks in the charge of gases from the carburetor. The compression stroke starts when the inlet valve closes and the piston is moving upward. The mixture is ignited while under pressure by a high-tension electric spark, which is produced across the points of the sparking plug. . . . The power stroke begins with a downward thrust of the piston. The burning of the gases gives off heat, which causes expansion. Pressure is thus exerted on the piston, forcing it downward as far as it will go. This completes the power stroke. The last movement of the piston is the exhaust stroke. The exhaust valve opens toward the end of the power stroke and most of the burned gases rush out. The piston moves upward, pushing the remaining gases out of the cylinder. The four-stroke cycle is now complete.

Summation of how the parts work together to perform the engine's function

—*Reprinted by permission* © *Encyclopedia Britannica,* 14th edition (1956).

An aspect of description sometimes called ''rendering'' is the act of showing or demonstrating, as opposed to telling, primarily through the advantageous use of images and details. Rendering makes writing more vivid and interesting by providing the reader with more details and a sense of immediacy. This technique is especially useful to technical people because it can help clarify and make more concrete the complex, abstract, and sometimes subtle subjects with which they must deal. For example, the following sentence is typical of much technical writing.

EXAMPLE   Computers can scan and read punched cards.

The complex thought contained in this sentence can be rendered clearly by the skillful use of images and details from everyday life, as the following example shows.

When information appears in print, as on this page, people like you and me are able to read and understand it. However, if information is to be processed by a machine, like a computer, then other ways must be found to put these same letters and words into the machine. Computers, of course, do not have eyes like humans—but they do have electrical sensing equipment that in certain ways does almost the same thing.

For example, most of us have walked into supermarkets through doors that open by themselves. These doors are con-trolled by electrical sensing equipment known as photoelectric cells, which act like eyes. Each door is controlled by two pho-toelectric cells that shine a beam of light to each other. As you walk through the door, the light beam is broken causing the photoelectric cells or ''eyes'' to sense that someone is beginning to enter. The electric eye reacts in a split second by sending a burst of electrical energy to the door's mechanical hinge, and this automatically swings the door open.

The photoelectric sensing idea can also be used to detect the presence of a hole in a card or piece of paper.

Thus, if holes are punched in paper to represent certain letters of the alphabet or words, it is possible for a machine to electrically sense or read this information.

—Joseph Becker, *The First Book of Information Science* (Washington, D.C.: The U.S. Atomic Energy Commission, 1973), p. 20.

Illustrations can be powerful aids in descriptive writing, especially when they show details too cumbersome to explain in writing.

despite/in spite of

Despite and *in spite of* (both meaning "notwithstanding") should not be blended into *despite of.*

CHANGE   *Despite of* our best efforts, the plan failed.
TO   *In spite of* (or *despite*) our best efforts, the plan failed.

Although there is no literal difference between *despite* and *in spite of, despite* suggests an effort to avoid blame.

EXAMPLES   *Despite* our best efforts, the plan failed. (We are not to blame for the failure.)
             In spite of our best efforts, the plan failed. (We did everything possible, but failure overcame us.)

diacritical marks

Diacritical marks are **symbols** added to letters to indicate their specific sounds. They include the phonetic symbols used by **dictionaries** as well as the marks used with foreign words. Some dictionaries place a quick reference list of common marks and sounds at the bottom of each page. Following is a list of common diacritical marks with their equivalent sound values. (See also **foreign words in English.**)

Name	Symbol	Example	Meaning
macron	¯	cāke	a "long" sound that signifies the standard pronunciation of the letter
breve	˘	brăcket	a "short" sound, in contrast to the standard pronunciation of the letter
dieresis	¨	coöperate	indicates that the second of two consecutive vowels is to be pronounced separately
acute	´	cliché	indicates a primary vocal stress on the indicated letter
grave	`	crèche	indicates a deep sound articulated toward the back of the mouth
circumflex	^	crêpe	indicates a very soft sound
tilde	~	cañon	a Spanish diacritical mark identifying the palatal nasal "ny" sound
cedilla	ç	garçon	a mark placed beneath the letter c in French, Portuguese, and Spanish to indicate an s sound

diagnosis/prognosis

Because they sound somewhat alike, these words are often confused with each other. *Diagnosis* means "an analysis of the nature of something" or "the conclusions reached by such analysis."

> EXAMPLE   The chairman of the board *diagnosed* the problem as the allocation of too few dollars for research.

Prognosis means "a forecast or prediction."

> EXAMPLE   He offered his *prognosis* that the problem would be solved next year.

dictation

If you write many short letters or **memorandums,** dictation may save you numerous hours each week. (See also **correspondence.**)

You should be aware, however, that dictation has certain limitations. First, although some people can compose **sentences** easily while speaking into a microphone or telephone, others find dictating a chore and can compose faster on paper. Only experience will tell whether dictation is for you. Second, certain kinds of material do not lend themselves to dictation. **Reports,** for example, or any material with **mathematical equations, tables, graphs,** or scientific terminology pose special problems for the person dictating as well as for the person transcribing. In general, dictation proves less effective with materials that have a complicated **format.** Finally, dictation sometimes causes a choppy and unconversational **style,** and so you may need to allow extra time to polish awkward phrases after you receive the transcription.

If you determine that dictation will save you time and effort, you may find the following tips helpful.

1. Once you learn how to use your firm's dictating equipment, you might want to make a few practice runs in order to become familiar with switches and buttons or the response time of a centralized dictating service.
2. Outline before you dictate. The length and complexity of the letter will determine how detailed your outline should be. However, even if you jot down only several key words in the order you wish to present your ideas, don't skip the outline—it will help prevent you from "freezing" when the machine is running.
3. Begin your dictation with any special **instructions** for the person transcribing: "This is a rough draft—please double-space" or "Prepare four carbon copies."
4. Speak in a normal voice and enunciate clearly. Avoid vocal pauses ("uh," "er," "hm"); simply stop speaking or turn off the machine when you cannot think of what to say next.
5. Spell out words that sound like other words *(ensure, insure);* personal or company names when used for the first time ("capital *M,* small *a, c,* capital *G,* small *i,l,l,i,s*"); **abbreviations,** and unusual terms.
6. Slow down for **numbers,** addresses, and difficult phrases (*"send six showroom* display cases").
7. Dictate as much **punctuation** as your transcriber needs; however,

always specify "paragraph," "quote," "unquote," "open parenthesis," "close parenthesis," and "underline."
8. Give a good secretary or transcriber the latitude to make corrections
   in punctuation, grammar, or **usage.** You, however, are responsible
   for the content and **tone.**
9. Proofread the transcribed dictation carefully, and make any necessary revisions—especially in important correspondence. Never skip
   this important stage, no matter how much you trust the person
   transcribing the dictation. (See also **proofreading.**)
10. Set a time each day for dictation, preferably early enough to allow
    time to review the correspondence before mailing it.

diction

The term *diction* is often misunderstood because it means both "the
choice of words used in writing and speech" *and* "the degree of distinct-
ness of enunciation of speech." In discussions of writing, *diction* applies to
choice of words. (See also **word choice.**)

dictionaries

A dictionary lists, in alphabetical order, a selection of the words in a
language. It defines them, gives their **spelling** and pronunciation, and
tells their function as a **part of speech.** In addition, it provides information on a word's origin and historical development. Often it provides a
list of **synonyms** for a word and, where pertinent, an **illustration** to help
clarify the meaning of a word.

The explanation of a word's meaning makes up the bulk of a dictionary entry. A dictionary gives primarily denotative rather than conno-
tative meanings. It also labels the field of knowledge to which a specific
meaning applies (grid, *electricity;* merger, *law*). The order in which a
word's meanings are given varies. Some dictionaries give the most
widely accepted current meaning first; others list the meaning in historical order, with the oldest meaning first and the current meaning last. If a
word has two or more fundamental meanings, they are listed in separate
paragraphs, with secondary meanings numbered consecutively within
each paragraph. A dictionary's preface normally indicates whether
current or historical meanings are listed first in an entry.

A dictionary entry also includes a word's spelling. It lists any variant spellings (align/aline, catalog/catalogue) and indicates which is most commonly used. Entries show if and where a word ought to be hyphenated—information especially helpful for **compound words**—and how the word can be abbreviated.

A word's pronunciation, indicated by phonetic symbols, is also given. The **symbols** are explained in a key in the front pages of the dictionary, and an abbreviated symbol key is often found at the bottom of each page as well. (See also **diacritical marks.**)

Information about the origin and history of a word (its etymology) is also given, usually in **brackets** at the end of the entry. This information can help clarify a word's correct meaning.

Entries also include information about a word's part of speech, and its inflected forms (how it forms plurals, or how it changes form to express comparative and superlative degree). Many entries include a list of synonyms in a separate paragraph following the main paragraph. Some dictionaries label words according to **usage** levels to indicate whether a word is considered **standard English, nonstandard English,** slang, and so on. Many dictionaries supplement a word's definition by providing illustrations in the form of **maps, photographs, tables, graphs,** and the like.

TYPES OF DICTIONARIES

The following unabridged dictionaries attempt to be exhaustive.

1. *The Oxford English Dictionary* (Oxford, England: The Clarendon Press, 1933) is the standard historical dictionary of the English language. It follows the chronological developments of over 240,000 words, from about the year 1000 to the early part of this century, providing numerous examples of uses and sources. *A Supplement to the Oxford English Dictionary,* Vol. 1, A–G, and Vol. 2, H–N (New York: Oxford University Press, 1972 and 1976), containing 17,000 and 13,000 entries respectively, is now available. Volumes 3 and 4 are still being compiled.

2. *Webster's Third New International Dictionary of the English Language, Unabridged* (Springfield, Mass.: G. & C. Merriam Company, 1976) contains over 450,000 entries. Since word meanings are listed in historical order, the current meaning is given last. This dictionary does not list personal and geographical names, nor does it include usage information.

3. *The Random House Dictionary of the English Language* (New York: Random House, 1973) contains 260,000 entries and is encyclopedic

in its use of examples. It gives a word's most widely used current meaning first and includes personal and geographical names.

Unabridged dictionaries provide complete and authoritative linguistic information, but they are impractical for desk use because of their size and expense. Desk dictionaries are often abridged versions of larger dictionaries. There is no single "best" dictionary, but there are several guidelines for selecting a good desk dictionary. Choose a recent edition. The older the dictionary, the less likely it is to have the up-to-date information you need. Select a dictionary with upward of 125,000 entries. Pocket dictionaries are convenient for checking spelling, but for detailed information the larger range of a desk dictionary is necessary. The following are considered good desk dictionaries.

1. *The Random House College Dictionary.* Rev. ed. New York: Random House, 1975.
2. *The American Heritage Dictionary of the English Language.* New York: American Heritage Publishing Co., Inc., 1973.
3. *Funk and Wagnalls Standard College Dictionary.* New, Updated ed. New York: Funk and Wagnalls, 1977.
4. *Webster's New World Dictionary of the American Language.* 2nd College ed. Cleveland: World Publishing Co., 1980.
5. *Webster's New Collegiate Dictionary.* 8th ed. Springfield, Mass.: G. & C. Merriam Company, 1973.

differ from/differ with

Differ from is used to suggest that two things are not alike.

EXAMPLE The vice-president's background *differs from* the general manager's background.

Differ with is used to indicate disagreement between persons.

EXAMPLE Our attorney *differed with* theirs over the selection of the last jury member.

different from/different than

In formal writing, the **preposition** *from* is used with *different*.

EXAMPLE The fourth-generation computer is *different from* the third-generation computer.

Different than is acceptable when it is followed by a **clause.**

EXAMPLE   The job cost was *different than* we had estimated it.

direct address

Direct address refers to a **sentence** or **phrase** in which the person being spoken or written to is explicitly named.

EXAMPLE   *John,* call me as soon as you arrive at the airport.

The person's name in a direct address is set off by **commas.**

direct objects

A direct object is a **noun** (or noun equivalent) that names a person or thing that receives the action of a **transitive verb.** A direct object normally answers the question *what* or *whom.*

EXAMPLES   John built a *business.* (noun)
I like *jogging.* (gerund)
I like *to jog.* (infinitive)
I like *it.* (pronoun)
I like *what I saw.* (noun clause)

See also **indirect objects, objects,** and **complements.**

discreet/discrete

Discreet means "having or showing prudent or careful behavior."

EXAMPLE   Since the matter was personal, he asked Bob to be *discreet* about it at the office.

Discrete means something is "separate, distinct, or individual."

EXAMPLE   The publications department was a *discrete* unit of the marketing division.

disinterested/uninterested

Disinterested means "impartial, objective, unbiased."

EXAMPLE   Like good judges, scientists should be passionately interested in the problems they tackle but completely *disinterested* when they seek to solve these problems.

Uninterested means simply "not interested."

EXAMPLE   Despite Jane's enthusiasm, her manager remained *uninterested* in the project.

division-and-classification method of development

One effective way to develop a complex subject is to divide it into logical parts and then go on to explain each part separately. This technique is especially well suited to subjects that can be readily broken down into units, and it can be used as the method of development in much technical writing. You might use this approach to describe a physical object, like a machine; to examine an idea, like the terms of a new labor-management contract; to explain a process, like the stages of an illness; or to give **instructions,** like the steps necessary to prepare a rusty metal surface for painting.

To explain the different types of printing processes currently in use, for example, you could divide the field into its major components and, where a fuller explanation is required, subdivide those (see Figure 1). The emphasis in division is on breaking down a complex whole into a number of like units, because it is easier to consider smaller units and to examine the relationship of each to the other.

The process by which a subject is divided is similar to the process by which a subject is classified. While division involves the separation of a whole into its component parts, classification is the grouping of a number

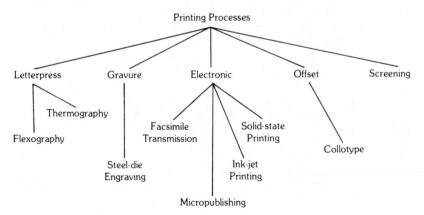

Figure 1

of units (people, objects, ideas, and so on) into related categories. Consider the following list:

Triangular file	Steel tape ruler
Needle-nose pliers	Vise
Pipe wrench	Keyhole saw
Mallet	Tin snips
C clamps	Rasp
Hacksaw	Plane
Glass cutter	Ball-peen hammer
Steel square	Spring clamp
Claw hammer	Utility knife
Crescent wrench	Folding extension ruler
Slip-joint pliers	Crosscut saw
Tack hammer	Utility scissors

To group the items in the list, you would first determine what they have in common. The most obvious characteristic they share is that they all belong in a carpenter's tool chest. With that observation as a starting point, you can begin to group the tools into related categories. Pipe wrenches belong with slip-joint pliers because both tools grip objects. The rasp and the plane belong with the triangular file because all three tools smooth rough surfaces. By applying this kind of thinking to all the items in the list, you can group each tool according to function (see Figure 2).

To divide or classify a subject, you must first divide the subject into its largest number of equal units. The basis for division depends, of course, on your subject and your purpose. For explaining the functions of tools, the classifications in Figure 2 (smoothing, hammering, measuring, gripping, and cutting) are excellent. For recommending which tools a new homeowner should buy first, however, the classifications are not helpful—they *all* contain tools that a new homeowner might want to purchase right away.

Once you have established the basis for the division or clasification, apply it consistently, putting each item in only one category. For example, it might seem logical to classify needle-nose pliers both as "tools that cut" and as "tools that grip" because most needle-nose pliers have a small section for cutting wires. However, the *primary* function of needle-nose pliers is to grip. So, listing them only under "tools that grip" would be most consistent with the basis used for listing the other tools.

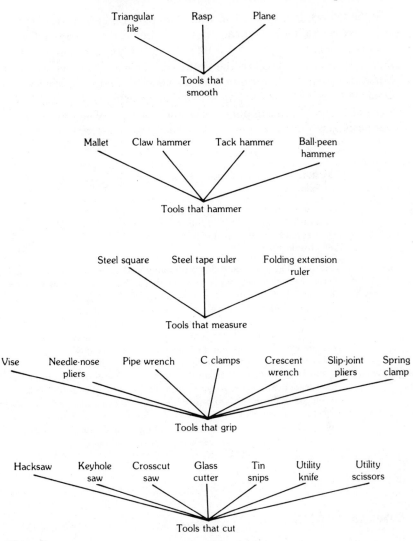

Figure 2

In the following example, two Canadian park rangers classify typical park users according to four catgories; the rangers then go on to discuss how to deal with potential rulebreaking by members of each group. The rangers could have classified the visitors in a variety of other ways, of course: as city and country residents, backpackers and drivers of recreational vehicles, United States and Canadian citizens, and so on. But for law-enforcement agents in public parklands, the size of a group and the relationships among its members were the most significant factors.

Dealing With Groups

First, recognize the various types of campers. They can be broken down as follows:

1. Family for weekend stay
2. Small groups ("a few of the boys")
3. Large groups or conventions
4. Hostile gangs

Persons in groups A and B can probably be dealt with on a one-to-one basis. For examples, suppose a member of the group is picking wild flowers, which is an offense in most park areas. Two courses of action are open. You could either issue a warning or charge the person with the offense. In this situation, a warning is preferable to a charge. First, advise the person that this action is an offense but, more importantly, explain why. Point out that the flowers are for all to enjoy and that most wild flowers are delicate and die quickly when picked.

For large groups, other approaches may be necessary. Every group has a leader. The leader may be official or, in informal or hostile groups, unofficial. If the group is organized, seek the official leader and hold this person responsible for the group's behavior. For informal groups, seek the person who assumes command and try to deal with this person.

—A.W. Moore and J. Mitchell, "Vandalism and Law Enforcement in Wildland Areas," in *Proceedings of the Wildland Recreation Conference*, Feb. 28–March 3, 1977, Banff Centre, Alberta, Canada.

documentation

The word *documentation* in the academic world refers to the giving of formal credit to sources used or quoted in a research paper or **report**—that is, the form used in a **bibliography, footnote,** or **reference.** In the business and industrial world, *documentation* may refer also to information that is recorded, or "documented," on paper. The **technical manuals** provided to customers by manufacturers are examples of such documentation. Even **flowcharts** and engineering **drawings** are considered documentation.

double negatives

A double negative is the use of an additional negative word to reinforce an expression that is already negative. It is an attempt to emphasize the negative, but the result is only awkward. Double negatives should never be used in writing.

> CHANGE   I *haven't* got *none.*
> TO   I have *none.*

> CHANGE   I *don't have* any.
> TO   I *haven't* any.

Barely, hardly, and *scarcely* cause problems because writers sometimes do not recognize that these words are already negative.

> CHANGE   I *don't hardly* ever have time to read these days.
> TO   I *hardly* ever have time to read these days.

Not unfriendly, not without, and similar constructions do not qualify as double negatives because in such constructions two negatives are meant to suggest the gray area of meaning between negative and positive. Be careful, however, how you use these constructions; they are often confusing to the **reader** and should be used only if they serve a purpose.

> EXAMPLES   He is *not unfriendly.* (meaning that he is neither hostile nor friendly)
> It is *not without* regret that I offer my resignation. (implying mixed feelings rather than only regret)

The **correlative conjunctions** *neither* and *nor* may appear together in a **clause** without creating a double negative, so long as the writer does not attempt to use the word *not* in the same clause.

> CHANGE   It was *not,* as a matter of fact, *neither* his duty *nor* his desire to fire the man.
> TO   It was *neither,* as a matter of fact, his duty *nor* his desire to fire the man.
> OR   It was *not,* as a matter of fact, *either* his duty *or* his desire to fire the man.

CHANGE He did *not neither* care about *nor* notice the error.
TO He *neither* cared about *nor* noticed the error.

Negative forms are full of traps that often entice inexperienced writers into errors of **logic,** as illustrated in the following example.

EXAMPLE There is *nothing* in the book which has *not* already been published in some form, but some of it is, I believe, very little known.

In this sentence, "some of it," logically, can only refer to "*nothing* in the book which has *not* already been published." The sentence can be corrected in one of two ways. The **pronoun** *it* can be replaced by a specific **noun.**

EXAMPLE There is nothing in the book which has not already been published in some form, but some of the *information* is, I believe, very little known.

Or the idea can be stated positively. (See also **positive writing.**)

EXAMPLE Everything in the book has been published in some form, but some of it is, I believe, very little known.

drawings

A drawing is useful when you wish to focus on details or relationships that a **photograph** cannot capture. A drawing can emphasize the significant part of a mechanism, or its function, and omit what is not significant. However, if the precise details of the actual appearance of an object are necessary to your **report** or document, a photograph is essential.

There are various types of drawings, each with unique advantages. The type of drawing used for an **illustration** should be determined by the specific purpose it is intended to serve. If your **reader** needs an impression of an object's general appearance or an overview of a series of steps or directions, a conventional drawing of the type illustrated by Figure 3 will suffice.

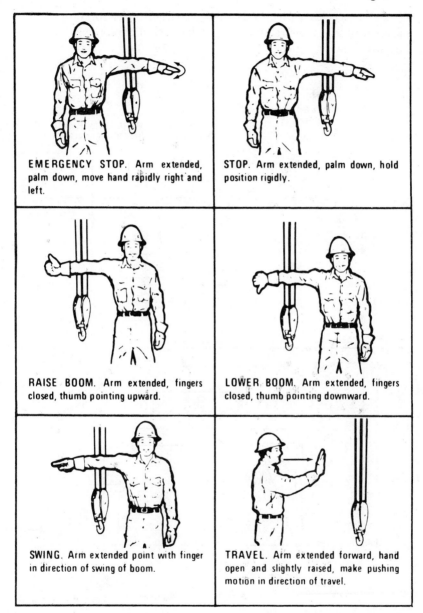

Figure 3.   Hand Signals for Crane Operation

Source: Harnischfeger Corporation

Where it is necessary to show the internal parts of a piece of equipment in such a way that their relationship to the overall equipment is clear, a cutaway drawing is necessary, as in Figure 4.

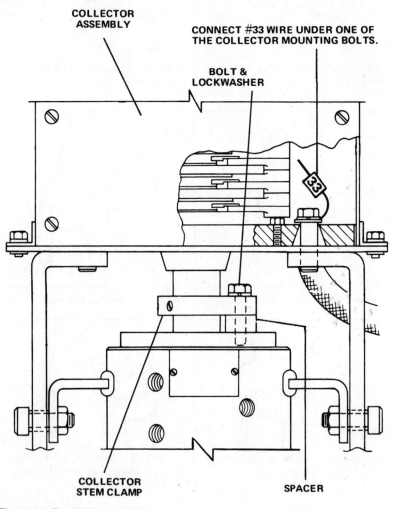

Figure 4.   Detail View—Collector
Source: Harnischfeger Corporation

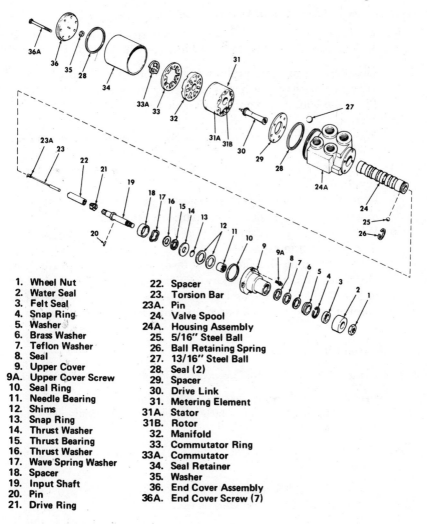

1. Wheel Nut
2. Water Seal
3. Felt Seal
4. Snap Ring
5. Washer
6. Brass Washer
7. Teflon Washer
8. Seal
9. Upper Cover
9A. Upper Cover Screw
10. Seal Ring
11. Needle Bearing
12. Shims
13. Snap Ring
14. Thrust Washer
15. Thrust Bearing
16. Thrust Washer
17. Wave Spring Washer
18. Spacer
19. Input Shaft
20. Pin
21. Drive Ring
22. Spacer
23. Torsion Bar
23A. Pin
24. Valve Spool
24A. Housing Assembly
25. 5/16" Steel Ball
26. Ball Retaining Spring
27. 13/16" Steel Ball
28. Seal (2)
29. Spacer
30. Drive Link
31. Metering Element
31A. Stator
31B. Rotor
32. Manifold
33. Commutator Ring
33A. Commutator
34. Seal Retainer
35. Washer
36. End Cover Assembly
36A. End Cover Screw (7)

Figure 5.   Power Steering Valve

Source: Harnischfeger Corporation

To show the proper sequence in which parts fit together, or when it is essential to show the details of each individual part, use an exploded-view drawing such as Figure 5.

TIPS FOR CREATING AND USING DRAWINGS

Many organizations have their own **format** specifications. In the absence of such specifications, the following tips will be helpful.

1. Give the drawing a clear title and a figure number (if figure numbers are used), both of which should be centered below the drawing.
2. Place the source or courtesy line, if necessary, in the lower left corner.
3. Show the equipment from the point of view of the person who will use it.
4. When illustrating a subsystem, show its relationship to the larger system of which it is a part.
5. Draw the different parts of an object in proportion to one another, unless you indicate that certain parts are enlarged.
6. Where a sequence of drawings is used to illustrate a process, arrange them from left to right.
7. Label parts in the drawing so that text references to them are clear.
8. Depending on the complexity of what is shown, labels may be placed on the parts themselves, or the parts may be given letter or number symbols, with an accompanying key (see Figure 5).

due to/because of

Due to (meaning "caused by") is acceptable following a **linking verb.**

EXAMPLES   His short temper was *due to* work strain.

His short temper, which was *due to* work strain, made him inefficient.

Due to is not acceptable, however, when it is used with a nonlinking **verb** to replace *because of.*

CHANGE   He went home *due to* illness.

TO   He went home *because of* illness.

E

each

When *each* is used as a **subject,** it takes a singular **verb** or **pronoun.**

EXAMPLES   *Each* of the reports *is* to be submitted ten weeks after *it* is assigned.
                Each worked as fast as *his* ability would permit.

When *each* occurs after a plural subject with which it is in apposition, it takes a plural verb or pronoun. (See also **agreement.**)

EXAMPLE   The reports *each have* white embossed titles on *their* covers.

each and every

Although the **phrase** *each and every* is commonly used in speech in an attempt to emphasize a point, the phrase is redundant and should be eliminated from your writing. Replace it with *each* or *every.*

CHANGE   *Each and every* part was accounted for.
TO   *Each* part was accounted for.
OR   *Every* part was accounted for.

economic/economical

Economic refers to the production, development, and management of material wealth. *Economical* simply means "not wasteful or extravagant."

EXAMPLES   The strike at General Motors had an *economic* impact on the
                country.
                Since gasoline will be in short supply, drivers should be as
                economical as possible.

e.g.

The **abbreviation** *e.g.* stands for the Latin *exempli gratia,* meaning "for example." Since a perfectly good English expression exists for the same use *(for example),* there is no need to use a Latin expression or abbreviation. In addition, many people confuse *e.g.* with i.e., which means "that is.'' The abbreviation *e.g.* does not save enough space to justify possible misunderstanding. Avoid *e.g.* in your writing.

CHANGE　Some terms of the contract (*e.g.*, duration and job classification) were settled in the first two bargaining sessions.

TO　Some terms of the contract (*for example,* duration and job classification) were settled in the first two bargaining sessions.

OR　Some terms of the contract, *such as* duration and job classification, were settled in the first two bargaining sessions.

elegant variation

The attempt to avoid repeating a **noun** in a **paragraph** by substituting other words (often pretentious **synonyms**) is called elegant variation. Avoid this practice—repeat a word if it says what you mean, or use a suitable **pronoun**. (See also **affectation** and **long variants**.)

CHANGE　The *use* of modules in the assembly process had increased production. *Modular utilization* has also cut costs.

TO　The *use* of modules in the assembly process has increased production. The *use* of modules has also cut costs.

OR　The use of modules in the assembly process has increased production. *It* has also cut costs.

ellipses

When you omit words in quoted material, use a series of three spaced **periods**—called ellipsis dots—to indicate the omission. However, such an omission should not detract from or alter the essential meaning of the **sentence.**

EXAMPLE　Technical material distributed for promotional use is sometimes charged for, particularly in high-volume distribution to schools, although prices for these publications are not uniformly based on the cost of developing them. (without omission)

Technical material distributed for promotional use is sometimes charged for . . . although prices for these publications are not uniformly based on the cost of developing them. (with omission)

When the omitted portion comes at the beginning of the sentence, begin the **quotation** with a lower-case letter.

EXAMPLE ''When the programmer has determined a system of runs, he must create a system flowchart to provide a picture of the data flow through the system.'' (without omission)

The letter states that the programmer ''. . . must create a system flowchart to provide a picture of the data flow through the system.'' (with omission)

When the omission comes at the end of a sentence and you continue the quotation following the omission, use four periods to indicate both the final period to end the sentence and the omission.

EXAMPLE In all publications departments except ours, publications funds—once they are initially allocated by higher management—are controlled by publications personnel. Our company is the only one to have nonpublications people control funding within a budgeted period. In addition, all publications departments control printing funds as well. (without omission)

In all publications departments except ours, publications funds—once they are initially allocated by higher management—are controlled by publications personnel. . . . In addition, all publications departments control printing funds as well. (with omission)

Use a full line of periods across the page to indicate the omission of one or more **paragraphs.**

EXAMPLE A computer system operates with two types of programs: the software programs supplied by the manufacturer and the programs created by the user. The manufacturer's software consists of a number of programs that enable the system to perform complicated manipulations of data from the relatively simple instructions specified in the user's program.

Programmers must take a systematic approach to solving any data processing problem. They must first clearly define the problem, and then they must define a system of runs that will solve the problem. For example, a given problem may require a validation and sort run to handle account numbers or employee numbers, a computation and update run to manipulate the data for the output reports, and a print run to produce the output reports.

> When a system of runs has been determined, the programmer must create a system flowchart to provide a picture of the data flow through the system. (without omission)

> A computer system operates with two types of programs: the software programs supplied by the manufacturer and the programs created by the user. The manufacturer's software consists of a number of programs that enable the system to perform complicated manipulations of data from the relatively simple instructions specified in the user's program.
>
> . . . .
>
> When a system of runs has been determined, the programmer must create a system flowchart to provide a picture of the data flow through the system. (with omission)

Do not use ellipsis dots for any purpose other than to indicate omission. Advertising copywriters, particularly those not skilled at achieving the effect they strive for with words, often use ellipsis dots to substitute for all marks of **punctuation**—and sometimes even use them where no punctuation is needed. This practice achieves no positive effect, and it is a poor one to emulate.

eminent/imminent/immanent

Someone or something that is *eminent,* is outstanding or distinguished.

EXAMPLE   He is an *eminent* scientist.

If something is *imminent,* it is about to happen.

EXAMPLE   He paced nervously, knowing that the committee's decision was *imminent.*

Immanent, meaning "inherent" or "indwelling," is used chiefly by theologians and philosophers to refer to the theory that a Deity pervades and sustains all material existence.

EXAMPLE   The theologian suggested that God is *immanent* in all life forms.

emphasis

Emphasis is the principle by which ideas in writing are stressed according to their importance. The first step in achieving emphasis is to write simply and concisely. If a **sentence** expresses the writer's meaning simply and directly, it has the proper emphasis. But there are also mechanical means of achieving emphasis: by position within a sentence, **paragraph,** or **report;** by the use of **repetition;** by selection of sentence type; by varying the length of sentences; by the use of climactic order within a sentence; by **punctuation** (use of the **dash**); by the use of **intensifiers;** by the use of mechanical devices, such as **italics** and **capital letters;** and by direct statement (using such terms as *most important* and *foremost*).

Emphasis can be achieved by position because the first and last words of a sentence stand out in the **reader's** mind.

CHANGE   Because they reflect geological history, moon craters are important to understanding the earth's history.

TO   Moon craters are important to understanding the earth's history because they reflect geological history.

OR   Because moon craters reflect geological history, they are important to understanding the earth's history.

Notice that the revised versions of the sentence emphasize moon craters simply because the term is in the front part of the sentence. Similarly, the first and last sentences in a paragraph and the first and last paragraphs in a report or paper tend to be the most emphatic to the reader.

EXAMPLE   Energy does far more than simply make our daily lives more comfortable and convenient. Suppose you wanted to stop—and reverse—the economic progress of this nation. What would be the surest and quickest way to do it? Find a way to cut off the nation's oil resources! Industrial plants would shut down, public utilities would stand idle, all forms of transportation would halt. The country would be paralyzed, and our economy would plummet into the abyss of national economic ruin. *Our economy, in short, is energy-based.*

—*The Baker World* (Los Angeles: Baker Oil Tools, Inc., 1964) p. 2.

Emphasis can be achieved by the repetition of key words and **phrases.**

EXAMPLE   Similarly, atoms *come and go* in a molecule, but the molecule *remains;* molecules *come and go* in a cell, but the cell *remains;* cells *come and go* in a body, but the body *remains;* persons *come and go* in an organization, but the organization *remains.*

—Kenneth Boulding, *Beyond Economics* (Ann Arbor: University of Michigan Press, 1968). p. 131.

Different emphasis can be achieved by the selection of a **compound sentence, complex sentence,** or **simple sentence.**

EXAMPLES   The report turned in by the police detective was carefully illustrated, and it covered five pages of single-spaced copy. (This
compound sentence carries no special emphasis because it
contains two coordinate **independent clauses.**)
The police detective's report, which was carefully illustrated,
covered five pages of single-spaced copy. (This complex sentence emphasizes the size of the report.)
The carefully illustrated report turned in by the police detective
covered five pages of single-spaced copy. (This simple sentence
emphasizes that the report was carefully illustrated.)

Emphasis can be achieved by making a sentence distinctly different in
length from the preceding one.

EXAMPLE   I believe that man is about to learn that the most practical life is
the moral life and that the moral life is the only road to survival.
He is beginning to learn that he will either share part of his
material wealth or lose all of it; that he will respect and learn to
live with other political ideologies if he wants civilization to go
on. This is the kind of argument that man's actual experience
equips him to understand and accept. *This is the low road to morality.
There is no other.*

—Saul Alinsky, *Rules for Radicals* (New York: Random House, 1971), p. 25.

Emphasis can be achieved by a climactic order of ideas or facts within a
sentence.

EXAMPLE   Over subsequent weeks the industrial relations department
worked diligently, management showed tact and patience, and
the employees finally accepted the new policy.

Emphasis can be achieved by setting an item apart with a **dash.**

EXAMPLE   Here is where all the trouble begins—in the American confidence that technology is ultimately the medicine for all ills.

Emphasis can be achieved by the use of intensifiers (*most, very, really*), but
this technique is so easily abused that it should be used only with caution.

EXAMPLE   The final proposal is *much* more persuasive than the first.

Emphasis can be achieved by such mechanical devices as italics (underlining) and capital letters. But this technique is also easily abused and should be used with caution.

EXAMPLE   When we add the commonplace situations that allow these systems to function for hundreds of researchers *simultaneously* and from all five continents, the power of this new information medium is even more remarkable.

The word *do* (or *does*) may be used for emphasis, but this is also easily overdone, so proceed with caution. Compare the following examples.

EXAMPLES   You believe weekly staff meetings are essential, don't you?
You *do* believe weekly staff meetings are essential, don't you?

Direct address may also be used for emphasis in **correspondence.**

EXAMPLE   John, I believe we should rethink our plans.

(See also **subordination.**)

English, varieties of

There are two broad varieties of written English: standard and nonstandard. These varieties are determined through **usage** by those who write in the English language. **Standard English** (also called American edited English) is used to carry on the daily business of the nation. It is the language of business, industry, government, education, and the professions. Standard English is characterized by exacting standards of **punctuation** and capitalization, by accurate **spelling,** by exact **diction,** by an expressive vocabulary, and by knowledgeable usage choices. **Nonstandard English,** on the other hand, is the language of those not familiar with the standards of written English. This form of English rarely appears in printed material except when it is used for special effect by fiction writers. Nonstandard English is characterized by inexact or inconsistent punctuation, capitalization, spelling, diction, and usage choices. Both standard and nonstandard English find their way into everyday speech and writing in the forms of the following subcategories.

COLLOQUIAL

Colloquial English is spoken standard English or writing that attempts to re-create the flavor of this kind of speech by using words and expressions common to casual conversation. Colloquial English is appropriate to some kinds of writing (personal letters, notes, etc.) but not to most technical writing.

VERNACULAR

Vernacular English is the spoken form of the language, as opposed to its written form. It is the form used by the majority of those who speak the language. To write "in the vernacular" is to imitate this kind of language. Ordinarily, such writing is confined to fiction. Vernacular can also refer to the manner of expression common to a trade or profession (one might speak of "the legal vernacular"), although **jargon** more clearly expresses this meaning.

DIALECT

Dialectal English is a social or regional variety of the language that is comprehensible to people of that social group or region but that may be incomprehensible to outsiders. Dialect, which is usually nonstandard English, involves distinct **word choices,** grammatical forms, and pronunciations. Technical writing, because it aims at a broad audience, should be free of dialect.

LOCALISM

A localism is a word or **phrase** that is unique to a geographical region. For example, the words *poke, sack,* and *bag* all denote "a paper container," each term being peculiar to a different region of the United States. Such words should normally be avoided in writing because knowledge of their meanings is too narrowly restricted.

SLANG

Slang refers to a manner of expressing common ideas in new, often humorous or exaggerated, ways. Slang often finds new use for familiar words ("He *crashed* early last night"—meaning that he went to bed early) or coins words ("He's a *kook*"—meaning that he is offbeat, unconventional). Slang expressions usually come and go very quickly. The fact that a fair number of words in the present standard English vocabulary were once considered slang indicates that when a slang word fills a legitimate need it is accepted into the language. *Skyscraper, bus,* and *date* (as in "to go on a date"), for example, were once considered slang expressions. Although slang may have a valid place in some writing, particularly in fiction, be careful not to let it creep into technical writing.

BARBARISM

A barbarism is any obvious misuse of standard words, grammatical forms, or expressions. Examples include *ain't, irregardless, done got, drownded.*

equal/unique/perfect

Logically, *equal* (meaning "having the same quantity or value as another"), *unique* (meaning "one of a kind"), and *perfect* (meaning "a state of highest excellence") are **absolute words** and therefore should not be compared. However, colloquial usage of *more* and *most* as **modifiers** of *equal, unique,* and *perfect* is so common that an absolute prohibition against such use is impossible.

   EXAMPLE   Yours is a *more unique* (or *perfect*) coin than mine.

Some writers try to overcome the problem by using *more nearly equal (unique, perfect).* Where **clarity** and preciseness are critical, the use of **comparative degrees** with *equal, unique,* and *perfect* can be misleading. The best rule of thumb is to avoid using the comparative degrees with absolute terms.

   CHANGE   Ours is a more *equal* percentage split than theirs.
      TO   Our percentage split is 51-49; theirs is 54-46.

-ese

The **suffix** *-ese* is used to designate types of **jargon** or certain languages or literary **styles** (official*ese*, journal*ese*, Chin*ese*, Pentagon*ese*).

essay questions, answering

One of the most practical applications for a systematic approach to writing is answering essay questions. The following is a procedure that has been used successfully by good students for years.

1. Read the question carefully, looking for key terms that help you determine precisely whether you are to define, describe, contrast, or trace the subject.
2. On a separate piece of paper, list the major points as your formulate your answer.

3. If time permits, develop a brief main idea statement. Quite often, the main idea statement will take the form of a restatement of the question.
4. Determine the **method of development** best suited to your subject.
5. Number the major points of your answer in the sequence best adapted to your method of development. This list of numbered points constitutes your **outline.**
6. Write a rough draft on your scratch sheet if time permits.
7. Correct any obvious errors in the rough draft and copy the draft onto your examination paper, continuing to revise as you do so.

Be careful to pace yourself so that you allow enough time for each question. If you must sacrifice one step of the process because time is short, omit the rough draft and write your final draft directly from your outline. The outline, however, is essential.

This method, which should take only a few minutes at most, will save you a great deal of writing time. Since it will provide a logical structure for your response, the instructor can evaluate your work more effectively—a factor that usually works to the student's advantage.

etc.

Etc. is an **abbreviation** for the Latin *et cetera,* meaning "and others"; therefore, *etc.* should not be used with *and.*

> CHANGE   He brought pencils, pads, erasers, a slide rule, *and etc.*
> TO   He brought pencils, pads, erasers, a slide rule, *etc.*

Do not use *etc.* at the end of a list or series introduced by the **phrases** *such as* or *for example* because these phrases already indicate that there are other things of the same category that are not named.

> CHANGE   He brought camping items, *such as* backpacks, sleeping bags, tents, *etc.*
> TO   He brought backpacks, sleeping bags, tents, *etc.*
> OR   He brought camping items, such as backpacks, sleeping bags, and tents.

In careful writing, *etc.* should be used only when there is logical progression (1, 2, 3, etc.) and when at least two items are named. It is often better to avoid *etc.* altogether, however, because there may be confusion as to the identity of the "class" of items listed.

EXAMPLE   He brought backpacks, sleeping bags, tents, and other camping items.

It is always best to be as specific as possible.

euphemism

A euphemism is a word that is an inoffensive substitute for one that could be distasteful, offensive, or too blunt.

EXAMPLES   *remains* for *corpse*
passed away for *died*
marketing representative for *salesman*
previously owned or *pre-owned* for *used*

Used judiciously, a euphemism might help you avoid embarrassing or offending someone. Overused, however, euphemisms can hide the facts of a situation (such as *incident* for *accident*). As a rule, call things by their right names. (See also **word choice.**)

everybody/everyone

Both *everybody* and *everyone* are considered singular and so take singular **verbs** and **pronouns.**

EXAMPLES   *Everybody is* happy with the new contract.
Everyone here *eats* at 11:30 a.m.
Everybody at the meeting made *his* proposals separately.
Everyone went *his* separate way after the meeting.

An exception to this is when use of singular verbs and pronouns would be offensive by implying sexual bias; in such a situation, it is better to use plural verbs and pronouns or to use the expression "his or her" than to offend.

CHANGE   *Everyone* went *his* separate way after the meeting.
TO   *They* all went *their* separate ways after the meeting.
OR   Everyone went *his or her* separate way after the meeting.

Although normally written as one word, *everyone* is written as two words when each individual in a group should be emphasized.

EXAMPLES *Everyone* here comes and goes as he pleases.
Every one of the team members contributed to this discovery.

(See also **indefinite pronouns.**)

ex post facto

The expression *ex post facto,* which is Latin for ''after the fact,'' refers to something that operates retroactively. The term is normally used to refer to laws and is best avoided in other contexts.

CHANGE The contract will be in effect *ex post facto* of the negotiations.
TO The contract will be in effect after the negotiations.

exclamation marks

The purpose of the exclamation mark (!) is to indicate the expression of strong feeling. It can signal surprise, fear, indignation, or excitement. It should not be used for trivial emotions or mild surprise. It cannot make an argument more convincing, lend force to a weak statement, or call attention to an intended irony—no matter how many are stacked like fence posts at the end of a **sentence.** It is a **punctuation** mark more common to fiction and personal letters than to technical writing.

USES

The most common use of an exclamation mark is after a word (**interjection**), **phrase, clause,** or sentence to indicate surprise. Interjections are words that express strong emotion.

EXAMPLES Ouch! Wow! Oh! Stop! Hurry!

An exclamation mark can also be used after a whole sentence, or even an element of a sentence.

EXAMPLES The subject of this meeting—please note it well!—is our budget deficit.
How exciting is Stravinsky's *Rite of Spring!*

An exclamation mark is sometimes used after a title that is an exclamatory word, phrase, or sentence.

EXAMPLES "Our Information Retrieval System Must Change!" is an article by Richard Moody.
The Cancer With No Cure! is a book by Wilbur Moody.

When used with **quotation marks,** the exclamation mark goes outside unless what is quoted is an exclamation.

EXAMPLE The boss yelled, "Get in here!" Then Ben, according to Ray, "jumped like a kangaroo"!

executive summaries

The executive summary is a shortened version of a **report** that presents an overview of the report's important ideas, conclusions, and recommendations. It is also known as a *brief,* a *digest,* a *review,* or a *synopsis.* The executive summary enables people who may not have time to read a lengthy report to scan its primary points quickly and then to decide whether it would be valuable to read it in full. Some executive summaries follow the organization of the report, but others highlight the important ideas, conclusions, and recommendations by placing them up front.

An executive summary is more complete than an **abstract;** in addition to summing up the major points, it contains a statement of the purpose and scope of the report as well as of the methods used. Whereas an abstract generally runs 200 words or less, an executive summary can be up to 1,000 words. (See also **formal reports.**)

The following piece about cooling towers is an example of an executive summary.

> Cooling towers are water-recycling devices that can reduce the quantity of waste heat discharged to surface waters by dissipating up to 95% of the heat directly to the atmosphere. Cooling towers are used for air conditioning large buildings and for cooling in various industries—increasingly, at steam-electric power plants.
>
> One essential component of wet cooling towers is the "fill," the material through which the hot water flows and by which the evaporative (cooling) process is enhanced. Fill can be composed of a variety of materials, including asbestos paper or asbestos cement. Release of asbestos fibers from these cooling towers into the ambient air or into surface waters is thus a possibility, and

may create a potential human health hazard. (Occupational exposure to asbestos is known to result in a number of diseases, including lung and digestive system cancers.)

The study included a review of the human health-effects literature on asbestos; collection of information from cooling-tower vendors and users; collection of water, sediment, and air samples at operating cooling towers and analysis for asbestos-fiber content using electron microscope techniques; and estimation of the quantities of fibers that could be released to the ambient air due to drift from cooling towers. Results from the study are discussed in this document, and have relevance to any site using cooling towers.

At 10 of 18 cooling-tower sites sampled, chrysotile-asbestos fibers were detected in cooling-tower basin and blowdown waters. Concentrations ranged up to 10^8 fibers/liter, with mass concentrations ranging to 37 μg/liter. At all but two sites, 53 to 100% of the chrysotile fibers were less than 5 μm long; aspect ratios ranged from 3.5 to 1700. The major source of these fibers, at least in nonurban areas, appears to be the components of the towers rather than the air drawn through the towers or the water taken into the towers to replace evaporative and other losses. Ash- or other material-settling ponds appear to reduce, by several orders of magnitude, asbestos-fiber concentrations in cooling-tower effluent.

Assuming a 100-fold dilution in a receiving stream, asbestos-fiber concentrations in public water supplies may range from "none detected" to 10^6 fibers/liter as a result of effluent from cooling towers using asbestos fill. This is the same range of asbestos-fiber concentrations reported for some beverages and drinking water.

The bulk of the literature reviewed during the course of this study did not support a causal relationship between the presence of asbestos fibers in food, beverages, or drinking water and the occurrence of human cancers. However, the uncertainties involved in the epidemiological studies preclude any conclusions regarding "safe" concentrations of asbestos fibers. Also, at least one study indicated a statistical association between certain human cancers and asbestos in drinking water.

In view of the lack of consistent evidence linking adverse health effects to drinking water containing asbestos concentrations on the order of 10^6 fibers/liter or less, as well as ignorance of the health risks and resource costs related to the manufacture and use of other types of cooling-tower fill, it is not presently suggested that the use of asbestos fill be discontinued. However, it is strongly recommended that the ratio of asbestos fibers in cooling-tower effluent to offsite receiving waters be kept as low as possible by adequate control of circulating water chemistry. Pilot studies should be initiated by industrial research groups on the feasibility of side-stream filtration as a means of reducing asbestos fibers in cooling-tower effluent to below detection limits. The effectiveness of asbestos-fiber removal by ash- or other material-settling ponds should be seriously investigated.

It is also recommended that monitoring programs for asbestos in cooling-tower waters be carried out at sites where asbestos fill is used. This recommendation is based on the opinion that caution and surveillance are dictated by the present uncertainties in evaluating health risks from long-term consumption of barely detectable concentrations of asbestos in drinking water.

On the basis of estimates made in this study, monitoring for asbestos fibers in drift from cooling towers, under conditions similar to those at the study sites, does not appear to be warranted. Estimates made using drift-deposition models indicated that asbestos concentrations in air at ground level due to drift are one to several orders of magnitude less than a proposed ambient air-quality standard of 30 ng/m^3. Exceptions might be sites where older, mechanical-draft cooling towers with draft rates equal to or greater than 0.01% of the circulating water flow are in operation; in these cases, the ground-level concentrations of asbestos in the air within 250 m downwind of the towers may exceed Occupational Safety and Health Administration standards.

TIPS ON WRITING AN EXECUTIVE SUMMARY

1. If you are writing an executive summary for your own report, you should write it last, after completing the report.
2. Summarize the most important ideas in the report. (You can find these quickly in the major and minor **heads** of your **outline.**)
3. Cover all the basic ideas, but do not include the details.
4. Begin with a statement of the topic.
5. Avoid using technical terms; your readers may include people who are not familiar with the topic.
6. Make your executive summary concise but not brusque. Leave out any unnecessary words or ideas, but do not delete **articles** *(a, an, the)* or transitional words and **phrases** *(however, moreover, therefore, for example, in summary,* etc.).
7. Do not include **illustrations, tables,** or bibliographic **references.**

explaining a process

Many kinds of technical writing explain a process. The explanation involves putting in the appropriate order the steps that a specific mechanism or system uses to accomplish a certain result. The process itself might range from the legal steps necessary to form a corporation to the steps necessary to develop a roll of film. At the least, it is necessary to divide the process into distinct steps and present them in their normal order.

In your opening **paragraph,** tell you **reader** why it is important to become familiar with the process you are explaining. Before you explain the steps necessary to form a corporation, for example, you could cite the tax savings that incorporation would permit. To provide your reader with a framework for the details that will follow, you might present a brief overview of the process. Finally, you might describe how the process works in relation to a larger whole. In explaining the air brake system of a large dump truck, you might note that the braking system is one part of the vehicle's air system, which also controls the throttle and transmission-shifting mechanisms.

A process explanation can be long or short, depending on how much detail is necessary. The following description of the way in which a camera controls light to expose a photographic film fits into one paragraph.

EXAMPLE   The camera is the basic tool for recording light images. It is simply a box from which all light is excluded except that which passes through a small opening at the front. Cameras are equipped with various devices for controlling the light rays as they enter this opening. At the press of a button, a mechanical blade or curtain, called a shutter, opens and closes automatically. During the fraction of a second that the shutter is open, the light reflected from the subject toward which the camera is aimed passes into the camera through a piece of optical glass called the *lens.* The lens focuses, or projects, the light rays onto the wall at the back of the camera. These light reflections are captured on a sheet of film attached to the back wall.

Many process explanations require more details and will, of course, be longer than a paragraph. The following example discusses several methods of surface mining for coal. The writer begins with an overview of the elements common to all the processes, defines the terms important to the explanation, and then describes each separately. Transitional words and **phrases** serve to achieve **unity** within paragraphs, and headings mark the **transition** from one process to the next.

Surface Mining of Coal

The process of removing the earth, rock, and other strata (called *overburden*) to uncover an underlying mineral deposit is generally referred to as surface mining. Strip mining is a specific kind of surface mining in which all the overburden is removed in strips, one cut at a time. Three types of strip-mining methods are used to mine coal: *area, contour,* and *mountain-top removal.* Which method is used depends upon the topography of the area to be mined.

AREA STRIP MINING

Area strip mining is used in regions of flat to gently rolling terrain, like those found in the Midwest and West. Depending on applicable reclamation laws, the topsoil may be removed from the area to be mined, stored, and later reapplied as surface material during reclamation of the mined land. Following removal of the topsoil, a trench is cut through the overburden to expose the upper surface of the coal to be mined. The length of the cut generally corresponds to the length of the property or of the deposit. The overburden from the first cut is placed on unmined land adjacent to the cut. After the first cut is completed, the coal is removed and a second cut is made parallel to the first. The overburden (now referred to as spoil) from each of the succeeding cuts is deposited in the adjacent pit from which the coal was just removed. The final cut leaves an open trench equal in depth to the thickness of the overburden plus the coal bed, bounded on one side by the last spoil pile and on the other side by the undisturbed soil. The final cut may be as far as a mile from the first cut. The overburden from all the cuts, unless graded and leveled, resembles the ridges of a giant washboard.

CONTOUR STRIP MINING

In areas of rolling or very steep terrain, such as in the eastern United States, contour strip mining is used. In this method, the overburden is removed from the mineral seam in a pattern that follows the contour line around the hillside. The overburden is then deposited on the downslope side of the cut until the depth of the overburden becomes too great for economical recovery of the coal. This method leaves a bench, or shelf, on the side of the hill, bordered on the inside by a highwall (30 to 100 feet high) and on the other side by a high ridge of spoil.

A method of mining that is often used in conjunction with contour mining is *auger mining*. This method is employed when the overburden becomes too thick and renders contour mining uneconomical and when extraction by underground mining would be too costly or unsafe. In auger mining an instrument bores holes horizontally into the coal seam. The coal can then be removed like the shavings produced by a drill bit. The exposed coal seam in the highwall is left with a continous series of bore holes from which the coal was removed.

MOUNTAIN-TOP REMOVAL

In areas of rolling or steep terrain an adaptation of area mining to conventional contour mining is used; it is called the mountain-top removal method. With this method entire mountain tops are removed down to the coal seam by a series of parallel cuts. This method is economical when the coal lies near the tops of mountains, ridges, or knobs. If there is excess overburden that cannot be stored on the mined land, it may be transported elsewhere.

expletives

An expletive is a word that fills the position of another word, **phrase,** or **clause.** *It* and *there* are the usual expletives.

EXAMPLE   *It* is certain that he will go.

In this example, the expletive *it* occupies the position of **subject** in place of the real subject, *that he will go.* Although expletives are sometimes necessary to avoid **awkwardness,** expletives are commonly overused and most **sentences** can be better stated without them.

CHANGE   *There are* several reasons why I did it.
TO   I did it for several reasons.

CHANGE   *There were* many orders lost for unexplained reasons.
TO   Many orders were lost for unexplained reasons.
OR   We lost many orders for unexplained reasons.

In addition to its usage as a grammatical term, the word *expletive* means an exclamation or oath, especially one that is profane.

explicit/implicit

As explicit statement is one expressed directly, with precision and **clarity.** An implicit meaning may be found within a statement even though it is not directly expressed.

EXAMPLES   His directions to the new plant were *explicit,* and we found it with no trouble.

Although he did not mention the nation's financial condition, the danger of an economic recession was *implicit* in the President's speech.

exposition

Exposition, or expository writing, informs the **reader** by presenting facts and ideas in direct and concise language that is usually not adorned with colorful or figurative words and **phrases.** Expository writing attempts to explain to the reader what its subject is, how it works, how it relates to something else, and so on. Exposition is aimed at the reader's understanding, rather than at his imagination or emotions; it is a sharing of the writer's knowledge with the reader. The most important function of exposition is to provide accurate and complete information to the reader and analyze it for him.

The following sample explains the use of software in a computer system by defining it, explaining how it works, and showing how it functions with the hardware equipment.

A major component of an electronic data processing system is the software—the programs supplied by the manufacturer to direct the computer's internal operation. The software supplied with the NCR Century 50 System is an extensive package that includes the operating system, utility routines, applied programs, and the NEAT/3 Compiler. These programs have been designed to perform specific functions but, at the same time, to interact with one another to increase processing efficiency and to avoid duplication.

To conserve valuable memory space, a large portion of the software package remains on disc; only the most frequently used portion resides in internal memory all of the time. The disc-resident software is organized into small modules that are called into memory as needed to perform specific functions.

The memory-resident portion of the operating system maintains strict control of processing. It consists of routines, subroutines, lists, and tables that are used to perform common program functions, such as processing input and output operations, calling other software routines from disc as needed, and processing errors.

The disc-resident portion of the operating system contains routines that are used less frequently in system operation, such as the peripheral-related software routines that are used for correcting errors encountered on the various units, and the log and display routines that record unusual operating conditions in the system log. The disc-resident portion of the operating system also contains Monitor, the software program that supervises the loading of utility routines and the user's programs.

—*NCR Century Elementary Systems Manual* (Dayton: NCR Corporation, 1974), p. 10.

Because it is the most effective **form of discourse** for explaining difficult subjects, exposition is widely used in technical writing. To write exposition, you must have a thorough knowledge of your subject. As with all writing, how much of that knowledge you pass on to your reader should depend upon the reader's needs and your **objective.**

F

fact

Expressions containing the word *fact* ("due to the *fact* that," "except for the *fact* that," "as a matter of *fact*," or "because of the *fact* that") are often wordy substitutes for more accurate terms.

> CHANGE   *Due to the fact that* the sales force has a high turnover rate, sales have declined.
>
> TO   *Because* the sales force has a high turnover rate, sales have declined.

The word *fact* is, of course, valid when facts are what is meant.

> EXAMPLE   Our research has brought out numerous *facts* to support your proposal.

Do not use the word *fact,* however, to refer to matters of judgment or opinion.

> CHANGE   *It is a fact that* sales are poor in the Midwest because of insufficient market research.
>
> TO   In my opinion, sales are poor in the Midwest because of insufficient market research.
>
> OR   Our analysis of the statistics indicates that sales are poor in the Midwest because of insufficient market research.
>
> OR   We infer from our statistics that sales are poor in the Midwest because of insufficient market research.

feasibility reports

When the management of an organization plans to undertake a new project—a move, the development of a new product, an expansion, or the purchase of new equipment—they try to determine the project's chances for success. A feasibility report is the study conducted to help them make this determination. This **report** presents evidence about the practicality of the proposed project: How much will it cost? Is sufficient manpower available? Are any legal or other special requirements necessary? Based on the evidence, the writer of the feasibility report recommends whether or not the project should be carried out. Management officials then consider the recommendation.

The most efficient way to begin work on a feasibility report is to state clearly and concisely the purpose of the study.

EXAMPLES   The purpose of this study is to determine what type and how many new vans should be purchased to expand our present delivery fleet.

This study will determine which of three possible sites should be selected for our new warehouse.

A clearly worded statement of purpose acts as an effective guide for gathering and organizing information for the study. It both states the objective and defines the **scope** of the study. In a feasibility study, the scope includes the alternatives for accomplishing the purpose and the criteria by which each alternative will be examined.

A firm needing a word processing system, for example, might conduct a feasibility study to determine which system would best suit its requirements. The firm's requirements would establish the criteria by which each alternative system is evaluated. The following example shows how the preliminary topic outline for such a study might be organized.

   I. Purpose: To determine which word processing system would best serve our office needs

  II. Alternatives: Word processing systems of various vendors

III. Criteria:

    A. Tasks the equipment must perform

        1. Ability to work with repetitive elements

        2. Ability to accommodate extensive revision

    B. Costs

        1. Base vs. rental

        2. Installation

        3. Maintenance

        4. Training time

In writing a feasibility report, you must first identify the alternatives and then evaluate each against your established criteria. After completing these analyses, summarize them in a **conclusion.** This summary of relative strengths and weaknesses usually points to one alternative as the best, or most feasible. Make your recommendation on the basis of this conclusion.

The structure of a feasibility report depends on its scope. Some studies are brief and informal; others are detailed and formal. Every feasibility report should contain the following sections: (1) an introduction, (2) a body, (3) a conclusion/summary, and (4) a recommendation.

INTRODUCTION

The introduction should state the purpose of the report, describe the problems that led to it, and include any pertinent background information. You may discuss the scope or extent of the report. You might also discuss procedures or methods used in the analyses of alternatives: Were any special techniques used to gather or analyze information? Was information obtained by interviews, an examination of financial records, computer analyses, or other means? Any limitations on the study should be noted here: Did a deadline have to be met? Were there any restrictions on how data could be obtained?

BODY

The body of the report should present a detailed evaluation of all alternatives under consideration. Evaluate each alternative according to your established criteria. Ordinarily, each evaluation would comprise a separate section of the report.

CONCLUSION/SUMMARY

The conclusion should summarize the evaluation of each alternative, usually in the order in which they are discussed in the body of the report. In your conclusion you may want to interpret the detailed evaluations.

RECOMMENDATION

This section must state the alternative which best meets the criteria.

SAMPLE FEASIBILITY REPORT

The sample feasibility report that follows opens with an **introduction** that states the purpose of the report, the problem that prompted the study, the alternatives that were examined, and the criteria used. The body presents a detailed discussion of each alternative, particularly in terms of the criteria stated in the introduction. The conclusion draws together and summarizes the details in the body of the study. The recommendation section suggests the course of action that the company should take.

INTRODUCTION

The purpose of this report is to determine which of two proposed computer processors would best enable the Jonesville Engineering and Manufacturing Branch to increase its data processing capacity and thus to meet its expanding production requirements.

Problem

In October 1978 the Information Systems and Support Group at Jonesville put the MISSION System into operation. Since then the volume of processing transactions has increased fivefold (from 1,000 to 5,000 updates per day). This increase has severely impaired system response time from less than 10 seconds in 1978 to 120 seconds on average at present. Degraded performance is also apparent in the backlog of batch-processing transactions. During a recent check 70 real-time and approximately 2,000 secondary transactions were backlogged. In addition, the ARC 98 Processor that runs MISSION is nine years old and frequently breaks down. Downtime caused by these repairs must be made up in overtime. In a recent 10-day period in January, processor downtime averaged 25% during working hours (7:30 A.M. to 6:00 P.M.). In February the system was down often enough that the entire plant production schedule was endangered.

Finally, because the ARC 98 cannot keep up with the current workload, the following new systems, all essential to increased plant efficiency and productivity, cannot be implemented: shipping and billing, labor collection, master scheduling, and capacity planning.

Scope

Two alternative solutions to provide increased processing capacity have been investigated: (1) purchase of a new ARC 98 Processor to supplement the first, and (2) purchase of a Landmark I Processor to replace the current ARC 98. The two alternatives will be evaluated primarily according to cost and, to a lesser extent, according to expanded capacity for future operations.

PURCHASING A SECOND ARC 98 PROCESSOR

This alternative would require additional annual maintenance costs, salary for an additional computer operator, increased energy costs, and a one-time construction cost for a new facility to house the processor.

Annual maintenance costs	$ 45,000
Annual salary for computer operator	16,000
Annual increased energy costs	7,500
Annual operating costs	$ 68,500
Facility cost (one-time)	$ 50,000
Total first-year cost	$118,500

These costs for the installation and operation of another ARC 98 Processor are expected to produce the following anticipated savings in hardware reliability and system readiness.

Hardware Reliability

A second ARC 98 would reduce current downtime periods from four to two per week. Downtime recovery averages 30 minutes and affects 40 users. Assuming that 50% of users require the system at a given time, we determined that the following reliability savings would result:

> 2 downtimes × 30 minutes × 40 users × 50% × $9.00/hour overtime × 52 weeks = $9,360 (annual savings)

System Readiness

Currently, an average of one day of batch processing per week cannot be completed. This gap prevents online system readiness when users report to work and affects all users at least one hour per week. Improved productivity would yield these savings:

> 40 users × 1 hour/week × $6.00/hour average wage rate × 52 weeks = $12,480 (annual savings)

Summary of Savings

Hardware reliability	$ 9,360
System readiness	12,480
Total annual savings	$21,840

Costs and Savings for ARC 98 Processor

Costs

Annual	$ 68,500
One-time	50,000
First-year total	$ 118,500
	– 50,000
Annual total	$ 68,500

Savings

Hardware reliability	$ 9,360
System readiness	12,480
Total annual savings	$21,840

Annual Costs Less Savings

Annual costs	$ 68,500
Annual savings	– 21,840
Net additional annual operating cost	$ 46,660

ARC 98 Capacity

By adding a second ARC 98 processor, current capacity will be doubled. Each processor could process 2,500 transactions per day while cutting response time from 120 seconds to 60 seconds. However, if new systems essential to increased plant productivity are added to the MISSION System, efficiency could be degraded to its present level in the next three to five years. This estimate is based on the assumption that the new systems will add between 250 and 500 transactions per day immediately. These figures could increase tenfold in the next several years if current rates of expansion continue.

PURCHASING A LANDMARK I PROCESSOR

This alternative will require additional annual maintenance costs, increased energy costs, and a one-time facility adaption cost.

Annual maintenance costs	$45,000
Annual energy costs	6,500
Annual operating costs	$51,500
Cost of adapting existing facility	15,300
Total first-year cost	$66,800

These costs for installation of the Landmark I Processor are expected to produce the following anticipated savings in hardware reliability, system readiness, and staffing for the Information Systems and Services Department.

Hardware Reliability

Annual savings will be the same as those for the ARC 98 Processor: $9,360.

System Readiness

Annual savings will be the same as those for the ARC 98 Processor: $12,480.

Wages for the Information Systems and Services Department

New system efficiencies would permit the following wage reductions in the department:

One computer operator (wages and fringe benefits)	$16,000
One-shift overtime premium (at $100/week × 52 weeks)	5,200
Total annual wage savings	$21,200

Summary of Savings

Hardware reliability	$ 9,360
System readiness	12,480
Wages	21,200
Total annual savings	$43,040

Costs and Savings for Landmark I Processor

Costs

Annual	$ 51,500
One-time	15,300
First-year total	$ 66,800
	– 15,300
Annual total	$ 51,500

Savings

Hardware reliability	$ 9,360
System readiness	12,480
Wages	21,200
Total annual savings	$43,040

Annual Costs Less Savings

Annual costs	$ 51,500
Annual savings	– 43,040
Net additional annual operating cost	$ 8,460

Landmark I Capacity

The Landmark I processor can process 5,000 transactions per day with an average response time of 10 seconds per transaction. Should the volume of future transactions double, the Landmark I could process 10,000 transactions per day without exceeding 20 seconds per transaction on average. This increase in capacity over the present system would permit implementation of plans to add four new systems to MISSION.

CONCLUSION

A comparison of costs for both systems indicates that the Landmark I would cost $36,380 less in first-year costs.

ARC 98 Costs

Net additional annual operating	$46,660
One-time facility	50,000
First-year total	$96,660

Landmark I Costs

Net additional annual operating	$ 8,460
One-time facility	15,300
First-year total	$23,760

Installation of a second ARC 98 Processor will permit the present information processing systems to operate relatively smoothly and efficiently. It will not, however, provide the expanded processing capacity that the Landmark I Processor would for implementing new subsystems essential to improved production and record keeping.

RECOMMENDATION

The Landmark I Processor should be purchased because of the initial and long-term savings and because its expanded capacity will allow the addition of essential systems.

female

Female is usually restricted to scientific, legal, or medical contexts (a *female* patient or suspect). Keep in mind that this term sounds cold and impersonal. The terms *girl, woman,* and *lady* are acceptable substitutes in other contexts; however, be aware that these substitute words have connotations involving age, dignity, and social position. (See also **male**.)

few/a few

In certain contexts, *few* carries more negative overtones than the **phrase** *a few* does.

EXAMPLES   They have *a few* scruples. (positive)
They have *few* scruples. (negative)
There are *a few* good things about your report. (positive)
There are *few* good things about your report. (negative)

fewer/less

Fewer refers to items that can be counted **(count nouns).**

EXAMPLES   A good diet can mean *fewer* colds.
Fewer members took the offer than we expected.

Less refers to mass quantities or amounts **(mass nouns).**

EXAMPLES   *Less* vitamin C in your diet may mean more, not *fewer,* colds.
The crop yield decreased this year because we had *less* rain than necessary for an optimum yield.

figuratively/literally

These two words are often confused. *Literally* means ''really'' and should not be used in place of *figuratively,* which means ''metaphorically.'' Do not say that someone ''literally turned green with envy'' unless that person actually changed color.

EXAMPLES   In the winner's circle the jockey was, *figuratively* speaking, ten feet tall.
When he said, ''Let's run it up the flagpole,'' he did not mean it *literally.*

Avoid the use of *literally* to reinforce the importance of something.

CHANGE   She was *literally* the best of the group.
TO   She was the best of the group.

figures of speech

A figure of speech is an imaginative **comparison,** either stated or implied, between two things that are basically unlike but have at least one thing in common. If a device is cone-shaped with an opening at the top, for example, you might say that it looks like a volcano.

Technical people may find themselves using figures of speech to clarify the unfamiliar by relating a new and difficult concept to one with which the **reader** is familiar. In this respect, figures of speech help establish a common ground of understanding between the specialist and the nonspecialist. Technical people may also use figures of speech to help translate the abstract into the concrete; in the process of doing so, figures of speech also make writing more colorful and graphic.

Although figures of speech are not used extensively in technical writing, a particularly apt figure of speech may be just the right tool when you must explain or describe a complex concept. A figure of speech must be appropriate, however, to achieve the desired effect.

CHANGE   Without the fuel of tax incentives, our economic engine would operate less efficiently. (It would not operate at all without fuel.)

TO   Without the fuel of tax incentives, our economic engine would sputter and die. (This is not only apt, but it states a rather dry fact in a colorful manner—always a desirable objective.)

A figure of speech must also be consistent to be effective.

CHANGE   We must get our research program back *on the track,* and we are counting on you to *carry the ball.* (inconsistent)

TO   We must get our research program back on the track, and we are counting on you to do it. (inconsistency removed)

A figure of speech should not, however, attract more attention to itself than to the point the writer is making.

EXAMPLE   The whine of the engine sounded like ten thousand cats having their tails pulled by ten thousand mischievous children.

Trite figures of speech, which are called **clichés,** defeat the purpose of a figure of speech—to be fresh, original, and vivid. A surprise that comes "like a bolt out of the blue" is not much of a surprise. It is better to use no figure of speech than to use a trite one.

TYPES OF FIGURES OF SPEECH

Analogy is a comparison between two objects or concepts that shows ways in which they are similar. It is very useful in technical writing, especially when you are writing to an educated but nontechnical audience. In effect, analogies say "A is to B as C is to D." The resemblance between these concepts is partial but close enough to provide a striking way of illuminating the relationship the writer wishes to establish.

EXAMPLE   Pollution (A) is to the environment (B) as cancer (C) is to the body (D).

Antithesis is a statement in which two contrasting ideas are set off against each other in a balanced syntactical structure.

EXAMPLES   Man proposes, but God disposes.
Art is long, but life is short.

Hyperbole is gross exaggeration used to achieve an effect or **emphasis**.

EXAMPLE   He *murdered* me on the tennis court.

Litotes are understatement, for emphasis or effect, achieved by denying the opposite of the point you are making.

EXAMPLES   Einstein was no dummy.
Fifty dollars is not a small price for a book.

Metaphor is a figure of speech that points out similarities between two things by treating them as though they were the same thing. Metaphor states that the thing being described *is* the thing to which it is being compared.

EXAMPLE   He is the sales department's *utility infielder.*

Metonymy is a figure of speech that uses one aspect of a thing to represent it, such as *the red, white, and blue* for the American flag, *the blue* for the sky, and *wheels* for an automobile. This device is common in everyday speech because it gives our expressions a colorful twist.

EXAMPLE   *The hard hat* area of the labor force was especially hurt by unemployment.

Simile is a direct comparison of two essentially unlike things, linking them with the word *like* or *as.*

EXAMPLE   His feelings about his business rival are so bitter that in recent conversations with his staff he has returned to the subject compulsively, *like a man scratching an itch.*

Personification is a figure of speech that attributes human characteristics to nonhuman things or abstract ideas. One characteristically speaks of the *birth* of a planet and the *stubborness* of an engine that will not start.

EXAMPLE   Early tribes of human beings attributed scientific discoveries to gods. To them, fire was not a *child of man's brain* but a gift from Prometheus.

fine

When used in expressions such as "I feel *fine*" or "a *fine* surf," *fine* is colloquial. The colloquial use of *fine*, like that of *nice*, is too vague for technical writing. In writing, the word *fine* should retain the sense of "refined," "delicate," or "pure."

EXAMPLES　A *fine* film of oil covered the surface of the water.
There was a *fine* distinction between the two possible meanings of the disputed passage.
Fine crystal is currently made in Austria.

finite verbs

A finite **verb** is the main verb of a **clause** or **sentence**. The nonfinite forms are the **verbals** (**infinitive, gerund,** and **participle**). Unlike verbals, finite verbs show **number, tense,** and person.

NUMBER　The punch press *stamps* 30 pieces a minute. (singular)
The punch presses *stamp* 25 pieces a minute. (plural)

TENSE　The plane *lifts* from the runway with the thrust of the engine. (present)
The plane *lifted* from the runway with the thrust of the engine. (past)
The plane *will lift* from the runway with the thrust of the engine. (future)

PERSON　*I give* to charity through a payroll deduction program. (first person)
You give to charity through the same program. (second person)
He gives to charity through the same program. (third person)

first/firstly

Firstly is an unnecessary attempt to add the *-ly* form to an **adverb.** *First* is an adverb in its own right, and sounds much less stiff than *firstly.*

CHANGE　*Firstly,* we should ask for an estimate.
TO　*First,* we should ask for an estimate.

flammable/inflammable/nonflammable

Both *flammable* and *inflammable* mean "capable of being set on fire."

EXAMPLES   The cargo of gasoline is *flammable*.
The cargo of gasoline is *inflammable*.

Since the *in-* **prefix** usually causes the word following to take its opposite meaning *(incapable, incompetent)*, *flammable* is preferable to *inflammable* because it avoids possible misunderstanding. *Nonflammable* is the opposite, meaning *"*not capable of being set on fire."

EXAMPLE   The asbestos suit was *nonflammable*.

flowcharts

A flowchart is a diagram of a process that involves stages, with the sequence of stages shown from beginning to end. The flowchart presents an overview of the process that allows the **reader** to grasp the essential steps of the process quickly and easily. The process being illustrated could range from the steps involved in assembling a bicycle to the stages by which bauxite ore is refined into aluminum ingots for fabrication.

Flowcharts can take several forms to represent the steps in a process. They can consist of labeled blocks (Figure 6), pictorial representations (Figure 7), or standardized symbols (Figure 8); the items in any flowchart are always connected according to the sequence in which the steps occur. The normal direction of flow in a chart is left to right or top to bottom. When the flow is otherwise, be sure to indicate it with arrows.

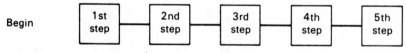

Figure 6.   Example of a Block Flowchart

Flowcharts that document computer programs and other information-processing procedures use standardized **symbols.** These standards are set forth in *U.S.A. Standard Flowchart Symbols and Their Usage in Information Processing,* published by the American National Standards Institute publication X3.5. When creating a flowchart, follow the guidelines on page 216.

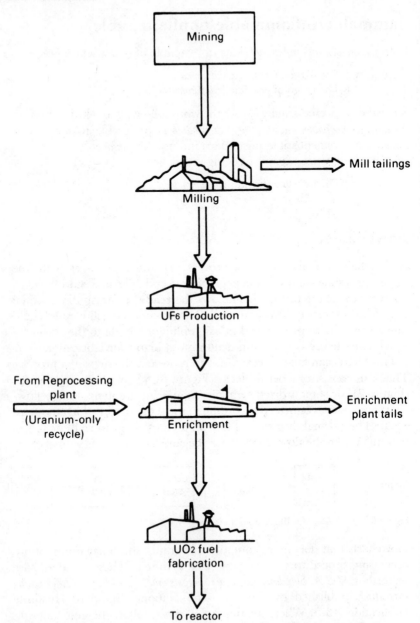

Figure 7.　Light-Water Reactor Uranium Fuel Cycle Front-End Operations

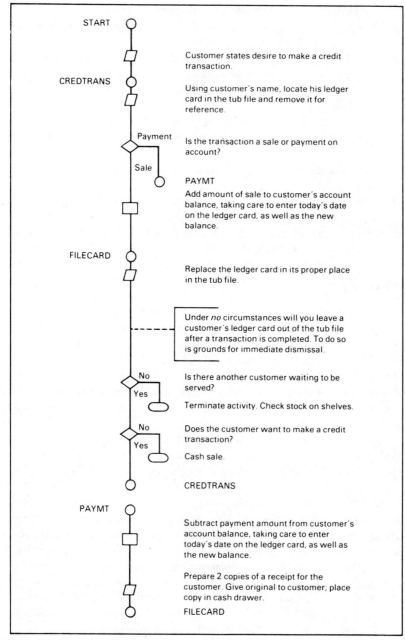

START

Customer states desire to make a credit transaction.

CREDTRANS

Using customer's name, locate his ledger card in the tub file and remove it for reference.

Payment

Is the transaction a sale or payment on account?

Sale

PAYMT

Add amount of sale to customer's account balance, taking care to enter today's date on the ledger card, as well as the new balance.

FILECARD

Replace the ledger card in its proper place in the tub file.

Under *no* circumstances will you leave a customer's ledger card out of the tub file after a transaction is completed. To do so is grounds for immediate dismissal.

No

Is there another customer waiting to be served?

Yes

Terminate activity. Check stock on shelves.

No

Does the customer want to make a credit transaction?

Yes

Cash sale.

CREDTRANS

PAYMT

Subtract payment amount from customer's account balance, taking care to enter today's date on the ledger card, as well as the new balance.

Prepare 2 copies of a receipt for the customer. Give original to customer; place copy in cash drawer.

FILECARD

Figure 8. Flowchart of a Credit Transaction

1. Label the flowchart clearly and concisely.
2. Assign the chart a figure number if it is being used in a document that contains five or more **illustrations.**
3. With labeled blocks and standardized symbols, use arrows to show the direction of flow only if the flow is opposite to the normal direction. With pictorial representations, use arrows to show the direction of all flow.
4. Label each step in the process, or identify it with a conventional symbol. Steps can also be represented pictorially or by captioned blocks.
5. Include a key if the flowchart contains symbols your reader may not understand.
6. Leave adequate white space on the page. Do not crowd your steps and directional arrows too close together.
7. As with other illustrations, place the flowchart as near as possible to that portion of the text that refers to it.

footnotes

Footnotes indicate the source of facts or ideas and tell the **reader** where to find additional information about the **topic.** Footnotes can also offer explanatory comments to support ideas mentioned in the main body of the text. Since footnotes are used to identify direct **quotations** or paraphrased material taken from other sources, they help writers avoid **plagiarism** as well. To decide what information should be acknowledged in a footnote, you must distinguish between those ideas that are considered common knowledge and those that are not.

Footnotes in scientific and technical writing differ in form from those used in other fields. To save space, footnotes in scientific writing are gathered together and placed in a **references** section at the end of an article or chapter.

For other kinds of research papers, footnotes are normally located either at the bottom of the appropriate page or at the end of a chapter or article. In either place, they are numbered consecutively, with superscript numerals, numbers that are raised slightly above the line. Within the text the footnote number, also raised, should follow the quotation it identifies, whether the quotation is within the text or is set off from the text. The number should appear at the end of the **sentence** whenever possible. When there are only a few footnotes, they are usually marked with asterisks or daggers rather than numbers.

Indent the first line of a footnote five spaces from the left margin, and type any additional lines of the footnote flush with the left margin. Single-space footnotes that take more than one line, and double-space between footnotes.

FIRST REFERENCE

For the first full footnote reference to a book, include the following information in this order: (1) author's full name; (2) complete title; (3) editor or compiler (if any); (4) title of the series in which the book appears, if it is part of a series, and its volume or number in the series; (5) edition, if it is other than the first; (6) number of volumes for a multivolume book; (7) publication information, including city, publisher, and year; (8) volume number of that specific book, if any; and (9) page number or numbers. Capitalize key words in the title and italicize (underline) the complete title.

For the first full reference to a **journal article,** include the following information in this order: (1) author's full name; (2) complete title of the article; (3) name of the journal; (4) volume and number of the journal; (5) date of the volume or issue; and (6) page number or numbers. Capitalize key words in the article and journal titles, enclose the article title in **quotation marks,** and italicize (underline) the complete journal title. Footnote forms in this section are based on those recommended in the University of Chicago Press *A Manual of Style* (12th ed.).

BOOK, ONE AUTHOR

[1]Wayne Hanley, <u>Natural History in America</u> (New York: Quadrangle Press/New York Times Books, 1977), p. 51.

BOOK, TWO OR MORE AUTHORS

[2]Herbert M. Jelley and Robert O. Hermann, <u>The American Consumer: Issues and Decisions</u> (New York: McGraw-Hill, 1973), p. 9.

BOOK EDITION, IF NOT THE FIRST

[3]S.R. Silverman and Hollowell Davis, <u>Hearing and Deafness</u>, 2nd ed. (New York: Holt, Rinehart and Winston, 1978), p. 35.

MULTIVOLUME BOOK

[4]John Bartholemew, ed. <u>Times Atlas of the World</u>, 5 vols. (London: Times Publishing Co., Ltd., 1955-1959), 3:27.

The volume from which the material cited was taken is noted at the end of the entry, along with the appropriate page number.

BOOK IN A SERIES

[5]E.L. Cooper, ed., <u>Invertebrate Immunology</u>, Contemporary Topics in Immunology, vol. 4 (New York: Plenum Publishing Corp., 1974), p. 33.

TRANSLATION

[6]Zhores A. Medvedov, <u>Nuclear Disaster in the Urals</u>, trans. George Saunders (New York: W.W. Norton, 1979), p. 13.

SYMPOSIUM OR CONFERENCE PAPER

[7]J. Higgonson, "The Role of Geographic Pathology in Environmental Carcinogenesis," in <u>Proceedings of the 1971 Symposium on Fundamental Cancer Research</u>, ed. R.L. Clarke (Baltimore: Williams and Wilkins, 1972), p. 170.

JOURNAL ARTICLE

[8]Willard Thomas, "Interrelated Video Factors and a Basic Model for System Selection," <u>Technical Communication</u> 27 (1980): 16-18.

MAGAZINE ARTICLE

[9]"The Assault on the Federal Trade Commission," <u>Consumer Reports</u>, March 1980, p. 151.

NEWSPAPER ARTICLE

[10]John H. Allan, "Fixed-Income Securities Ease," <u>The New York Times</u>, Late City Ed., 15 June 1979, Sec. D, p. 70.

ENCYCLOPEDIA ARTICLE

[11]"Electricity," Encyclopaedia Britannica, 14th
ed.

Because encyclopedia articles are arranged
alphabetically, page and volume numbers are
unnecessary unless the reference is to a page in
an article that runs longer than one page. Then
it is necessary to give the volume and page
number or numbers.

CORPORATE AUTHOR

[12]U.S. Controller General, Waste Disposal
Practices - A Threat to Health and the Nation's
Water Supply, A Report to Congress (Washington,
D.C.: U.S. General Accounting Office, 1978,
p. 12.

REPORT

[13]Office of Noise Abatement and Control,
Potential Effectiveness of Barriers Toward
Reducing Highway Noise Exposure on a National
Scale (Washington, D.C.: U.S. Environmental
Protection Agency, 1978), p. 33.

PAMPHLET OR BOOKLET

[14]Laurence Hursh, Coronary Heart Disease:
Risk Factors and the Diet Debate (Rosemont, Ill.:
National Dairy Council, 1980), p. 7.

THESIS OR DISSERTATION

[15]M.G. Johnson, "Infiltration Capacities and
Surface Erodibility Associated with Forest
Harvesting Activities in the Oregon Cascades"
(Master's Thesis, Oregon State University, 1978),
p. 38.

PERSONAL CORRESPONDENCE

[16]Dr. Robert T. Brady, Salvo Corp., June 4,
1980, letter to author.

INTERVIEW

[17]Interview with Virgil Denlinger, Assistant Chief of Police, Anytown, Mass., March 15, 1980.

SUBSEQUENT REFERENCES

In subsequent references to the same book or article, list only the author's last name and the relevant page number or numbers. For a work that has more than one author, give the last name of each author and the appropriate page number.

[18]Hursh, p. 5.

[19]Jelley and Hermann, p. 15.

If you cite more than one work by the same author, subsequent references should give (1) the author's last name, (2) a shortened title, and (3) the page number or numbers.

[20]Higgonson, "Geographic Pathology," p. 170.

The author's name may be omitted from a footnote when his or her full name is given in the text near the reference to the work cited, as in the example at the top of page 221.

Conventional footnote forms can differ in varying degrees from one field of study to another and at times even within the same field. Be alert to the forms used in the organization or publication for which your material is being written. In the absence of specific guidelines, use the forms presented here, remembering to include enough accurate information to enable a reader wishing to do so to find your source. Once you have adopted a form, however, be consistent.

> In a recent article on magnetic fluids, John Free explains that they are made by crushing magnetite (Fe_3O_4), an iron oxide, into tiny particles that are mixed with an oil-like substance called a diester.[1]
>
> ---
>
> [1]"Magnetic Fluids Put New Highs in Hi-Fi," Popular Science, June 1979, p. 57.

forceful/forcible

Although *forceful* and *forcible* are both **adjectives** meaning "characterized by or full of force," *forceful* is usually limited to persuasive ability and *forcible* to physical force.

EXAMPLES   The thief made a *forcible* entry into my apartment.
John made a *forceful* presentation at the committee meeting.

foreign words in English

The English language has a long history of borrowing words from other languages. Most of these borrowings occurred so long ago that we seldom recognize the borrowed terms (also called "loan words") as being of foreign origin. To check the origin of a word, consult the etymology portion of its **dictionary** entry.

EXAMPLES   whiskey (Gaelic), animal (Latin), church (Greek)

GUIDELINES FOR USE OF FOREIGN WORDS IN ENGLISH
Foreign expressions should be used only if they serve a real need. The overuse of foreign words in an attempt to impress your **reader** or to be

elegant is **affectation**. Your goal of effective communication can be accomplished only if your reader understands what you write; choose foreign expressions, therefore, only when they serve the purpose of making an idea clearer.

Words not fully assimilated into the English language are set in **italics** if printed (underlined in typed manuscript).

EXAMPLES   *sine qua non, coup de grâce, in res, in camera*

Words that have been fully assimilated need not be italicized. When in doubt, consult a current dictionary.

EXAMPLES   cliché, etiquette, vis-à-vis, de facto, résumé

As foreign words become current in English, their plural forms give way to English plurals.

EXAMPLES   *formulae* becomes *formulas*
memoranda becomes *memorandums*

In addition, accent marks tend to be dropped from words (especially from French words) the longer they are used in English. But because not all foreign words shed their accent marks with the passage of time, consult a dictionary whenever you are in doubt.

foreword/forward

Although the pronunciation is the same, the spellings and meanings of these two words are quite different. The word *foreword* is a **noun** meaning "introductory statement at the beginning of a book or other work."

EXAMPLE   The department chairman was asked to write a *foreword* for the professor's book.

The word *forward* is an **adjective** or **adverb** meaning "at or toward the front."

EXAMPLES   Move the lever to the *forward* position on the panel. (adjective)
Turn the dial until the needle begins to move *forward*. (adverb)

foreword/preface

The terms *foreword* and *preface* are sometimes used interchangeably, but are increasingly differentiated. A foreword is usually an introductory statement about a book or **formal report** written by someone other than the author. A preface is a statement by the author about the purpose, background, or **scope** of the book or report. The preface of this book, as an example, is on page *v.*

formal reports

A formal report is the written account of a major project. Projects that are likely to generate formal reports include **research** into new developments within a field, a study on the advisability of launching a new product, or an end-of-year developmental review. The **scope** and complexity of the project will determine the length and complexity of the **report.** Most formal reports—certainly those that are long and complex—require a carefully planned structure. Aids such as the **table of contents,** lists of **illustrations** and **tables,** and the **abstract** serve to make the information in the report accessible to the **reader.** Often, formal reports address more than one audience. The **executive summary,** for example, mainly interests managers who need an overview of the report, while the **appendix** will offer detailed technical data to specialists.

Most formal reports are divided into three major parts—front matter, body, and back matter—each of which contains a number of elements. Just how many elements are needed for a particular report depends on a combination of features, including the subject, the length of the report, and the kind of material the report contains.

ORDER OF ELEMENTS IN A FORMAL REPORT

The arrangement of the elements in a formal report may vary. Many companies and institutions have a preferred style for formal reports and furnish guidelines that staff members must follow. If your employer has prepared a set of style guidelines, follow it; if not, use the **format** recommended here.

The following list includes all the elements a formal report might contain in their normal order. Not all formal reports will include every element.

FRONT MATTER
Title Page
Abstract
Table of Contents
List of Figures
List of Tables
Preface
List of Abbreviations and Symbols

BODY
Executive Summary
Introduction
Explanatory Footnotes
Heads
References

BACK MATTER
Bibliography
Appendixes
Glossary

FRONT MATTER

The front matter serves several purposes: It gives a general idea of what the report is about and of the author's purpose in writing it; it gives the reader a chance to determine whether the report contains the kind of information he or she is looking for; and it lists the location of chapters, heads, illustrations, and tables. Front matter pages are numbered with lower-case Roman numerals. Following are the seven elements that normally make up front matter, though not all formal reports require every element.

Title Page. The title page, which is unnumbered but counted as page *i*, should include the following information.

1. Full **title** of the report.
2. Writer, principal investigator, compiler, and so on.
3. Date or dates of the report. For one-time reports, list the date the report is distributed. For periodical reports, which may be issued monthly or quarterly, list the time period that the present report covers, as well as the date the report is issued.
4. Name of the organization for which the writer works.

These categories are standard on most title pages, though some organizations may require additional information.

```
         Card Uniqueness:  A Solution to the

               Credit Card Problems

                  of Tomorrow



                 Jason T. Smiley



                  April 19--



         Fidelity Financial Consultants, Inc.
```

Sample Title Page

Abstract. The abstract provides a brief summary of the contents of the full report. Usually in two hundred words or less, it states the subject and scope of the report and notes its findings, conclusions, or recommendations. Going through an **abstract** helps the reader decide whether it will be worthwhile to read the report in full. It always starts on a new page.

Table of Contents. A table of contents lists all the headings or sections of the report in their order of appearance, along with their page numbers. It includes a listing of all front matter and back matter elements, except the title page and the table of contents itself. The table of contents always begins on a new page.

```
                    TABLE OF CONTENTS
```

Along with the abstract, the table of contents gives the reader an idea of what the report discusses and of whether it is worth reading. It also aids readers with interest only in certain sections of the report. For this reason the table of contents lists chapter and section titles and their page numbers exactly as they are given in the text.

List of Figures. Figures include all **drawings, photographs, maps,** charts, and **graphs** contained in the report. When a report contains more than five figures, they should be listed, along with their page numbers, in a separate section immediately following the table of contents. The section should be entitled "List of Figures" and should begin on a new page. Figure numbers, titles, and page numbers should be identical to those in the text. In most reports figures are numbered consecutively with Arabic numbers.

List of Tables. When a report contains more than five tables, they should be listed, along with their titles and page numbers, in a separate section entitled "List of Tables," immediately following the list of figures (if there is one). Tables are numbered consecutively with Arabic numbers throughout most reports.

Preface. The preface is the author's statement of the purpose, background, or scope of the report. The preface may contain acknowledgment of help received on the project or the report; it may also cite permission obtained for the use of copyrighted works. For an example, see the preface to this text on page *v*.

The preface follows the table of contents (and any lists of figures or tables) and begins on a separate page entitled "Preface."

List of Abbreviations and Symbols. When a report uses many **abbreviations** and **symbols** that the reader may not be able to interpret, the front matter should include a list of all abbreviations and symbols, with an explanation of what they stand for in the report. Such a list, which follows the preface, is particularly necessary in technical reports that are not limited to technicians. If it is a long list, it usually starts on a new page.

BODY

The first page of text is unnumbered but is considered page 1. If there is no introduction, the report begins with the first major text heading.

Executive Summary. Many organizations require reports to contain a summary of the report's major findings, **conclusions,** and recommendations. Usually called an **executive summary,** it allows managers who may not have time to read the full report to review its main points quickly and then decide whether they should read the entire work.

Usually 500 to 1,000 words, an executive summary is more complete than an abstract: besides summarizing the major points, it states the purpose and scope of the longer report as well as describing the methods used. An executive summary starts on a new page.

Introduction. An **introduction** should state the scope and limitations of the report. This statement should answer two questions: What was the **objective** of the study, and why was the study done? Not every report needs an introduction, however. Background information on scope and objective that can be stated briefly may appear in the preface instead.

Explanatory Footnotes. Sometimes it is necessary to give additional explanation of something that appears in the text. If such an explanation would interrupt the text discussion, it should be given in a note at the foot of the page. Explanatory **footnotes** are marked with a single or a double asterisk (*, **) or a single or a double dagger (†, ‡).

Heads. In most reports the body of the discussion is divided into logical parts or segments. These parts are often labeled with heads, or headings. Besides dividing the discussion into manageable parts, heads call attention to the main topics and signal changes of topic.

Long and complicated reports may require several levels of heads to indicate major divisions and subdivisions. For a discussion of how to use heads in formal reports, see **heads.**

References. If a report quotes or refers to material in other published works or sources, it must name the works and sources in a separate references, or notes, section. If your employer or instructor has a preferred reference style, follow that format; otherwise, follow the guidelines provided in the **references** entry in this book.

```
                       References

     1.  Beating the new cards.  Business Week.
         120-121; 1973 August.

     2.  Frank, C. A.  New jobs ahead for bank
         cards.  Banking.  115-117; 1973
         September.

     3.  Richardson, Dennis W.  Electric money.
         Cambridge, MA:  MIT Press; 1973.

     4.  Financial Systems Marketing Newsletter.
         Dayton, OH:  NCR Corp.; 1975.

     5.  Walker, Gerald M.  Electronic funds
         transfer systems.  Electronics.
         13-19; 1975 July.
```

In relatively short reports the references may all go at the end on a new page. In longer reports with a number of sections or chapters, each section or chapter should have a references section. In either case, every references section should be clearly labeled as such. If a particular reference must be listed in more than one section or chapter, it should be repeated in full in each one. If a reference must be repeated within a references section, follow the style for subsequent references explained in the **references** entry of this book.

BACK MATTER

The back matter of a formal report contains supplemental information— details that may not fit directly in the text but that are helpful for understanding the report or give additional details.

Bibliography. The **bibliography** usually appears in the back matter and lists alphabetically by author's last name all the sources of your information. Like other elements in the front and back matter, the bibliography starts on a new page and is labeled by name. If a bibliography is necessary, it should be consistent in format with the **references.**

Appendix. The appendix includes such materials as long charts, graphs, or tables; sample **questionnaires;** transcripts of **interviews; correspondence;** and explanations that are too long for explanatory footnotes but that may be helpful to the reader.

A report may have one or more appendixes; generally, each appendix contains one type of material. For example, a report with two appendixes might include charts in one and graphs in the other. The first appendix goes on a new page directly after the bibliography. Additional appendixes also begin on a new page. Each should be given a title (for example, Appendix A) and a heading (for example, Questionnaire).

Glossary. A **glossary** lists selected terms on a particular subject that are defined or explained. It is necessary to include a glossary in reports that use many words and expressions that may be unfamiliar to the audience. The terms are arranged alphabetically, with each entry on a new line and followed by its definitions, dictionary style. The glossary, labeled as such, appears on a new page directly after any appendixes.

GRAPHIC AND TABULAR MATTER

Formal reports often contain illustrations and tables to clarify and support the text as well as to compress a lot of data into a small space. Numbering and sequencing of illustrations and tables may vary. The following guidelines show one conventional system for integrating such material smoothly into the text.

Figures. Beneath each figure put a title and a number, in Arabic numerals. Number figures sequentially throughout the report. For long reports, number figures by chapter or section according to this system: Make the first figure in Chapter 1 Figure 1.1 (or Figure 1-1), the second figure Figure 1.2 (or Figure 1-2), the first figure in Chapter 2 Figure 2.1 and so on. In the text refer to figures by number rather than by location ("Figure 2," rather than "the figure below").

Tables. Above each table give a title and a number, in Arabic numerals. Number the tables sequentially throughout the report. For long reports, number tables with Arabic numerals by chapter or by section according to the system described for numbering figures. Refer to tables in the text by number rather than by location ("Table 4" rather than "the above table").

formal writing style

Formal writing style is difficult to define because there is no clear dividing line between formal and **informal writing style**—elements of both appear in almost all writing. Formal writing style can perhaps best be defined by pointing to certain material that is clearly formal, such as scholarly and scientific articles in professional journals, lectures read at meetings of professional societies, and legal documents. Although **technical writing style** is not as formal as it once was, technical writing is nonetheless formal writing.

Any particular sample of writing is judged to be formal or informal depending on a number of distinctive qualities. Material written in formal style is usually the work of a specialist in a particular field writing to other specialists. As a result, the vocabulary is specialized. The writer's **tone** is impersonal and objective because the subject matter looms larger in the writing than the author's personality (see **point of view**). Unlike informal writing style, formal writing style does not use **contractions,** slang, or dialect (see **English, varieties of**). **Sentences** are often elaborate because complex ideas are generally being examined.

Formal writing need not be dull and lifeless, however. By using such techniques as the active **voice,** variety in sentences, and **subordination,** formal writing can be made lively if the subject matter is inherently interesting to the **reader.**

EXAMPLE   Although a knowledge of the morphological and chemical constitution of cells is necessary to the proper understanding of living things, in the final analysis it is the activities of their cells that distinguish organisms from all other objects in the world.

Many of these activities differ greatly among the various types of living things, but some of the basic sorts are shared by all, at least in their essentials. It is these fundamental actions with which we are concerned here. They fall into two major groups—those that are characteristic of the cell in the *steady state,* that is, in the normally functioning cell not engaged in reproducing itself, and those that occur during the process of *cellular reproduction.*
—Lawrence S. Dillon, *The Principles of Life Sciences* (New York: Macmillan, 1964), p. 32.

Whether you should use a formal style in a particular instance depends on your reader and your **objective.** When writers attempt to force a formal style where it should not be used, they are likely to fall into **affectation, awkwardness,** and **gobbledygook.**

format

Format is the physical arrangement and general appearance of a finished writing project. A good general appearance is important to any writing project because it puts your **readers** in a favorable frame of mind toward your work and lets them know that you have put much planning and thought into it. An unpleasing appearance, on the other hand, is likely to put your readers in a negative frame of mind toward your work.

The type of writing project normally determines the format the writer uses. Such writing projects as **formal reports, memorandums, questionnaires, progress reports, proposals, technical manuals, correspondence, trip reports, résumés, instructions, test reports,** and **laboratory reports** have fairly standard formats. Some may, for example, include a **foreward** or preface, a **glossary,** a **table of contents,** an **index,** or a **bibliography.** Some may use such devices as **heads,** leaders (see **periods**), **headers and footers, illustrations, indentation, references,** or **footnotes,** and others may not. Format is important because any way you can make your finished product more pleasing and accessible to your reader will help you achieve your ultimate **objective**—the clear communication of information.

former/latter

Former and *latter* should be used to refer to only two items in a **sentence** or **paragraph.**

EXAMPLE   The president and his trusted aide emerged from the conference, the *former* looking nervous and the *latter* looking downright glum.

Because these terms make the **reader** look back to previous material to identify the reference, they impede reading and are best avoided.

CHANGE   Model 190432 had the necessary technical modifications, while model 129733 lacked them, the *former* being this year's model and the *latter* last year's.

TO   Model 190432, this year's model, had the necessary technical modifications; model 129733, last year's model, did not.

forms

Forms provide a time-saving, efficient, and uniform way to record data. It is easier and quicker to supply information by filling out a form than by writing a **memorandum,** letter, or **report.** Also, it is far easier to read, tabulate, and evaluate information on a form than in more individual forms of writing.

PREPARING A FORM

To be effective, a form should make it easy for one person to supply information and for another person to retrieve and interpret the information. Ideally, a form should be self-explanatory, even to a person who is seeing it for the first time. If you are preparing a form, plan carefully. Determine what kind of information you will be seeking and arrange the questions in a logical order.

Writing Instructions and Captions. Instructions should go at the beginning of the form and are often preceded by a **head** designed to attract the attention of the **reader,** as in the following example.

Instructions for Completing This Form

1. Complete the applicable blue shaded portions on the front of pages 1, 2, and 3.
2. Mail page 1 to the Securi-Med Insurance Company at the address shown above.
3. Give page 2 to your doctor.
4. If services are rendered in a hospital, give page 3 to the hospital.
5. Use the back of page 1 to itemize bills that are to go toward your major medical deduction.

Instructions for distributing the various copies of multiple-copy forms are normally placed at the bottom of the form, and are repeated on every copy of the form.

Requests for information on the form are normally worded as captions. Keep captions brief and to the point; avoid unnecessary **repetition** by combining related information under an explanatory heading.

CHANGE   What make of car do you drive? _____

What year was it manufactured? _____

What model is it? _____

What is the body style? _____

TO   Vehicle Information

Make _____   Year _____

Model _____   Body Style _____

Planning for Responses. Forms should ask questions that can be answered simply and briefly. The best responses are check marks, circles, or underlining; next best are numbers, single words, or brief **phrases. Sentence** responses are the least effective, since they take the most time to write and to read.

Make captions as specific as possible. For example, if a requested date is other than the date on which the form is being filled out, make the caption read, "Effective date," "Date issued," or whatever it may be, rather than simply "Date." As in all technical writing, put yourself in your reader's place and try to imagine what sort of requests would be clear to you.

Sequencing Data. Try to arrange requests for information in an order that will be most helpful to the person filling out the form. At the top of the form, include preliminary information, such as the name of your organization, the title of your form, and any file number or reference number. In the main portion of the form, include the entries you need in order to obtain the necessary data. At the end of the form, include space for a signature and a date.

The arrangement of the entries depends upon several factors. First, the subject matter of the entries will frequently determine the most logical order. A form requesting reimbursement for travel expenses, for instance, would logically begin with the first day of the week (or month) and end with the last day of the appropriate period. Second, if the response to one item is based upon the response to another item, be sure that the items appear in the correct order. Third, group requests for related information together whenever possible. Finally, entries should, in general, be arranged on the form from left to right and from top to bottom, since that is the way we are accustomed to reading.

DESIGNING A FORM

When you prepare the final version of a form, pay particular attention to design details—especially to the placement of entry lines and the amount of space allowed for responses.

Entry Lines. The form can be laid out so that the person filling it out supplies information on a writing line, in a writing block, or in square boxes. The *writing line* is simply a rule with a caption.

| (Name) | (Telephone) |

(Street Address)

| (City) | (State) | (Zip Code) |

The *writing block* is essentially the same except that each entry is enclosed in a ruled block, making it impossible for the person filling out the form to associate a caption with the wrong line.

Name	Telephone
Street Address	
City	State	Zip Code

On some forms, captions are set horizontally.

Destination	Source	Supplier	Method of Shipment

On other forms, they are set vertically.

Destination	
Source	
Supplier	
Method of Shipment	

When it is possible to anticipate all likely responses, you can make the form easy to fill out by writing the question on the form, supplying a labeled box for each possible answer, and asking people to put an *X* in the appropriate box. Such a plan also makes it easy to retrieve the data. Be sure that your questions are both simple and specific.

EXAMPLE   Would you buy another Whapo?   ☐ Yes   ☐ No

Spacing. Be sure to provide enough space to enable the person filling out the form to enter the data. Insufficient writing space makes it difficult for people to respond—and thus causes responses that are hard to read. Reading responses that are too tightly spaced or that snake around the side of the form can cause headaches and errors.

Forms may be filled out either in longhand or on a typewriter. Always allow sufficient space to accommodate both. If you think that some people will fill out the form in longhand, provide adequate space for relatively large handwriting. It is especially important to plan for type-written responses if you intend to have your form typeset. If you do not inform the typesetter that the printed form must be easy to answer on a typewriter, those who do may have difficulty aligning the form vertically in their machines. The reason is simple: On a typewriter, there are 6 lines to the inch, but on typesetting equipment there are 6.0386 lines to the inch—there's a hairline less space between typeset lines than between typewritten lines—and any typewriter will be off the writing line by the time it nears the bottom of a page-long form. Remind the typesetter to set *exactly* 6 lines to the inch. And if the form will contain more than one response per line, align the column entries vertically whenever possible, so that the person filling out the form can set tabs on the typewriter.

Look on pages 235 and 236 for two sample forms.

Annual Reappointment Form   July 1, 19__ to June 30, 19__

CHILDREN'S MEDICAL CENTER

1735 Chapel Street

Dayton, Ohio 45404

NAME:

List appointments or offices held, teaching positions, independent studies in medical or dental societies or other medical organizations, and any other professional recognitions you would like to have included in your file:

_____

_____

_____

_____

_____

_____

_____

Do you wish a change in your privileges? If so, specify:_____

_____

Have there been any changes in your board specialties?   Yes [ ]   No [ ]

    If yes: Date_____

             Specialty Board_____

Signature_____

Date_____

                              Return to: Chairman
                                          Credentials & Nominating Com...

Sample Business Form

CHILDREN'S MEDICAL CENTER

PHYSICIAN:	DEPARTMENT:
OFFICE ADDRESS:	SECTION:
OFFICE PHONE:	OHIO LICENSE NO:
STAFF STATUS:	BIRTHDATE:

****

1978-79		1979-80		1980-81		1981-82	
Satisfactory Health Status		Satisfactory Health Status		Satisfactory Health Status		Satisfactory Health Status	
Satisfactory Meeting Attendance		Satisfactory Meeting Attendance		Satisfactory Meeting Attendance		Satisfactory Meeting Attendance	
Satisfactory Medical Record Completion		Satisfactory Medical Record Completion		Satisfactory Medical Record Completion		Satisfactory Medical Record Completion	
No Disciplinary Action		No Disciplinary Action		No Disciplinary Action		No Disciplinary Action	
COMMITTEE APPOINTMENTS		COMMITTEE APPOINTMENTS		COMMITTEE APPOINTMENTS		COMMITTEE APPOINTMENTS	
APPROVED	DATE	APPROVED	DATE	APPROVED	DATE	APPROVED	DATE
Chairman, Credentials Comm.		Chairman, Credentials Comm.		Chairman, Credentials Comm.		Chairman, Credentials Comm.	
Chief of Staff		Chief of Staff		Chief of Staff		Chief of Staff	
Secretary, Board of Trustees		Secretary, Board of Trustees		Secretary, Board of Trustees		Secretary, Board of Trustees	

Sample Business Form

forms of discourse

There are four forms of discourse: exposition, description, persuasion, and narration. **Exposition** is the straightforward presentation of facts and ideas with the **objective** of informing the **reader; description** is an attempt to re-create an object or situation with words so that the reader can visualize it mentally; **persuasion** attempts to convince the reader that the writer's point of view is the correct or desirable one; and **narration** is the presentation of a series of events in chronological order. These types of writing rarely exist in pure form; rather, they usually appear in combination.

formula/formulae

The plural form of *formula* is either *formulae* (a Latin derivative) or *formulas*. *Formulas* is more common, however, since it is the more natural English plural. (See also **foreign words in English.**)

EXAMPLE   You may present the underlying theory in your introduction, but save proofs or *formulas* for the body of your report.

fortuitous/fortunate

When an event is *fortuitous,* it happens by chance or accident and without plan. Such an event may be lucky, unlucky, or neutral.

EXAMPLE   My encounter with the general manager in Denver was entirely *fortuitous;* I had no idea he was there.

When an event is *fortunate,* it happens by good fortune or happens favorably.

EXAMPLE   Our chance meeting had a *fortunate* outcome.

functional shift

Many words shift easily from one **part of speech** to another, depending on how they are used. When they do, the process is called a functional shift, or a shift in function.

EXAMPLES   It takes ten minutes to *walk* from the sales office to the accounting department. (verb)
The long *walk* from the sales office to the accounting department reduces efficiency. (noun)

Let us *run* the new data through the computer. (verb)
May I see the last computer *run?* (noun)
I talk to the Chicago office on the *telephone* every day. (noun)
Sometimes I have to call from a *telephone* booth. (adjective)
He will *telephone* the home office from London. (verb)
After we discuss the project, we will begin work. (conjunction)
After lengthy discussions, we began work. (preposition)
The partners worked well together forever *after.* (adverb)

Do not arbitrarily shift the function of a word in an attempt to shorten a **phrase** or expression. Such attempts at achieving **conciseness** can result in the worst kind of **jargon,** as in the following examples.

EXAMPLES   In medical jargon, an *attending physician* becomes an "attending."
(a shift from an adjective to a noun)
In computer jargon *to pass a signal through a computer logic gate*
becomes "to gate." (a shift from a noun to a verb)
In nuclear energy jargon, a *reactor containment building* becomes a
"containment." (a shift from an adjective to a noun)

Avoid these shortcut versions; they are unnecessary. Linguistically acceptable ways for saying the same things already exist. If you are unsure about how to use similar words and expressions, consult a **dictionary** for both the word's meaning and its part of speech. Identifying the word's part of speech will allow you to determine quickly whether or not you are using it properly. (See also **new words.**)

G

garbled sentences

A garbled **sentence** is one that is so tangled with structural and grammatical problems that it cannot be repaired by simply replacing words or rewriting **phrases.** The following sentence appeared in an actual **job description.**

EXAMPLE My job objectives are accomplished by my having a diversified background which enables me to operate effectively and efficiently, consisting of a degree in mechanical engineering, along with twelve years of experience, including three years in Staff Engineering-Packaging sets a foundation for a strong background in areas of analyzing problems and assessing economical and reasonable solutions.

A garbled sentence often results from an attempt to squeeze too many ideas into one sentence. Do not try to patch such a sentence; rather, analyze the ideas it contains, list them in a logical sequence, and then construct one or more entirely new sentences. An analysis of the previous example yields the following five ideas:

EXAMPLES My job requires that I analyze problems and find economical and workable solutions to them.
My diversified background helps me accomplish my job.
I have a mechanical engineering degree.
I have twelve years of job experience.
Three of these years have been in Staff Engineering-Packaging.

Using these ideas, the writer might have described his job as follows.

EXAMPLE My job requires that I analyze problems and find economical and workable solutions to them. Both my training and my experience help me achieve this goal. Specifically, I have a mechanical engineering degree and twelve years of job experience, three of which have been in the Staff Engineering-Packaging Department.

Notice that the revised job description contains three sentences with logical development and clear **transitions.** Such analysis and rewriting is useful for any sentence or **paragraph** that you cannot repair with minor editing. (See also **clarity.**)

gender

In English grammar, gender is a term for the way words are formed to designate sex. The English language provides for recognition of three genders: masculine, feminine, and neuter (to designate those objects that have no definable sex characteristics). The gender of most words can be identified only by the choice of the appropriate **pronoun** *(he, she, it)*. Only these pronouns and a select few **nouns** *(waiter/waitress)* reflect gender.

Gender is important to writers because they must be sure that nouns and pronouns within a grammatical construction agree in gender. A pronoun, for example, must agree with its noun antecedent in gender. We must refer to a woman as *she* or *her,* not as *it;* to a man as *he* or *him,* not as *it;* to a barn as *it,* not *he* or *she.*

An antecedent that includes both sexes, such as *everyone* and *student,* has traditionally taken a masculine pronoun. Since this **usage** might be offensive by implying sexual bias, however, it is better to use plural nouns and pronouns.

EXAMPLES   Every *employee* should be aware of *his* insurance benefits.
All *employees* should be aware of *their* insurance benefits.

It is best (though not always possible) to avoid the awkward "his or her," "he or she" type of construction.

general-to-specific method of development

In a general-to-specific **method of development** you begin with a general statement and then provide facts or examples to develop and support that statement. For example, if you begin a **report** with the general statement "Companies that diversify are more successful than those that do not," the remainder of the report would offer examples and statistics that prove to your **reader** that companies that diversify are, in fact, more successful than companies that do not.

A **memorandum** or report organized in a general-to-specific sequence discusses only one point—the point made in the opening general statement. All other information in the memo or report supports the general statement, as in the following example.

Locating Additional Circuit Suppliers

General
statement

Based on information presented at the Supply Committee meeting on April 14, we recommend that the company locate additional suppliers of integrated circuits. Several related events make such an action necessary.

Our current supplier, ABC Electronics, is reducing its output. Specifically, we can expect a reduction of between 800 and 1,000 units per month for the remainder of this fiscal year. The number of units should stabilize at 15,000 units per month thereafter.

Domestic demand for our calculators continues to grow. Demand during the current fiscal year is up 25,000 units over the last fiscal year. Sales Department projections for the next five years show that demand should peak next year at 50,000 units and then remain at that figure for at least the following four years.

Finally, our overseas expansion into England and West Germany will require additional shipments of 5,000 units per quarter to each country for the remainder of this fiscal year. Sales Department projections put calculator sales for each country at double this rate, or 20,000 units in a fiscal year, for the next five years.

gerunds

A gerund is a nonfinite **verb** ending in -*ing* and used as a **noun.**

EXAMPLE *Smelting* is a technique used to extract metal from ore.

A gerund may be used as a **subject,** a **direct object,** an object of a **preposition,** a **subjective complement,** or an **appositive.**

EXAMPLES *Estimating* is an important managerial skill. (subject)
I find *estimating* difficult. (direct object)
We were unprepared for their *coming.* (object of a preposition)
Seeing is *believing.* (subjective complement)
My primary departmental function, *programming,* occupies about two-thirds of my time on the job. (appositive)

Only the possessive form of a noun or *pronoun* should precede a gerund.

EXAMPLES *John's* working has not affected his grades.
His working has not affected his grades.

Do not confuse a gerund with a present **participle,** which has the same form, but functions as an **adjective.**

EXAMPLES *Operating* the machine is difficult. (gerund, used as a noun)
Please provide clearer *operating* instructions. (participle, used as an adjective)

glossary

A glossary is a selected list of terms defined and explained for a particular field of knowledge. An alphabetical glossary, such as is often found at the end of a textbook or a **formal report,** can be helpful for quick reference.

If you are writing a report that will go to people who are not familiar with many of your technical terms, you may want to include a **glossary.** This does not relieve you of the responsibility of defining in the text any terms you are certain your **reader** will not know, however. If you include a glossary, keep the entries concise and be sure they are so clear that any reader can understand the definitions. (See also **defining terms.**)

EXAMPLES   *Amplitude Modulations:* Varying the amplitude of a carrier current with an audio-frequency signal.

Carbon Microphone: A microphone that uses carbon granules as a means of varying resistance with sound waves.

Overmodulation: Distortion created when the amplitude of the modulator current is greater than the amplitude of the carrier current.

gobbledygook

Gobbledygook is writing that suffers from an overdose of traits guaranteed to make it stuffy, pretentious, and wordy. These traits include the overuse of big and mostly **abstract words,** inappropriate **jargon,** stale expressions, **euphemisms, jammed modifiers,** and deadwood. Gobbledygook is writing that attempts to sound official (officialese), legal (legalese), or scientific; it tries to make a ''natural elevation of the geosphere's outer crust'' out of a molehill. Consider the following statement from an auto repair release form.

CHANGE   I hereby authorize the above repair work to be done along with the necessary material, and hereby grant you and/or your employees permission to operate the car or truck herein described on streets, highways, or elsewhere for the purpose of testing and/or inspection. An express mechanic's lien is hereby acknowledged on above car or truck to secure the amount of repairs thereto.

TO   You have my permission to do the repair work listed on this work order and to use the necessary material. You may drive my vehicle to test its performance. I understand that you will keep my vehicle until I have paid for all repairs.

Translated into straightforward English, the statement gains in **clarity** what it loses in pomposity without losing its legal meaning.

Gobbledygook (also called "bombast" and "pedantic writing") is packed with unusual words when more common ones would be clearer; it is heavy with foreign words when their English equivalents would be more appropriate; and it generally stresses trivial matters in the hope that the **reader** will be impressed with the writer's mental agility. George Orwell parodies such writing in the essay "Politics and the English Language," from which the following excerpt is taken.

EXAMPLE   Objective consideration of contemporary phenomena compels the conclusion that success or failure in competitive activities exhibits no tendency to be commensurate with innate capacity, but that a considerable element of the unpredictable must invariably be taken into account.

The following revision of Orwell's passage demonstrates the clarity of direct writing.

EXAMPLE   Success or failure today depends as much on chance as on your capabilities.

See also **affectation, conciseness/wordiness,** and **word choice.**

good/well

The confusion about the use of *good* and *well* can be cleared up by remembering that *good* is an **adjective** and *well* is an **adverb.**

EXAMPLES   John presented a *good* plan.
The plan was presented *well.*

However, *well* can also be used as an adjective to describe someone's health.

EXAMPLES   She is not a *well* woman.
Jane is looking *well.*

government proposals

A **proposal** is a company's offer to provide goods or services to a potential buyer within a certain amount of time and at a specified cost. Government proposals are usually prepared as a result of either an Invitation for Bids or a Request for Proposals that has been issued by a government agency.

INVITATION FOR BIDS

An Invitation for Bids is inflexible; the rules are rigid and the terms are not open to negotiation. Any proposal prepared in response to an Invitation for Bids must adhere strictly to its terms.

The Invitation for Bids clearly defines the quantity and type of an item that a government agency intends to purchase. It is prepared at the request of the agency that will use the item, such as the Department of Defense. An advertisement indicating that the government intends to purchase the item is published in an official government publication, such as the *Commerce Business Daily* (available from the Superintendent of Documents, Washington, D.C. 20402). This publication is required reading for government-sales oriented organizations. The goods or services to be procured are defined in the Invitation for Bids by references to performance standards called **specifications.** If, for instance, the item to be purchased is a truck, the important characteristics of that truck (height, weight, speed, carrying capacity, and so on) will be listed in the machine specification. More than one specification may apply to a single purchase; that is, one specification may apply to the item, another to the manuals, another to welding procedures, and so forth. Bidders must be prepared to prove that their product will meet all requirements of all specifications.

A specification is restrictive, binding the bidder to production of an item that meets the exact requirements of the specification. For example, if a truck is the specified item, a company may not bid to furnish a cargo helicopter, even though the helicopter might do the job more efficiently than the truck. An Invitation for Bids requires the bidder to furnish a specific product that meets the specification parameters, and nothing else will be accepted or considered. An alternate suggestion will render the bid "nonresponsive," and it will be rejected.

Bearing in mind that the product will be tested, measured, and evaluated to see that it does, in fact, meet the requirements of the specification, price is the main criterion that enters into the selection of the vendor. However, an organization with superior engineering skills can often supply a product that meets the specification at a price well below that of a less sophisticated competitor. When a high degree of technological competence, unusual facilities, or other valid requirements make it necessary, the procuring agency may require the bidder to possess certain minimum qualifications in order to be considered; a paper clip manufacturer employing ten people, for instance, could not qualify to bid on a procurement of military aircraft.

REQUEST FOR PROPOSALS

In contrast to the Invitation for Bids, the Request for Proposals is flexible. It is open to negotiation, and it does not necessarily specify exactly what goods or services are required. Often, a Request for Proposals will define a problem and allow those who respond to it to suggest possible solutions. As an example, if a procuring agency wanted to develop a way to make foot soldiers more mobile (capable of covering difficult terrain at high speed), a Request for Proposals would normally be the means used to find the best method and select the most qualified vendor. In some instances, Requests for Proposals are presented in two or more stages, the first being development of a concept, the second being a "prototype" machine or device, and the third being the manufacture of the device selected. The procedure for preparing a proposal in response to a Request for Proposals is as follows.

The procuring agency defines the problem and publishes it as a Request for Proposals in one or more business journals. Companies interested in government business scan the appropriate publications daily. Upon finding a project of interest, the sales department of such a company obtains all available information from the procuring agency. It then presents the data to management for a decision on whether the company is interested in the project. If the decision is positive, the corporate technical staff is assigned the task of developing a "concept" to accomplish the task posed by the Request for Proposals. The technical staff normally considers several alternatives, selecting one that combines feasibility and price. The staff's concept is presented to management for a decision on whether the company wishes to present a proposal to the requesting government agency.

Assuming that the decision is to proceed, preparation of a proposal is the next step. At a minimum, the proposal should provide a clear-cut statement of the problem and the proposed solution. It should include data to show that the company is financially and technologically sound (this may require a **résumé** of the qualifications of the people in charge and an **organizational chart** to show the chain of command). A discussion of the company's manufacturing capabilities and any other advantages it may have is also advisable. However, in dealing with government agencies, bear in mind that unnecessarily elaborate and costly proposals may be construed as a lack of cost-consciousness.

In many instances, the final cost to the government is determined by negotiation after the best overall concept has been selected. However, if your concept is one that provides a simple and inexpensive solution to the problem, cost is an advantage that you should point out. If your concept is expensive, explain the benefits of advanced engineering, speed, increased capacity, and so forth.

Further information about the government procurement process can be obtained from the *Armed Service Procurement Regulations* (available from the U.S. Government Printing Office, Washington, D.C. 20402).

The following excerpt comes from a proposal that was prepared in response to a Request for Proposals.

Advantages of the Proposed 5,000 H.P. Design

Statement of problem

The Navy Department has indicated to Burdlorn Manufacturing Company a need for main diesel propulsion engines to be used in conjunction with gas turbines on a program with the code name of CODAG. We have been advised that the power requirement for this engine is 5,000 horsepower, plus 10 percent overload. This engine must be contained with a space 24 feet long, 11 feet high, and 7 feet wide and must not exceed 60,000 pounds in weight. In addition, at a 5,000 horsepower operating level, it is not expected that the engine will require replacement or renewal of wearing parts or other major components more frequently

Brief statement of solution

than every 5,000 hours. The proposed design changes will result in a very light weight engine (12 pounds/BHP) but one with durability characteristics that will not be equaled by any engine throughout the world in this compact high-horsepower output.

5,000 HORSEPOWER ENGINE

Company's technical capability

Burdlorn Manufacturing Company has completed shop tests on NOBS 72393, addenda 12 and 13. During these tests, the Navy engine serial GR-1047-0836 was operated at loads up to 4,950 brake horsepower (just 1 percent short of the 5,000 horsepower goal). As a result of these tests, we are satisfied that the 16-cylinder, 11″ bore, 12″ stroke V-engine, with a rating of 5,000 horsepower at 1,000 rpm, is well within practical limits. On the basis of an intensive review of the service history of the four (4) engines installed on the LST-1176 and all of the research and development work conducted at the Burdlorn factory, coupled with a substantial amoung of analytical design work, Burdlorn Manufacturing Company recommends that certain redesigns be effected, based on the above information.

As a result of the recent tests at loads up to 4,950 horsepower that have been run in our factory, Burdlorn Manufacturing Company has successfully dealt with the thermodynamic problems of air/fuel requirements, air flow, heat transfer, and other related problems. In proposing design criteria, which are conservative to insure the desired reliability, we have applied sufficient analysis to the individual components to make specific recommendations.

In order to summarize findings that we feel the Navy needs in order to evaluate our proposal, we will outline in numerical sequence the improvements and advantages based on our total experience with the 11″ bore, 12″ stroke, 16-cylinder V-engine.

1. Frame. Burdlorn Manufacturing Company recommends that the present 17 inches center distance from one cylinder to the adjacent cylinder be increased by 3 inches to provide 20 inches center distance from cylinder to cylinder in a given bank. The additional 3 inches per cylinder will mean an overall increase in engine length of 24 inches. This will mean an engine length of approximately 22 feet with the engine-driven auxiliary equipment. This will be well within the 24 foot maximum covered by the Navy specifications. The additional 3 inches will provide the following advantages.

   a. The principal load carrying member in the frame, which is the athwartship plate, will be increased in thickness from ½ inch to ¾ inch.
   b. Additional clearance is provided for welding, and machining, particularly on the intermediate deck.
   c. Additional length is provided for the main and crankpin journals. This will increase the bearing areas for the main and crankpin bearings and thus gain the advantage of reduced bearing loads.

The results of our experimental stress analysis have indicated a substantial bending movement in the top deck of the present design. Accordingly, Burdlorn Manufacturing Company proposes to increase the thickness of the top deck. The present design uses forgings. We are proposing to use controlled quality steel castings of substantially heavier construction to reduce the stress levels.

Burdlorn Manufacturing Company has done a substantial amount of stress analysis on a similar frame on our Model FS-13½″ × 16½″ V-engine. On the basis of this experience, we recommend a minor redesign of the main bearing saddles to create a more even stress distribution in transmitting the load through the athwartship plates to the main bearing saddles. These main bearing saddles would be made from steel castings such as our present 13½ ″ × 16½ ″ engines currently in use.

To provide for reduction in stress levels in the sidesheets of the engine, we propose increasing the plate thickness from ½ inch to ⅝ inch. In addition, we have found through extensive testing of the Navy engine and our 13½ ″ × 16½ ″ V-engine and also by exhaustive photoelastic stress analysis techniques that the present "hourglass" design results in a stress concentration factor at the "waist" of the frame. Accordingly, we propose a modification in this shape that will reduce the stress concentration factor due to this shape effect (see Figure 9).

One problem experienced repeatedly on the LST-1176 was associated with the bosses for the handhole covers. Stress analysis work has shown a high stress concentration factor around these bosses because of unfused weld roots and also the increased localized stiffness caused by these bosses. Accordingly, we have designed handhole covers held in place by clamping action, which do not impose additional stress in the sidesheet due to the torquing down of the bolts against the bosses. We therefore gain the advantage by this direct means of eliminating a superimposed stress of 5,000 psi (see Figure 10).

An additional problem experienced on earlier frames was cracking in the air manifold. The present design provides for

Introduction to illustration

Another problem and proposed solution with illustration

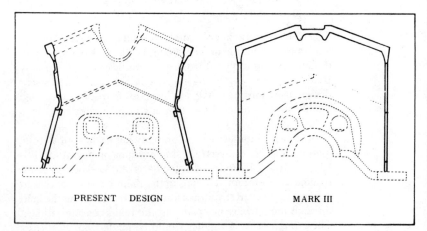

PRESENT DESIGN   MARK III

Figure 9.   Frame Profiles

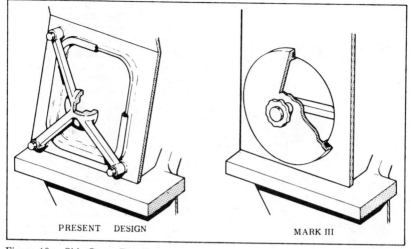

PRESENT DESIGN

MARK III

Figure 10.   Side Cover Details

the air header to be welded in as an integral stress member of the frame. A number of problems are associated with this type of design, some of which are stress concentration factors in the welding together of the top deck, the air header, and the athwartship plate. A second problem associated with this type of construction is the thermogradient experienced because the air manifold temperature is regulated to approximately 100°F., whereas other heat sinks at higher temperature levels create thermal stresses in the frame that are difficult to manage. Accordingly, Burdlorn proposes a separate and independent air header that will be bolted to the frame in such a manner as not to contribute to the overall stresses or stress concentration factors of the frame weldment.

2. Valves. During the 1,000 rpm test of the engine under Navy contract NOBS 72393, a wear rate of approximately .050 inch per 1,000 hours was experienced on the inlet valves. The wear rate on the exhaust valves was approximately .010 inch per 1,000 hours. A total wear of .100 inch would be considered acceptable for this size valve and insert. Accordingly, it will be appreciated that a wear life of more than 5,000 hours could be expected on the exhaust valve, but a much shorter life was experienced on the inlet valve. A careful analysis of all availa-

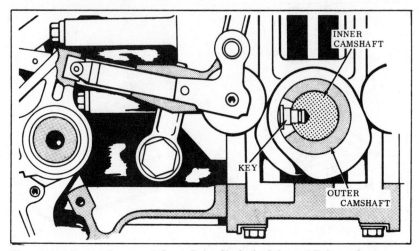

Figure 11.   Present Inner and Outer Drive Shafts for Inlet and Exhaust Cams

ble information has revealed only two differences between the intake valve and the exhaust valve that would contribute to this difference in wear rates. These two differences are as follows:

a. The exhaust cam is driven through a hollow shaft that is 4 inches O.D. by 2½ inches I.D. The inlet cam is driven by a solid shaft that runs within the hollow exhaust camshaft and is 2½ inches in diameter. It will be recognized that the relative stiffness of these two shafts in torsion is in the ratio of 5½:1 as revealed by our calculations (see Figure 11).

b. The lift profiles of the intake and exhaust cams are identical; however, there is a greater dwell on the exhaust cam than there is on the inlet cam. The inlet cam has 15 degrees dwell, whereas the exhaust cam has 50 degrees dwell. The longer dwell on the exhaust cam, which cannot be incorporated in the inlet cam due to timing requirements, provides additional time in which the vibratory amplitudes in the valve train dampen or attenuate to a greater extent than the 15 degrees dwell provides on the inlet cam (see Figure 12). . . .

The rest of the proposal continues to explain the design of the engine's cylinder heads, bearings, crankshaft, pistons, cylinder liners, connecting rod, and reversing mechanism.

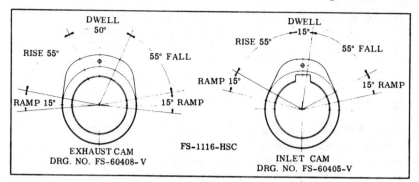

Figure 12. Comparison of Inlet and Exhaust Cam Profiles

grammar

Grammar is the systematic description of the way words work together to form a coherent language; in this sense, it is an explanation of the structure of a language. However, grammar is popularly taken to mean the set of "rules" that governs how a language ought to be spoken and written; in this sense, it refers to the **usage** conventions of a language.

These two meanings of grammar—how the language functions and how it ought to function—are easily confused. To clarify the distinction, consider the expression "ain't." Unless used purposely for its colloquial flavor, *ain't* is unacceptable to careful speakers and writers because a convention of usage prohibits its use. Yet taken strictly as a **part of speech,** the term functions perfectly well as a **verb;** whether it appears in a declarative **sentence** ("I ain't going.") or an interrogative sentence ("Ain't I going?"), it conforms to the normal pattern for all verbs in the English language. Although we may not approve of its *use* in a sentence, we cannot argue that it is *ungrammatical.*

To achieve **clarity,** writers need a knowledge of both grammar (as a description of the way words work together) and the conventions of usage. Knowing the conventions of usage helps writers select the appropriate over the inappropriate word or expression. A knowledge of grammar helps them diagnose and correct problems arising from how words and **phrases** function in relation to one another. For example, knowing that certain words and phrases function to modify other words and phrases gives the writer a basis for correcting those **modifiers** that are not doing their job. Understanding **dangling modifiers** helps the writer avoid or correct a construction that obscures the intended meaning. In short, an understanding of grammar and its special terminology is valuable for writers chiefly because it enables them to recognize problems and thus to communicate clearly and precisely.

graphs

A graph presents numerical data in visual form. This method has several advantages over presenting data in **tables** or within the text. Trends, movements, distributions, and cycles are more readily apparent in graphs than they are in tables. By providing a means for ready comparisons, a graph often shows a significance in the data not otherwise immediately apparent. Be aware, however, that although graphs present statistics in a more interesting and comprehensible form than tables do, they are less accurate. For this reason, they are often accompanied by tables giving the exact figures. There are many different kinds of graphs, most notably line graphs, bar graphs, pie graphs, and picture graphs.

LINE GRAPHS

The line graph, which is the most widely used of all graphs, shows the relationship between two sets of numbers by means of points plotted in relation to two axes drawn at right angles. The points, once plotted, are connected to one another to form a continuous line. In this way, what was merely a set of dots having abstract mathematical significance becomes graphic, and the relationship between the two sets of figures can easily be seen.

The line graph's vertical axis usually represents amounts, and its horizontal axis usually represents increments of time, as in Figure 13.

Line graphs with more than one line are common because they allow for comparisons between two sets of statistics for the same period of time. In creating such graphs, be certain to identify each line with a label or a legend, as shown in Figure 14. The difference between the two lines can be emphasized by shading the space between them.

Tips on Preparing Line Graphs

1. Give the graph a title that describes the data clearly and concisely. Either center the title below the graph or align it with the left margin. If the title is too long to fit within the margins of the graph on one

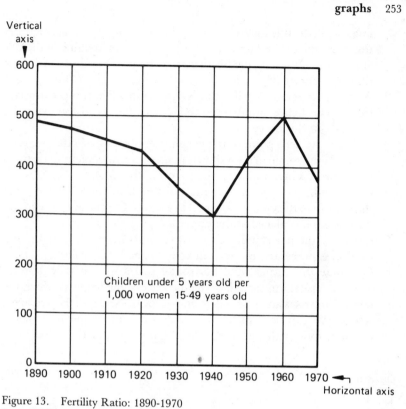

Figure 13.   Fertility Ratio: 1890-1970
Source: U.S. Bureau of the Census

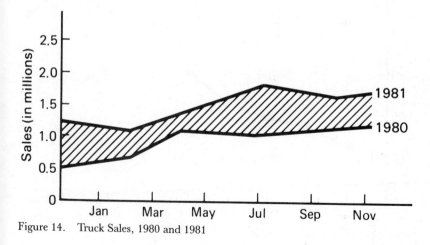

Figure 14.   Truck Sales, 1980 and 1981

line, break the title into two or more lines of roughly equal length. In this case, the figure number should precede the first line of the title and the other lines should begin directly beneath the first word of the title. See Figure 24.

2. Assign a figure number if your **report** includes five or more **illustrations.** The figure number precedes the title, as in the examples throughout this entry.

3. Indicate the zero point of the graph (the point where the two axes meet). If the range of data shown makes it inconvenient to begin at zero, insert a break in the scale as in Figure 15.

4. Graduate the verticle axis in equal portions from the least amount at the bottom to the greatest amount at the top. Ordinarily, the caption for this scale is placed at the upper left.

5. Graduate the horizontal axis in equal units from left to right. If a caption is necessary, center it directly beneath the scale.

6. Graduate the vertical and horizontal scales so that they give an accurate visual impression of the data, since the angle at which the curved line rises and falls is determined by the scales of the two axes. The curve can be kept free of distortion if the scales maintain a constant ratio with each other. See Figures 16 and 17.

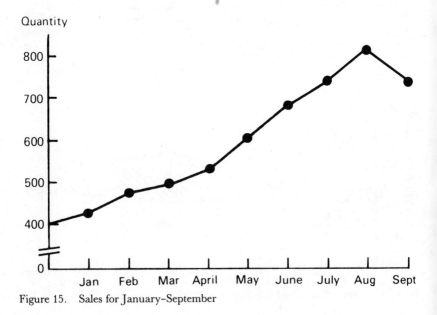

Figure 15.   Sales for January–September

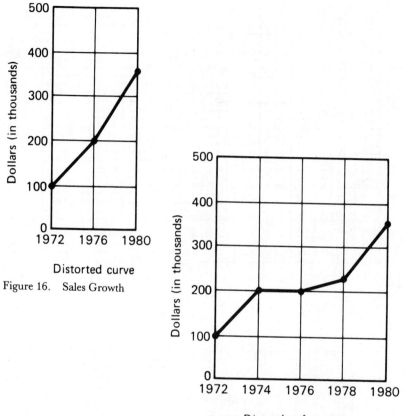

Distorted curve

Figure 16.   Sales Growth

Distortion-free curve

Figure 17.   Sales Growth

7. Hold grid lines to a minimum so that curved lines stand out. Since precise values are usually shown in a table of data accompanying a graph, detailed grid lines are unnecessary. Note the increasing clarity of the three graphs in Figures 18, 19, and 20.

8. Include a key (which lists and explains **symbols**) when necessary, as in Figure 19. At times a label will do just as well, as in Figure 20.

9. If the information comes from another source, include a source line below and on the same left margin as the caption, as in Figure 13 in this entry.

10. Place explanatory **footnotes** below the graph, to the lower left.

11. Place all lettering horizontally if possible.

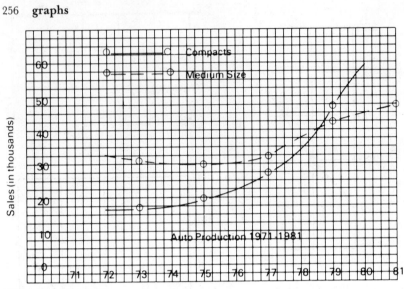

Figure 18. Production

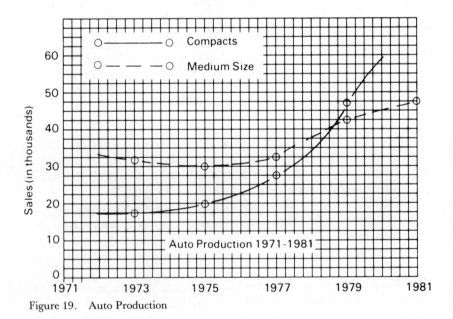

Figure 19. Auto Production

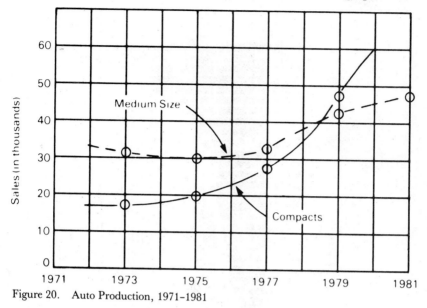

Figure 20. Auto Production, 1971–1981

BAR GRAPHS

Bar graphs consist of horizontal or vertical bars of equal width but scaled in length to represent some quantity. They are commonly used to show (1) quantities of the same item at different times, (2) quantities of different items for the same time period, or (3) quantities of the different parts of an item that make up the whole.

Figure 21 is an example of a bar graph showing varying quantities of the same item at the same time. Here each bar, which represents a different quantity of the same item, begins at the left scale. The left scale also provides additional information in the form of the percentage of the whole population each bar (and therefore each state) represents.

Some bar graphs show the quantities of different items for the same period of time. See Figure 22. (A bar graph with vertical bars is also called a column graph.)

Bar graphs can also show the different portions of an item that make up the whole. Here the bar is equivalent to 100 percent. It is then divided according to the appropriate proportions of the item sampled. This type of graph can be constructed vertically or horizontally and can indicate more than one whole where comparisons are necessary. See Figures 23 and 24.

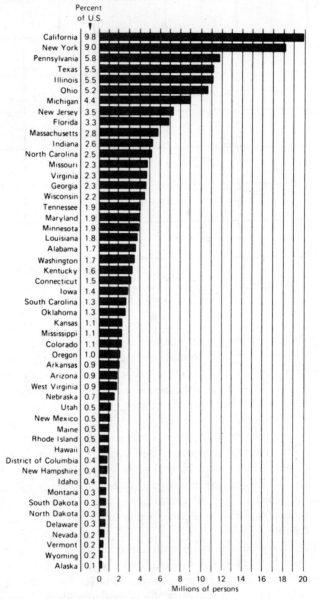

Figure 21.   States Ranked by Total Population: 1970
Source: U.S. Bureau of the Census

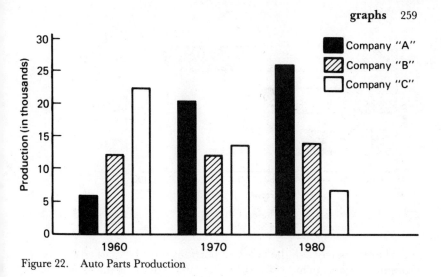

Figure 22. Auto Parts Production

Figure 23. Your Municipal Tax Dollar

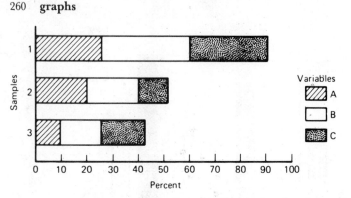

Figure 24.   Example of 100 Percent Bar Graph Showing Proportions of Three Variables in Three Samples

If the bar is not labeled, the different portions must be marked clearly by shading or crosshatching. Include a key that identifies the various subdivisions.

PIE GRAPHS

A pie graph presents data as wedge-shaped sections of a circle. The circle equals 100 percent, or the whole, of some quantity (a tax dollar, a bus fare, the hours of a working day), with the wedges representing the various ways in which the whole is divided. In Figure 25, for example, the circle stands for a city tax dollar, and it is divided into units equivalent to the percentage of the tax dollar spent on various city services.

Pie graphs provide a quicker, more striking way of presenting the same information that can be presented in a table; in fact, a table often accompanies a pie graph with a more detailed breakdown of the same information.

Tips on Preparing Pie Graphs

1. The complete 360° circle is equivalent to 100 percent; therefore, each percentage point is equivalent to 3.6°.
2. To make the relative percentages as clear as possible, begin at the 12 o'clock position and sequence the wedges clockwise, from largest to smallest.
3. If you shade the wedges, do so clockwise and from light to dark.
4. Keep all labels horizontal and, most important, give the percentage value of each wedge.
5. Finally, check to see that all wedges, as well as the percentage values given for them, add up to 100 percent.

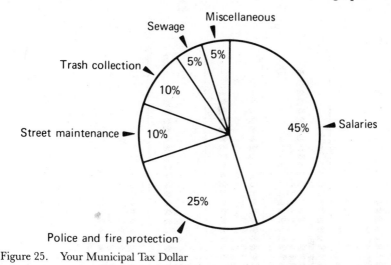

Figure 25. Your Municipal Tax Dollar

Although pie graphs have strong visual impact, they also have draw-backs. If more than five or six items are presented, the graph looks cluttered. Also, since they usually present percentages of something, they must often be accompanied by a table listing precise statistics. Further, unless percentages are shown on the sections, the **reader** cannot compare the values of the sections as accurately as with a bar graph.

PICTURE GRAPHS

Picture graphs are modified bar graphs that use picture symbols of the item presented. Each symbol corresponds to a specified quantity of the item. See Figure 26. Note that precise figures are included since the graph can present only approximate figures.

Tips on Preparing Picture Graphs

1. Make the symbol self-explanatory.
2. Have each symbol represent a single unit.
3. Show larger quantities by increasing the number of symbols rather than by creating a larger symbol. (It is difficult to judge relative sizes accurately.)
4. Consult a good graphics manual for details about preparing graphs. See **reference books** for a list of useful manuals.

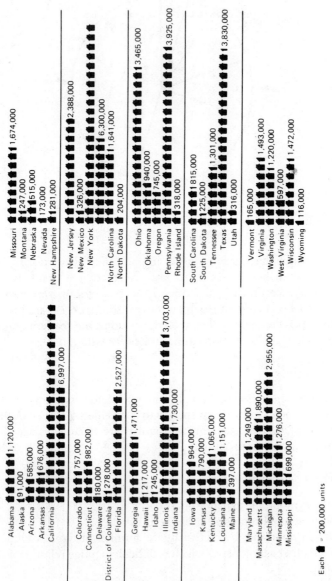

Each ▐ = 200,000 units

Figure 26.   Number of Housing Units, by States: 1970
Source: Bureau of the Census

H

half

Half a, half an, and *a half* are all correct idiomatic uses of the word *half*.

EXAMPLES   It takes *a half* second for the signal to reach your receiver.
Wait *half a* minute before stirring the solution.
Allow *half an* hour for the compound to solidify.

A half a (or *a half an*) is colloquial, however, and should be avoided in writing.

CHANGE   The subroutine is executed in *a half a* minute.
TO   The subroutine is executed in *half a* minute.

he/she

Since there is no singular **personal pronoun** in English that refers to both sexes, the word *he* has traditionally been used when the sex of the antecedent is unknown.

EXAMPLE   Whoever is appointed [either a man or woman] will find *his* task difficult.

Because use of a masculine **pronoun** might be offensive, however, it is better to rewrite the sentence in the plural (or avoid use of a pronoun altogether) than to offend.

EXAMPLES   *Employees* should take advantage of *their* insurance benefits.
Whoever is appointed will find *the* task difficult.

You should also use the **phrase** *he or she*.

EXAMPLE   *Whoever* is appointed will find *his or her* task difficult.

Unfortunately, *he, or she* and *his or her* are clumsy when used repeatedly; the best advice is to reword the **sentence** to use a plural pronoun. (See also **gender** and **Ms./Miss/Mrs.**)

headers and footers

A header in a **report, technical manual,** or **specification** is identifying information carried at the top of each page. The header normally

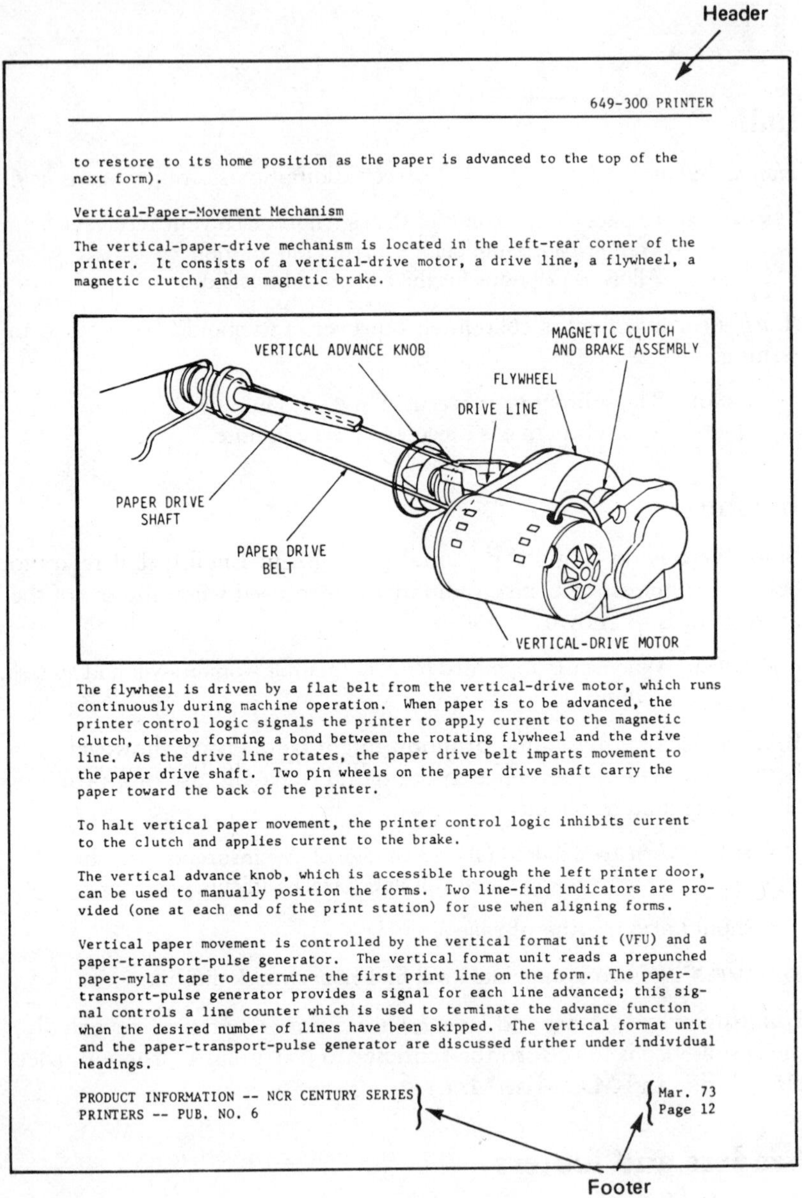

264 headers and footers

to restore to its home position as the paper is advanced to the top of the next form).

Vertical-Paper-Movement Mechanism

The vertical-paper-drive mechanism is located in the left-rear corner of the printer. It consists of a vertical-drive motor, a drive line, a flywheel, a magnetic clutch, and a magnetic brake.

MAGNETIC CLUTCH
AND BRAKE ASSEMBLY

VERTICAL ADVANCE KNOB

FLYWHEEL

DRIVE LINE

PAPER DRIVE
SHAFT

PAPER DRIVE
BELT

VERTICAL-DRIVE MOTOR

The flywheel is driven by a flat belt from the vertical-drive motor, which runs continuously during machine operation. When paper is to be advanced, the printer control logic signals the printer to apply current to the magnetic clutch, thereby forming a bond between the rotating flywheel and the drive line. As the drive line rotates, the paper drive belt imparts movement to the paper drive shaft. Two pin wheels on the paper drive shaft carry the paper toward the back of the printer.

To halt vertical paper movement, the printer control logic inhibits current to the clutch and applies current to the brake.

The vertical advance knob, which is accessible through the left printer door, can be used to manually position the forms. Two line-find indicators are provided (one at each end of the print station) for use when aligning forms.

Vertical paper movement is controlled by the vertical format unit (VFU) and a paper-transport-pulse generator. The vertical format unit reads a prepunched paper-mylar tape to determine the first print line on the form. The paper-transport-pulse generator provides a signal for each line advanced; this signal controls a line counter which is used to terminate the advance function when the desired number of lines have been skipped. The vertical format unit and the paper-transport-pulse generator are discussed further under individual headings.

Figure 27

contains the **topic** (or topic and subtopic) dealt with in that section of the report, manual, or specification (specifications also include the identification numbers of sections and **paragraphs**).

A footer is identifying information carried at the bottom of each page. The footer generally contains the date of the document, the page number, and sometimes the manual name and section title.

Although the types of information included in headers and footers vary greatly from one organization to the next, the example in Figure 27 is fairly typical.

heads

Heads (also called headings) are **titles** or subtitles within the body of a long piece of writing that serve as guideposts for the **reader.** They divide the material into manageable segments, call attention to the main topics, and signal changes of topic. If a **formal report, proposal,** or other document you are writing is long or complicated, you may need several levels of heads to indicate major divisions, subdivisions, and even smaller units of those. In extremely technical material one occasionally sees as many as five levels of heads, but as a general rule it is rarely necessary (and usually confusing) to use more than three. The following example, greatly shortened, is intended only to illustrate the use of the heads—in this case, three levels (beyond the title of the entire report).

Interim Report of the Committee to Investigate New Factory Locations

The committee initially considered thirty possible locations for the proposed new factory. Of these, twenty were eliminated almost immediately for one reason or another (unfavorable tax structure, remoteness from rail service, inadequate labor supply, etc.). Of the remaining ten locations, the committee selected for intensive study the three that seemed most promising: Chicago, Minneapolis, and Salt Lake City. These three cities we have now visited, and our observations on each of them follow.

CHICAGO

Of the three cities, Chicago presently seems to the committee to offer the greatest advantages, although we wish to examine these more carefully before making a final recommendation.

Location

Though not at the geographical center of the United States, Chicago is centrally located in an area that contains more than three-quarters of the U.S. population. It is within easy reach of our corporate headquarters in New York. And it is close to several of our most important suppliers of components and raw materials—those, for example, in Columbus, Detroit, and St. Louis.

Transportation

Rail Transportation. Chicago is served by the following major railroads. . . .
Sea Transportation. Except during the winter months when the Great Lakes are frozen, Chicago is an international seaport. . . .
Air Transportation. Chicago has two major airports (O'Hare and Midway) and is contemplating building a third. Both domestic and international air cargo service are available. . . .
Transportation by Truck. Virtually all of the major U.S. carriers have terminals in Chicago. . .

As you can infer from this example, the heads you use grow naturally from the divisions and subdivisions created in **outlining.** If more of the preceding example were given, the heads would look like this:

CHICAGO
Location
Transportation
Rail Transportation
Sea Transportation
Air Transportation
Truck Transportation
Labor Supply
Engineering Personnel
Manufacturing Personnel
Outside and Consulting Services
Tax Structure
Living Conditions
Housing
Education
Recreation
Climate
MINNEAPOLIS
Location
Transportation

There is no one correct format for heads. Sometimes a company settles on a standard format, which everyone within the company is then expected to follow. Or a customer for whom a report or proposal is being prepared may specify a particular format. In the absence of such guidelines, the system used in the example should serve you well. Note the following formal characteristics: (1) The first-level head is in all **capital letters,** typed flush to the left margin on a line by itself, and separated by a line space from the material it introduces. (2) The second-level head is in capital and lower-case letters, also typed flush to the left margin on a line by itself, and also separated by a line space from the material it introduces. (3) The third-level head is in capital and lower-case letters, but it is "run in" right on the same line with the first sentence of the material it introduces. Therefore, it is followed by a period to set it apart from what follows, and it is underlined or italicized so that it will stand out clearly on the page.

Following are some important things to keep in mind about heads.

1. They should signal a shift to a new topic (or, if they are lower-level heads, a new "subtopic" within the larger topic).
2. Within the larger unit they subdivide, all heads at any one level should be consistent in their relationship to the topic of the larger unit. For instance, notice in the extended list of heads on page 266 that the first-level heads ("CHICAGO" and "MINNEAPOLIS") both bear the same relationship to the larger topic of the report (possible factory locations). When the discussion narrows to Chicago, all of the second-level heads within that unit bear the same relationship to the first-level head ("CHICAGO"): they refer to Chicago's location, its transportation, its labor supply, its tax structure, and its living conditions. And the third-level heads conform to the same principle within their units.
3. The fact that one unit at a particular level is subdivided by lower-level heads does not mean that *every* unit at that particular level must also include lower-level heads. In the example, notice that under the first-level head "CHICAGO" some of the second-level units (for instance, "Transportation") are further subdivided by third-level heads and some (for instance, "Location") are not.
4. All heads at any one level within the same next-larger unit should be parallel with one another in structure (see **parallel structure**). For instance, note in the example that all of the heads are **nouns** or noun **phrases.**
5. Too many heads, or too many levels of heads, can be as bad as too few. A highway map that showed only New York and San Francisco

would not be very helpful, even to the traveler driving from New York to San Francisco. But a map that tried to show every street and every building in every town along the way would not be much better. Keep in mind the needs of your reader.

6. If your document needs a **table of contents,** create it from your heads and subheads. Be sure that the wording in your table of contents is the same as the wording in your text.

healthful/healthy

If something is *healthful* it promotes good health. If something is *healthy* it has good health. Carrots are *healthful;* people can be *healthy.*

EXAMPLES   Yogurt is *healthful.*
We provide *healthful* lunches in the cafeteria because we want our employees to be *healthy.*

helping verbs

A helping verb (also called an auxiliary verb) is a **verb** that is added to a main verb to help it indicate **mood, tense,** and **voice.** In doing so, it also forms a **verb phrase.**

EXAMPLES   I *am* going.
I *will* go.
I *should have* gone.
I *should have been* gone.
The order *was* given.

All helping verbs *(be, have, do, shall, will, may, can, must, ought)* can function alone as main verbs, but they usually function as helping verbs in verb phrases. Notice that the principal helping verbs are the various forms of the main verbs *be, have,* and *do.*

herein/herewith

The words *herein* and *herewith* are a type of business **jargon** that should be avoided.

CHANGE   I have *herewith* enclosed a copy of the contract.
TO   I have enclosed a copy of the contract.

house organ articles

A house organ is a company publication or an employee publication that is produced by the employer. Its primary purpose is to keep employees informed about the company and its operations and policies. Many different types of house organs are found in business and industry—from the gossip sheet that carries little more than bowling scores and who got married or had a baby to the more sophisticated instruments of company policy that increase the employee's understanding of the company, its purpose, and the business in which it is engaged. If your company publishes a house organ of the latter type, you may be asked to contribute an article on a subject that you are especially qualified to write about. The editor may offer some general advice but can give you little more help until you submit a draft.

Before beginning to write, consider the traditional *who, what, where, when,* and *why* of journalism (who did it? what was done? where was it done? when was it done? and why was it done?)—and then add *how,* since the *how* of your subject may be of as much interest to your fellow employees as any of the five *w*'s.

Before going any further, determine whether there is an official company policy or position on your subject. If so, adhere to it as you prepare your article. If there is no company policy on your subject, determine as nearly as you can what management's attitude is toward your subject.

Gather several fairly recent issues of your company house organ and study the **style** and **tone** of the writing and the approach the company has taken to various kinds of subjects in past issues. Try to emulate these as you work on your own article. Ask yourself the following questions about your subject: What is its significance to the company? What is its significance to employees? The answers to these questions should help you establish the style, tone, and approach for your article. These answers should also heavily influence the **conclusion** you write for your article.

Research for a house organ article frequently consists of **interviewing.** Interview everyone concerned with your subject. Get all available information and all points of view. (Be sure to give maximum credit to the maximum number of people.)

Writing a house organ article requires a little more imagination than writing **reports.** With a house organ, you do not have a captive audience; therefore, you may want to use your imagination to provide four helpful ingredients to ensure a successful article: (1) an intriguing **title** to catch **readers'** attention (a **rhetorical question** is often effective); (2) eye-catching **photographs** or **illustrations** that entice them to read your lead **paragraph** to see what the subject is really about; (3) a lead, or first paragraph, that is designed to encourage further reading into your article (this paragraph generally makes the **transition** from the **title** to the down-to-earth treatment of the subject); and (4) a well-developed presentation of your subject to hold the reader's interest all the way to the conclusion.

Your conclusion should emphasize the significance of your subject to your company and its employees. Because of its strategic location to achieve **emphasis,** your conclusion should include the thoughts that you want your reader to retain about your subject.

In preparing your house organ article you will find it helpful to refer to the Checklist of the Writing Process and to follow the steps listed there. Following is an article written for a house organ. The original included photographs and boxed quotations.

Author! Author!

It was a number of centuries back that the Roman satirist Persius said, "Your knowing is nothing, unless others know you know."

Today at Allen-Bradley some of our engineers are subscribing to Persius's philosophy. They contend it takes more than the selling of a product for a corporation to enjoy continuing success.

"We must sell our knowledge as well," stressed Don Fitzpatrick, Commercial Chief Engineer. "We have to get our customers to think of us as knowledge experts."

Fitzpatrick's point is well made. In today's high technology business climate, it is essential to present a vanguard image as well as produce quality products. A corporation can be considered a "knowledge expert" when it "not only has a product for sale, but is the acknowledged leader in the application of that product," Fitzpatrick explained.

How does a corporation acquire an image as a "knowledge expert"?

Customer Roundtables, distributor schools and participation by our engineers in professional socieites, such as the Institute of Electrical and Electronics Engineers (IEEE), have helped Allen-Bradley to enhance its image in the industrial control market, noted Fitzpatrick. Yet, one subliminal selling tool that Fitzpatrick believes A-B engineers could sharpen to foster more influence is the writing and presentation of technical papers.

"I'm talking about technical writing. Not the kind of technical writing that explains our products or how to use them safely, but the kind of writing that is done as a method of sharing knowledge among companies who have similar concerns," Fitzpatrick explained. Such writings discuss new methods being used in the industry, or a company's difficulties working with an electrical phenomenon. They introduce state-of-the-art technology, new theories, or are tutorials on the developments in the industry to date, he added.

FITZPATRICK'S PAPER CITED

Fitzpatrick knows about technical papers. The former Purdue University professor has been preparing and presenting topics for Allen-Bradley Roundtables and distributor schools for several years. He also has been active in the presentation of papers for IEEE conferences. This past September in Houston, Fitzpatrick received a second place award for a paper he presented at a Petrochemical Industrial Conference (PCIC) in San Diego in 1979. (PCIC is an organization of the Industry Application Society, which is part of the IEEE.) The award winning paper, entitled "Transient Phenomena in the Motor Control Distribution Systems," also was published in IEEE's renowned annual publication "IAS Transactions."

Only a few papers are accepted annually for presentation at professional conferences, noted Fitzpatrick. In turn, just a few of those presentations are published in IEEE related "Transactions" catalogs. Occasionally a paper's topic also may generate an article in a trade publication. For these reasons preparing such papers can be plums—for both company recognition and the author's esteem among colleagues.

Cultivating peer recognition is as important as nurturing an expert image for your company in today's technological world, said Steven Bomba, Director of Advanced Technology Development. "In the area of advanced technology the only way to test if your work is current is to share it with people outside of the company," he said. "When you publish, you don't reveal proprietary information. You publish information that will be economically beneficial to your organization. Our employees should be anxious to test their sword against their competitors'."

GROWING FRATERNITY HERE

Bomba and Fitzpatrick noted that there is a "new spirit for Allen-Bradley" in the area of technical writing and presentation. The number of engineers participating as guest speakers at national engineering, application and technical conferences is a small, but growing fraternity. Participation by A-B personnel in these professional conferences spurs a "psyche tune-up" and is critical to "innovation, staying current and invention. Without it, a company lacks a reference point. Encouraging your people to write these papers provides a measure of value and stimulation," Bomba contends.

Technical papers are always noncommercial. They don't mention specific products or the company's name. The papers are by-lined, however, usually with an editor's note containing information about the writer's background and employment. "If you word a paper right, you can sell your company. You can make the papers work as a subliminal public relations tool for you. One payoff for the extra time it often takes to prepare a paper is the direct influence in the public eye that is gained. It is impossible to measure the exact influence, but the exposure generally generates positive reaction," Fitzpatrick notes.

CONTACTS AT CONFERENCES

"Participation in conferences through writing of technical papers results in an employee becoming more active in professional organizations. That in turn leads to more business contacts. We're trying to broaden our market base, enter new industries. Roundtables, publishing and conference programs are by-products, or spin-offs, that work for us. They help us to sell our ability," said Fitzpatrick.

They also work for the engineers and technicians who get involved in the paper's preparation. Often a paper may be co-authored or researched as a team. Engineers or technicians who take the time to research a possible topic, prepare an abstract for query, and finally write a paper upon abstract acceptance, are taking those small steps of knowledge that lead to leaps in industry technology.

"Getting published places the author in the elite 'Invisible College of Knowledge Workers,' which is a collection of technical people who transcend the boundaries of industry, cities and nations," said Bomba. Only 1% of degreed engineers write technical papers that are published, he added.

"Peer fear" may be the greatest barrier for an engineer to pick up the pen—or word processor, as the case may be today. Walt Maslowski, Principal Engineer, Dept. 756, was an engineer for 12 years before he decided to give technical writing a whirl three years ago. He joined A-B in 1979. This past September he presented his second paper as an A-B employee at an Industry Application Society conference in Cincinnati. His latest paper, which addressed the techniques in controlling three phase induction motor drives, received a second place award from IAS. The article will be published in IEEE's next issues of IAS Transactions. "There was a lot of positive response to the paper, and motor drives recognition is an area that A-B has been pursuing for some time," Maslowski noted.

MANY BENEFITS INTANGIBLE

"The fear of possibly writing substandard papers held me back from writing for years," Maslowski continued. "What I've discovered, though, is that when you do write a qualified paper, your technical expertise becomes known. Many of your personal gains are intangible, but you meet people, open contacts and absorb more knowledge."

How does one decide to write a paper? "First you should review enough of a field to get a pulse for it. Read key papers. Review key results that people are talking about. Analyze the unsolved areas—or holes—of these writings and then figure out how those holes might relate to your job," suggested Dave Linton, a project engineer in Dept. 756. Linton presented a tutorial on computer software design at a Digital Equipment Computer User Society symposium in early November.

Another barrier for some engineers is getting an abstract accepted, since all papers are reviewed by published professionals, or "referees." "Making that jump from the inside to the outside is hard without some middle ground to test your ideas. Some companies active in publishing have internal review systems where your co-workers give you some input and encouragement. It would be nice if Allen-Bradley had a similar system of internal publishing," Linton suggested. Bomba, Fitzpatrick and others involved in technical writing here, concurred.

"Industry does not make the 'publish or perish' threat that is prevalent in the academic world. At A-B we really encourage our employees to contribute to our industry: to get involved," said Bomba.

Being recognized in your field builds self-confidence and generates a sense of pride and accomplishment. Writing technical papers is a good way to let "others know you know."

Reprinted with permission from the 1980 Nov.-Dec. issue of the employee publication of the Allen-Bradley Company, Milwaukee.

hyperbole

Hyperbole is exaggeration to achieve an effect or **emphasis.** It is a device used to emphasize or intensify a point by going beyond the literal facts. It distorts to heighten effect.

EXAMPLES   They *murdered* us at the negotiating session.
The hail was like *boulders.*

Used cautiously, hyperbole can magnify an idea without distorting it; therefore, it plays a large role in advertising.

EXAMPLE   A box of Clenso detergent can clean a *mountain of clothes.*

Everyday speech abounds in hyperbole.

EXAMPLES   I'm *starved.* (meaning "hungry")
I'm *dead.* (meaning "tired")

Because technical writing needs to be as accurate and precise as possible, always avoid hyperbole. (See also **figures of speech.**)

hyphens

Although the hyphen functions primarily as a **spelling** device, it also functions to link and to separate words; in addition, it occasionally replaces the **prepositions** *to* (0-100 for 0 *to* 100). The most common use of the hyphen, however, is to join **compound words.**

EXAMPLES   able-bodied, self-contained, carry-all, brother-in-law

A hyphen is used to form compound **numbers** from twenty-one to ninety-nine and fractions when they are written out.

EXAMPLES   twenty-one, one forty-second

When in doubt about whether or not to hyphenate a word, check your **dictionary.**

HYPHENS USED WITH MODIFIERS

Two-word and three-word unit **modifiers** that express a single thought are frequently hyphenated when they precede a **noun** (an *out-of-date* car, a *clear-cut* decision). If each of the words could modify the noun without the aid of the other modifying word or words, however, do not use a hyphen (a *new digital* computer—no hyphen). Also, if the first word is an **adverb** ending in -*ly*, do not use a hyphen (a *hardly* used computer, a *badly* needed micrometer).

A modifying **phrase** is not hyphenated when it follows the noun it modifies.

EXAMPLE   Our office equipment is *out of date.*

A hyphen is always used as part of a letter or number modifier.

EXAMPLES   5-cent, 9-inch, A-frame, H-bomb

In a series of unit modifiers that all have the same term following the hyphen, the term following the hyphen need not be repeated throughout the series; for greater smoothness and brevity, use the term only at the end of the series.

CHANGE   The third-floor, fourth-floor, fifth-floor rooms have recently been painted.

TO   The third-, fourth-, and fifth-floor rooms have recently been painted.

HYPHENS USED WITH PREFIXES AND SUFFIXES

A hyphen is used with a **prefix** when the root word is a **proper noun**.

EXAMPLES   pre-Sputnik, anti-Stalinist, post-Newtonian

A hyphen may optionally be used when the prefix ends and the root word begins with the same vowel. When the repeated vowel is *i,* a hyphen is almost always used.

EXAMPLES   re-elect, re-enter, anti-inflationary

A hyphen is used when *ex-* means ''former.''

EXAMPLES   ex-partners, ex-wife

A hyphen may be used to emphasize a prefix.

EXAMPLE   He was *anti-everything.*

The **suffix** -*elect* is hyphenated.

EXAMPLES   president-elect, commissioner-elect

OTHER USES OF THE HYPHEN

To avoid confusion, some words and modifiers should always be hyphenated. *Re-cover* does not mean the same thing as *recover,* for example; the same is true of *re-sent* and *resent, re-form* and *reform, re-sign* and *resign.*

Hyphens should be used between letters showing how a word is spelled.

EXAMPLE   In his letter he misspelled *believed* b-e-l-e-i-v-e-d.

Hyphens identify prefixes, suffixes, or written syllables.

EXAMPLE   *Re-, -ism,* and *ex-* are word parts that cause spelling problems.

A hyphen can stand for *to* or *through* between letters and **numbers.**

EXAMPLES   pp. 44-46
the Detroit-Toledo Expressway
A-L and M-Z

Finally, hyphens are used to divide words at the end of a line. Words are divided on the basis of their syllables, which can be determined with a **dictionary.** Do not divide a word, however, if only one or two letters remain, and never divide a word at the end of a page in a typewritten manuscript. A good rule of thumb on dividing a word is to determine whether each section is pronounceable. If a word is spelled with a hyphen, divide it only at the hyphen break unless to divide the word there would confuse the **reader** because the hyphen is essential to the word's meaning. It is generally better, however, to avoid dividing words unless this would leave a large gap at the end of a line. (See also **syllabication.**)

I

idea

The word *idea* means the "mental representation of an object of thought." Use *impression* or *intention* where they convey more precisely what you mean to say.

CHANGE We started out with the *idea* of returning early.

TO We started out with the *intention* of leaving early.

CHANGE She left me with the *idea* that he was in favor of the proposal.

TO She left me with the *impression* that he was in favor of the proposal.

idioms

Idioms are groups of words that have a special meaning apart from their literal meaning. Someone who "runs for office" in the United States, for example, need not be a track star. The same person would "stand for office" in England. In both nations, the individual is seeking public office; only the idioms of the two countries differ. This difference indicates why idioms give foreign writers and speakers trouble.

EXAMPLES The judge *threw the book* at the convicted arsonist.

The company's legal advisers debated whether to *stand fast* or to *press forward* with the lawsuit.

The foreigner must memorize such expressions; they cannot logically be understood. The native writer has little trouble understanding idioms and need not attempt to avoid them in writing, provided the **reader** is equally at home with them. Idioms are often helpful shortcuts, in fact, that can make writing more vigorous and natural.

If there is any chance that your writing might be translated into another language or read in other English-speaking countries, eliminate obvious idiomatic expressions that might puzzle readers. For example, the owner's manual for a Volvo automobile was obviously translated into English with British drivers in mind. Where the American expression would be "filling up," there are references to "topping up" various fluids; where Americans would expect "windshield wipers," they find "windscreen wipers"; and the rearview mirror has an "antidazzle knob" instead of an "antiglare adjustment" for night driving. Such idiomatic expressions can be an annoying hindrance to clear understanding.

Idioms are the reason that certain **prepositions** follow certain **verbs, nouns,** and **adjectives.** Since there is no sure system to explain such **usages,** the best advice is to check in a **dictionary.** Following are some common pairings that give writers trouble.

absolve from (responsibility)
absolved of (crimes)
accede to
accordance with
according to
accountable for (actions)
accountable to (a person)
accused by (a person)
accused of (a deed)
acquaint with

adapt for (a purpose)
adapt to (a situation)
adapt from (change)
adept in
adhere to
adverse to
affinity between, with
agree on (terms)
agree to (a plan)
agree with (a person)

angry with (a person)
angry at, about (a thing)
approve of
apply for (a position)
apply to (contact)
argue for, against (a policy)
argue with (a person)
arrive in (a city, country)
arrive at (a specific location,
    conclusion)
based on, upon

blame for (an action)
blame on (a person)

comply with
concur in (consensus)
concur with (a person)
conform to
consist of
convenient for (a purpose)
convenient to (a place)
correspond to, with (a thing)

correspond with (a person)
deal with
depend on
deprive of
devoid of
differ about, over (an issue)
differ from (a thing)
differ with (a person)
differ on (amounts, terms)
different from

disagree on (an issue, plan)
disagree with (a person)
disappointed in
disapprove of
disdain for
divide between, among
divide into (parts)
engage in
exclude from
expect from (things)

expect of (people)
expert in
identical with, to
impatient for (something)
impatient with (someone)

imply that
impose on
improve on
inconsistent with
independent of

infer from
inferior to
inseparable from
mediate between, among
necessary for (an action)
necessary to (a state of being)
occupied by (things, people)
occupied with (actions)
opposite of (qualities)
opposite to (positions)
part from (a person)

part with (a thing)
proceed with (a project)
proceed to (begin)
proficient in

profit by (things)
profit from (actions)
prohibit from
qualify as (a person)
qualify by (experience, actions)
qualify for (a position, award)

rely on
responsibility for, of
reward for (an action)
reward with (a gift)
similar to
surrounded by (people)
surrounded with (things)
talk to (a group)
talk with (a person)
wait at (a place)
wait for (a person, event)
wait on (a customer)
yield to

i.e.

The **abbreviation** *i.e.* stands for the Latin *id est,* meaning "that is."
Since *that is* is a perfectly good English expression, there is no need to use
a Latin expression or its abbreviation. In addition, many people confuse
i.e. with **e.g.,** never being certain which means what. This abbreviation
does not save enough space to justify having one person misunderstand
your meaning; avoid *i.e.* in your writing.

> CHANGE   We were a fairly heterogeneous group; *i.e.,* there were managers,
> foremen, and vice-presidents at the meeting.
>
> TO   We were a fairly heterogeneous group; *that is,* there were man-
> agers, foremen, and vice-presidents at the meeting.

If you must use *i.e.,* punctuate it as follows. If *i.e.* connects two **independent clauses,** precede it with a **semicolon** and follow it with a **comma,** as
in the above example. If *i.e.* connects a **noun** and **appositives,** a comma
should precede and follow it.

> EXAMPLE   We were a fairly heterogeneous group, *i.e.,* managers, foremen,
> and vice-presidents.

illegal/illicit

If something is *illegal,* it is prohibited by law. If something is *illicit,* it is prohibited by law or *custom.* *Illicit* behavior may or may not be *illegal,* but it does violate custom or moral codes and therefore usually has a clandestine or immoral **connotation.**

EXAMPLES   Their *illicit* behavior caused a scandal, and the district attorney charged that *illegal* acts were committed.

Explicit sexual material may be *illicit* without being *illegal.*

illustrations

The objective of using an illustration is to help your **reader** absorb the facts and ideas you are presenting. When used well, an illustration can convey an idea that words alone could never really make clear. Notice how the following description depends upon its illustration.

EXAMPLE   The 682-100 Card Reader, a serial Reader that uses photoelectric sensing for data detection, can read a maximum of 300 cards per minute, with the actual rate dependent upon the processor's demand for input. Each time the processor requests that a card be input, the Reader completes one cycle of the following concurrent steps: (1) the card in the read station moves past the photoelectric cells to the output hopper (the photoelectric cells sense the data in the card for transmission to the processor's main memory), (2) the card in the ready

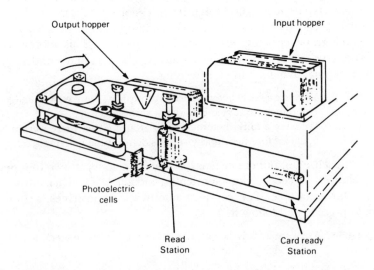

Output hopper

Input hopper

Photoelectric
cells

Read
Station

Card ready
Station

station moves to the read station, and (3) a card feeds from the input hopper to the ready station.

Illustrations should never be used as ornaments, however; they should always be functional working parts of your writing. Be careful not to overillustrate; use an illustration only when it makes a direct contribution to your reader's understanding of your subject. When creating illustrations, consider your objective and your reader. You would use a different illustration of an X-ray machine, for example, for a high-school science class than you would for a group of medical doctors. Many of the attributes of good writing—simplicity, **clarity, conciseness,** directness—are equally important in creating and using illustrations.

The most common types of illustrations are **photographs, graphs, tables, drawings, flowcharts, organizational charts, schematic diagrams,** and **maps.** Your material will normally suggest one of these types when an illustration is needed.

TIPS FOR CREATING AND USING ILLUSTRATIONS

Each type of illustration has its unique strengths and weaknesses, and these are discussed in the specific text entry for each type. The guidelines presented here apply to most visual material you use to supplement the information in your text. Following these tips should help you create and present your visual material to good effect.

1. Keep the information as brief and simple as possible.
2. Try to present only one type of information in each illustration.
3. Label or caption each illustration carefully.
4. Include a key that identifies all **symbols,** when necessary.
5. Specify the proportions used or include a scale of relative distances, when appropriate.
6. Make the lettering horizontal for easy reading if possible.
7. Keep terminology consistent. Do not refer to something as a "proportion" in the text and as a "percentage" in the illustration.
8. Allow enough white space around and within the illustration for easy viewing.
9. Position the illustration as close as possible to the text that refers to it; however, an illustration should normally not appear ahead of the first text reference to it.
10. Be certain that the significance of each illustration is clear from the text.
11. If figure numbers or table numbers are used, as they should be if five or more illustrations or tables are used, number the illustrations or tables consecutively.

12. If more than five illustrations or tables appear in a **report,** list them by title, together with figure and page numbers, under a separate heading ("List of Figures" or "List of Tables") following the **table of contents.**

Presented with clarity and consistency, illustrations can help your reader focus on key portions of your report. Be aware, though, that even the best illustration only supplements the text. Your writing must carry the major burden of providing the context for the illustration and pointing out its significance.

imply/infer

If you *imply* something, you hint or suggest it. If you *infer* something, you reach a conclusion on the basis of evidence.

EXAMPLES   His memo *implied* that the project would be delayed.
The general manager *inferred* from the memo that the project would be delayed.

in/into

In means "inside of "; *into* implies movement from the outside to the inside.

EXAMPLE   The equipment was *in* the test chamber, so she sent her assistant *into* the chamber to get it.

in order to

The **phrase** *in order to* is sometimes essential to the meaning of a **sentence.**

EXAMPLE   If the vertical scale of a graph line would not normally show the zero point, use a horizontal break in the graph *in order to* include the zero point.

The phrase *in order to* also helps control the **pace** of a sentence even though it is not essential to the meaning of the sentence.

EXAMPLE   The committee must know the estimated costs *in order to* evaluate the feasibility of the project.

Most often, however, the phrase *in order to* is just a meaningless filler phrase that is dropped into a sentence without thought.

CHANGE *In order to* start the engine, open the choke and throttle and then press the starter.

TO To start the engine, open the choke and throttle and then press the starter.

Search for these thoughtless uses of the phrase *in order to* in your writing and eliminate them.

in terms of

When used to indicate a shift from one kind of language, or terminology, to another, the **phrase** *in terms of* can be useful.

EXAMPLE *In terms of* gross sales, the year has been relatively successful; however, *in terms of* net income, it has been discouraging.

When simply dropped into a **sentence** because it comes easily to mind, however, the phrase *in terms of* is meaningless **affectation.**

CHANGE She was thinking *in terms of* subcontracting much of the work.

TO She was thinking *about* subcontracting much of the work.

OR She was thinking *of* subcontracting much of the work.

inasmuch as/insofar as

Inasmuch as, meaning "because of the fact that," is a weak **connective** for stating causal relationships.

CHANGE *Inasmuch as* the heavy spring rains delayed construction, the office building will not be completed on schedule.

TO *Because* the heavy spring rains delayed construction, the office building will not be completed on schedule.

Insofar as, meaning "to the extent that," should be reserved for that explicit **usage** rather than being used to mean *since.*

EXAMPLE *Insofar as* the report deals with causes of highway accidents, it will be useful in preparing future plans for highway construction.

CHANGE *Insofar as* you are here, we will review the material.

TO *Since* you are here, we will review the material.

increasing-order-of-importance method of development

When you want the most important of several ideas to be freshest in your **reader's** mind at the end of your writing, organize your information by **increasing order of importance.** This sequence begins with the least important point or fact, then moves to the next least important, and builds finally to the most important point at the end.

Writing organized by increasing order of importance has the disadvantage of beginning weakly, with the least important information, which can cause your reader to become impatient before reaching your main point. But when you build a case in which the ideas lead, point by point, to an important **conclusion,** increasing order of importance is an effective method of **organization. Reports** on production or personnel goals are often arranged by this method, as are **oral reports.**

In the following example, the writer begins with the least productive source of applicants and builds up to the most productive source.

```
                        MEMORANDUM

        To:  William D. Vane, Vice President for
             Operations
        From: Harry Matthews, Personnel Department
        Date: December 3, 19--

        Subject:  Recruiting Qualified Electronics
                  Technicians

        To keep our company staffed with qualified
        electronics technicians, we will have to redirect
        our recruiting program.

        In the past dozen years, we have relied heavily
        on the recruitment of skilled veterans.  An end
        to the military draft, as well as attractive
        reenlistment bonuses for skilled technicians
        now in uniform, has all but eliminated veterans
        as a source of trained employees.  Our attempts
        to reach this group through ads in service
        newspapers and daily newspaper ads has not been
        successful recently.  I think that the want ads
        should continue, although each passing month
        brings fewer and fewer veterans as job
        applicants.

        Our in-house apprentice program has not provided
        the needed personnel either.  High school enroll-
```

ments in the area are continually dropping. Each year, fewer high school graduates, our one source of trainees for the apprentice program, enter the shop. Even vigorous Career Day recruiting has yielded disappointing results. The number of students interested in the apprentice program has declined proportionately as school enrollment goes down.

The local and regional technical schools produce the greatest number of qualified electronics technicians. These graduates tend to be highly motivated, in part because many have obtained their education at their own expense. The training and experience they have received in the tech programs, moreover, are first-rate. Competition for these graduates is keen, but I recommend that we increase our recruiting efforts and hire a larger share of this group than we have been doing.

I would like to meet with you soon to discuss the details of a more dynamic recruiting program in the technical schools. I am certain that with the right recruitment campaign we can find the skilled personnel essential to our expanding role in electronics products and service.

incredible/incredulous

Something that is unbelievable or nearly unbelievable is *incredible*.

EXAMPLE   The precision of the new instrument is truly *incredible*.

Incredulous means that a person is skeptical or unbelieving about some thing or some event.

EXAMPLE   He was *incredulous* at the reports of the instrument's precision.

indefinite adjectives

Indefinite adjectives are so called because they do not designate anything specific about the **nouns** they modify.

EXAMPLES   *some* circuit boards, *any* branches, *all* aircraft engines

Notice that the **articles** *a* and *an* are included among the indefinite adjectives. (See also **a, an.**)

> EXAMPLE　It was *an* hour before the needle of the gauge finally approached the danger area.

For recommendations on the placement and **usage** of adjectives, see **adjectives.**

indefinite pronouns

Indefinite pronouns do not specify a particular person or thing; they specify any one of a class or group of persons or things. *Any, another, anyone, anything, both, each, either, everybody, few, many, most, much, neither, none, several, some,* and *such* are indefinite pronouns. Most indefinite **pronouns** require singular **verbs.**

> EXAMPLES　If *either* of these switches *detects* the absence of paper, the printer's control logic places the printer in a not-ready state.
> *Neither* of them *was* available for comment.
> *Each* of the writers *has* a unique style.
> *Everyone* in the department *has* completed the form.

Some indefinite pronouns require plural verbs.

> EXAMPLES　*Many are* called, but *few are* chosen.
> *Several* of them *are* aware of the problem.

A few indefinite pronouns may take either plural or singular verbs, depending on whether the **nouns** they stand for are plural or singular.

> EXAMPLES　*Most* of the employees *are* pleased, but *some are* not.
> *Most* of the oil *is* imported, but *some is* domestic.

indentation

Type that is indented is set in from the margin. The most common use of indentation is at the beginning of a **paragraph,** where the first line is usually indented five spaces in a typed manuscript and one inch in a longhand manuscript.

> EXAMPLE　　A hydraulic crane is not like a lattice boom friction crane in one very important way. In most cases, the safe lifting capacity of a lattice boom crane is based on the weight needed to tip the machine. Therefore, operators of friction machines sometimes

depend on signs that the machine might tip to warn them of impending danger.

   This is a very dangerous thing to do with a hydraulic crane. Hydraulic crane ratings are based on the strength of the material of the boom (and other components). Therefore, the hydraulic crane operator who waits for signs of tipping to warn him of an overloaded condition will often bend the boom or cause severe damage to his machine before any signs of tipping occur.

Another use of indentation is in **outlining,** where each subordinate entry is indented under its major entry.

EXAMPLE   II. Types of Outlines
           A. The Sentence Outline
              1. A sentence outline provides order and establishes the relationship of topics to one another to a greater degree than a topic outline.

Finally, a block **quotation** may be indented within a manuscript instead of being enclosed in **quotation marks;** it must, in this case, be single-spaced. Set off the indented material from the text by indenting five spaces from the left margin and five spaces from the right margin and by double-spacing above and below the passage.

EXAMPLE   The Operators Manual states:

   This is a very dangerous thing to do with a hydraulic crane. Hydraulic crane ratings are based on the strength of the material of the boom (and other components) . . .
   A hydraulic crane is not like a lattice boom friction crane in one very important way. In most cases, the safe lifting capacity of the lattice boom crane is based on the weight needed to tip the machine. Therefore, operators of friction machines sometimes depend on signs that the machine might tip to warn them of impending danger.

   This warning was given because many operators were accustomed to . . .

independent clauses

A independent clause is a group of words containing a **subject** and a **predicate** that expresses a complete statement and that could stand alone as a separate **sentence.**

EXAMPLE   *Production was not as great as we had expected,* although we met our quota.

In the example the independent **clause** is italicized and could clearly stand alone as a complete sentence. The second, unitalicized clause is a **dependent clause**; although it contains a subject and predicate, it does not express a complete thought and could not stand alone as a sentence.

indexing

An index is an alphabetical list of all the major topics discussed in a written work. It cites the pages where each topic can be found and thus allows **readers** to find information on particular topics quickly and easily. The index falls at the very end of the work.

The key to compiling a useful index is selectivity. Instead of listing every possible reference to a topic, select references to passages where the topic is discussed fully or where a significant point is made about it. For actual index entries choose those words or **phrases** that best represent a topic. For example, the key terms in a reference to the development of legislation about environmental impact statements would probably be *legislation* and *environmental impact statement,* not *development.* Key terms are those that a reader would most likely look for in an index. In selecting terms for index entries, use chapter or section titles only if they include such key words. For index entries on **tables** and **illustrations,** use the key words in their titles.

COMPILING AN INDEX

Do not attempt to compile an index until the final manuscript is completed, because terminology and page numbers will not be accurate before then. The best way to compile your list of topics is with 3 × 5-inch index cards. Read through your written work from the beginning; each time a key term appears in a significant context, enter the term and its page number on a separate index card. An index entry can consist solely of a main entry and its page number.

EXAMPLE   Aquatic monitoring programs, 42

An index entry can also include a main entry, subentries, and even subsubentries, as in the following example.

```
Monitoring programs, 27-44  ─────────── Main entry
    aquatic, 42
    ecological, 40                        Subentries
    meteorological, 37
        operational, 39
        preoperational, 37                Sub-subentries
    radiological, 30
    terrestrial, 41, 43-44                Subentries
    thermal, 27
```

A subentry indicates pages where a specific subcategory or subdivision of a main topic can be found. When compiling index entries on cards, enter a subentry on its own separate card, but include the main entry as well. Also put a sub-subentry on a separate index card, but include the entry and subentry to which it is related. You must include all this information so that you will be able to tell where each card belongs when you arrange the cards in alphabetical order.

Monitoring Programs
Meteorological, 37

Monitoring Programs
Meteorological
operational, 39.

Sample Index Cards

When you have completed this process for the entire work, arrange the cards alphabetically and type the index, two columns to a page, from the cards. Be sure to arrange subentries and sub-subentries alphabetically beneath the proper main entry.

WORDING INDEX ENTRIES

The first word of an index entry should be the principal word because the reader will look for topics alphabetically by their main words. Selecting the right word to list first is of course easier for some topics than for others. For instance, *Tips on repairing electrical wire* would not be a suitable index entry because a reader looking for information on electrical wire would not look under the word *tips*. Whether you select *electrical wire; wire, electrical;* or *repairing electrical wire* depends on the purpose and **scope** of your report. Ordinarily, an entry with two key words, like *electrical wire,* should be indexed under each word. (When entering words on index cards you make out one card for *electrical wire* and another for *wire, electrical.*) An index entry should be written as a **noun** or a noun **phrase** rather than as an **adjective** alone.

CHANGE   Electrical
            grounding, 21
            insulation, 20
            repairing, 22
            size, 21
            wire, 20–22
     TO   Electrical wire, 20-22
            grounding, 21
            insulation, 20
            repairing, 22
            size, 21

CROSS-REFERENCING

Cross-references in an index guide the reader to other pertinent topics in the text. A reader looking up *technical writing,* for example, might find a cross-reference to *memorandum.* Cross-references do not include page numbers; they merely direct the reader to another index entry, which of course gives pages. There are two kinds of cross-references: *see* references and *see also* references.

See references are most commonly used with topics that can be identified by several different terms. Listing the topic page numbers by only one of the terms, the indexer then lists the other terms throughout the index as *see* references.

EXAMPLE   Economic costs. *See* Benefit-cost analyses

See references also direct readers to index entries where a topic is listed as a subentry.

EXAMPLE   L-shaped fittings. *See* Elbows, L-shaped fittings

See also references indicate other entries that include additional information on a topic.

EXAMPLE   Ecological programs, 40–49
                    See also Monitoring programs

STYLING THE INDEX

Capitalize the first word of a main entry and all other terms normally capitalized in the text. Do not capitalize the first words in subentries and sub-subentries unless they appear that way in the text. Italicize (underline) the cross-reference terms *see* and *see also.*

Place each subentry in the index on a separate line, indented from its main entry. Indent sub-subentries from the preceding subentry. These indentations allow readers to scan a column quickly for pertinent subentries and sub-subentries.

Separate entries from page numbers with **commas.** Type the index in a double-column format, as is done in the index to this book.

indirect objects

An indirect object is a **noun** or noun equivalent that occurs with a **direct object** after certain kinds of **transitive verbs,** such as *give, wish, cause, tell,* and their **synonyms** or **antonyms.** The indirect object is usually a person or persons and answers the question "to whom or what?" or "for whom or what?" The indirect object always precedes the direct object.

EXAMPLES   Wish *me* success.
                    It caused *him* pain.
                    Their attorney wrote *our firm* a follow-up letter.
                    Give *the car* a push.

The indirect object is one of the four kinds of **complements,** along with the direct object, the **objective complement,** and the **subjective complement.** (See also **objects.**)

indiscreet/indiscrete

Indiscreet means "lacking in prudence or sound judgment."

EXAMPLE   His public discussion of the proposed merger was *indiscreet.*

Indiscrete means "not divided or divisible into parts."

EXAMPLE   The separate units, once combined, become *indiscrete.*

See also **discreet/discrete.**

individual

Avoid using *individual* as a **noun** if *person* is more appropriate.

CHANGE   Several *individuals* on the panel did not vote.
TO   Several *persons* on the panel did not vote.
OR   Several *people* on the panel did not vote.
CHANGE   She is an *individual* who says what she things.
TO   She is a *person* who says what she thinks.
OR   She says what she thinks.

Individual is most appropriate when used as an **adjective** to distinguish a single person from a group.

EXAMPLE   The *individual* employee's obligations to the firm are detailed in the booklet that describes company policies.

See also **persons/people.**

infinitives

An infinitive is the bare, or uninflected, form of a **verb** *(go, run, fall, talk, dress, shout)* without the restrictions imposed by **person** and **number.** Along with the **gerund** and **participle,** it is one of the nonfinite verb forms. The infinitive is generally preceded by the word *to,* which, although not an inherent part of the infinitive, is considered to be the sign of an infinitive.

EXAMPLES   It is time *to go* to work.
We met in the conference room *to talk* about the new project.

An infinitive is a **verbal** and may function as a **noun,** an **adjective,** or an **adverb.**

EXAMPLES   *To expand* is not the only objective. (noun)
These are the instructions *to follow.* (adjective)
The company struggled *to survive.* (adverb)

The infinitive may reflect two **tenses**: the present and (with a **helping verb**) the present perfect.

EXAMPLES   to go (present tense)
to have gone (present perfect tense)

The most common mistake made with infinitives is to use the present perfect tense when the simple present tense is sufficient.

CHANGE   I should not have tried *to have gone* so early.
TO   I should not have tried *to go* so early.

Infinitives formed with the root form of **transitive verbs** can express both active and (with a helping verb) passive **voice.**

EXAMPLES   to hit (present tense, active voice)
to have hit (present perfect tense, active voice)
to be hit (present tense, passive voice)
to have been hit (present perfect tense, passive voice)

SPLITTING INFINITIVES

A split infinitive is one in which an adverb is placed between the sign of the infinitive, *to,* and the infinitive itself. Because they make up a grammatical unit, the infinitive and its sign are better left intact than separated by an intervening adverb. Although it may be better to split an infinitive than allow a sentence to become awkward or incoherent, only rarely is a sentence improved by splitting an infinitive.

CHANGE   *To* initially *build* the table in the file, you could input transaction records containing the data necessary to construct the record and table.
TO   *To build* the table in the file initially, you could input transaction records containing the data necessary to construct the record and table.

infinitive phrases

An infinitive phrase consists of an **infinitive** (usually with *to*) and any **modifiers** or **complements.**

EXAMPLE   The product must be able *to resist repeated hammer blows.*

Do not confuse a **prepositional phrase** beginning with *to* with an infinitive phrase. In the infinitive phrase, the *to* is followed by a **verb;** in the prepositional phrase, *to* is followed by a **noun** or **pronoun.**

EXAMPLES   We went *to the building site.* (prepositional phrase)
Our firm tries *to provide a comprehensive training program.* (infinitive phrase)

An infinitive phrase may function as a noun, an **adjective,** or an **adverb.**

EXAMPLES   His goal is *to become sales manager.* (noun)
The need *to increase sales* should be obvious. (adjective)
We must work *to increase sales.* (adverb)

An infinitive phrase that functions as a noun may serve within the **sentence** as **subject, object,** complement, or **appositive.**

EXAMPLES   *To form her own company* was her lifelong ambition. (subject)
They want *to know when we can begin the project.* **(direct object)**
His objective is *to live well.* **(subjective complement)**
His ambition, *to form his own company,* is soon to be realized. (appositive)

The implied subject of an introductory infinitive phrase should be the same as the subject of the sentence. If it is not, the **phrase** is a **dangling modifier.** In the following example, the implied subject of the infinitive is *you,* or *one,* not *practice.*

CHANGE   *To learn shorthand,* practice is needed.
TO   *To learn shorthand,* you must practice.
OR   *To learn shorthand,* one must practice.

informal writing style

Informal writing style is a relaxed and colloquial way of writing **standard English.** It is the **style** found in most private letters, nonfiction books of general interest, many mass-circulation magazines, and some business **correspondence.** There is less distance between writer and **reader** because the **tone** is more personal than in **formal writing style. Contractions** and elliptical constructions are commonplace.

The vocabulary of an informal writing style is made up of generally familiar rather than unfamiliar words and expressions, although slang and dialect are usually avoided. Informal style approximates the cadence and structure of spoken English, while conforming to the grammatical conventions of written English.

EXAMPLE   All you need is a talent for spotting the idiocies now built into the system. But you'll have to give up being an administrator who loves to run others and become a manager who carries water for his people so they can get on with the job. And you'll have to keep a suspicious eye on the phonies who cater to your uncertainties or feed your trembling ego on press releases, office perquisites, and optimistic financial reports. You'll have to give substance to such tired rituals as the office party. And you'll certainly have to recognize, once you get a hunk of your company's stock, that you aren't the last man who might enjoy the benefits of shareholding. These elegant simplicities require a sense of justice that won't be easy to hang on to.

—Robert Townsend, *Up the Organization* (New York: Alfred A. Knopf, 1970), p. 11

Most technical writing tends to be more formal than informal. (See also **technical writing style** and **English, varieties of.**)

ingenious/ingenuous

Ingenious means "marked by cleverness and originality," and *ingenuous* means "straightforward" or "characterized by innocence and simplicity."

EXAMPLES   Wilson's *ingenious* plan, which streamlined production in Department L, was the beginning of his rise within the company.
I believe that the *ingenuous* co-op students bring some freshness into the company.

inquiry letters and responses

An inquiry letter may be as simple as requesting a free brochure or as complex as asking a financial consultant to define the specific requirements for floating a multimillion-dollar bond issue. Regardless of your purpose, you need to state clearly what it is you want, why you want it, and what you are going to do with it. If the information you are requesting is at all sensitive, promise your **reader** that you will keep it confidential.

WRITING AN INQUIRY LETTER

Since your primary **objective** in writing an inquiry letter is to get a response, make it as easy as possible for your reader to respond. Keep the number of questions to a minimum, and make them as clear and concise as possible.

There are two broad categories of inquiry letters. One kind provides a benefit (or a potential benefit) to the reader; the other benefits only the writer.

Suppose, for example, that you are inquiring about a product or service you have seen advertised recently. Such an inquiry letter benefits the reader, who has a built-in motivation to respond—you represent a potential customer. With this kind of inquiry letter, get directly to the point. Identify the product or service you are interested in and ask for the brochure or other item you want the reader to send you. Letter 11 is typical of this type of request letter.

The other kind of inquiry letter benefits only the writer. It asks for a special favor from someone who has no apparent motivation to respond. For example, you might ask a trade journal to send you information on the number of subscribers in a geographical area. This kind of inquiry letter requires **persuasion,** and it is generally best to start with an inducement designed to get your reader to respond. Such an inducement might be the recognition of your recipient as an expert in the field, thereby appealing to his or her ego. Another inducement often used to encourage people to complete and return **questionnaires** is the offer of some form of reward (for example, a publisher might offer the reader a complimentary book). Or, you might emphasize the intrinsic value of what you are doing as a means of inducing the reader to contribute to your effort.

Letter 12 is typical of an inquiry letter that benefits only the writer. He is writing to get a recommendation from a job applicant's former employer. As an inducement to get the reader to respond, the writer asks him to help *the former employee.* The writer also makes it easy for his reader to respond by asking a series of *specific questions* about the applicant rather than simply asking the employer to speak about the person in general terms.

To save yourself and your reader time and trouble, plan your letter carefully. Phrase your questions carefully and precisely; nothing is more frustrating than receiving an answer to the wrong question. Be sure to include your address—surprisingly, this vital piece of information is

```
┌─────────────────────────────┐
│ HOWARD ENTERPRISES          │
│ P.O. Box 113                │
│ Clemson, IL 49301           │
└─────────────────────────────┘
```

March 11, 19--

Ms. Jane Metcalf
Engineering Services
Miami Valley Power Company
P.O. Box 1444
Clemson, IL 49305

Dear Ms. Metcalf:

We are presently designing an all-electric, energy-efficient,
middle-priced home which we plan to market in your service area.
The house, which contains 2,000 square feet of living space
(17,600 cubic feet), meets all the requirements stipulated in
your company's brochure "Insulating for Efficiency." However,
we need some additional information on heating systems.
Specifically, we need the following information.

   1. The proper size heat pump to use in this climate
      for such a home;

   2. The wattage of the supplemental electrical furnace
      that would be required for this climate; and

   3. The estimated power consumption, and current rates,
      of these units for a calendar year.

We will be happy to send you a copy of our preliminary design
report for your information.

Thank you very much.

               Sincerely yours,

               Kathryn J. Parsons

               Kathryn J. Parsons

Letter 11

SCAIBORN CORPORATION 19 WATERFORD WAY MINNEAPOLIS, MN 94321

January 19, 19--

Mr. Jason Jaborski
Manager, Engineering Department
Glendon Products, Inc.
2660 Willow Avenue
Birmingham, AL 49326

Dear Mr. Jaborski:

Ms. Kimberly Thomas worked for you as a mechanical engineer
from 19-- to 19--. We are considering her for a position as
a mechanical engineer. She has given your name as a reference,
and we are interested in your comments about her and her work.
Would you mind taking a few minutes to assist her in her quest
for a position with our firm?

Specifically, we would like your responses to the following
questions.

　　　1.　Our engineers must be extremely accurate in their
　　　　　work. Can you comment on Ms. Thomas' accuracy?

　　　2.　Did she demonstrate the kind of stamina that
　　　　　would enable her to work long hours when
　　　　　necessary?

　　　3.　She would have to supervise a small drafting
　　　　　section. Can you comment on her leadership
　　　　　abilities?

　　　4.　The job requires some traveling. Do you believe
　　　　　she would function well when separated from her
　　　　　home and family?

We would greatly appreciate your answers to these questions
and any other information that would help us reach a
decision. Of course, any information you provide will be
kept confidential.

Sincerely,

Henry Winton

Henry Winton
Chief Engineer

HW:ls

Letter 12

often forgotten. In fact, it is usually a good idea to include a stamped, self-addressed return envelope. End your letter with an expression of appreciation for your reader's anticipated help.

RESPONDING TO AN INQUIRY LETTER

If you cannot respond to an inquiry letter immediately, do at least acknowledge receipt of the letter, indicate the reason for the delay, and indicate when you will be able to respond.

If you cannot respond to an inquiry letter at all, find someone who can and forward the letter to that person. Then send a note to the sender saying that you have forwarded the letter to this person, who will respond shortly. Be sure to send a copy of this letter to the person who will eventually answer the inquiry.

Begin and end a response to an inquiry letter with positive statements. If it is necessary to include negative facts, cushion them in the body of your letter—but do not hide them. Be sure to answer all questions, even those that seem obvious or unimportant. If you must ask for additional information or point out a misunderstanding, do so politely. Take care not to be sarcastic or condescending.

End your letter by offering to answer any further questions the reader may have. Letters 13 and 14 show two responses to inquiry letters.

inside/inside of

In the **phrase** *inside of,* the word *of* is redundant and should be omitted.

> CHANGE   The switch is just *inside of* the door.
> TO   The switch is just *inside* the door.

Using *inside of* to mean "in less time than" is colloquial but should be avoided in writing.

> CHANGE   They were finished *inside of* an hour.
> TO   They were finished *in less than* an hour.

insoluble/unsolvable

Although *insoluble* and *unsolvable* are sometimes used interchangeably to mean "incapable of being solved," careful writers distinguish between

MIAMI VALLEY POWER COMPANY
P.O. BOX 1444
CLEMSON, IL 49305

(513) 264-4800

March 15, 19--

Ms. Kathryn J. Parsons
Howard Enterprises
P.O. Box 113
Clemson, IL 49301

Dear Ms. Parsons:

The project you mentioned in your letter sounds interesting. Unfortunately, I cannot give you an answer that would apply specifically to the house you are designing unless I have additional information about particular design features of the house (one or two stories, number of windows, number of entrances, for example).

I can, however, estimate the insulation requirement of a typical home of 17,600 cubic feet.

1. We would generally recommend, for such a home, a heat pump capable of delivering 40,000 BTUs. Our model AL-42 (17 kilowatts) meets this requirement.

2. With the efficiency of the AL-42, you would not need a supplemental electrical furnace.

3. Depending on usage, the AL-42 unit averages between 1,000 and 1,500 kilowatt hours from December through March. To determine the current rate for such usage, you will need to check with the Chicago Power and Light Company. The Clemson area is not serviced by CP&L, and therefore I do not know what its rates are.

I am sorry that I cannot give you more specific answers. If you could send me more detailed information, however, I would be happy to provide you with more precise figures.

Sincerely,

Jane E. Metcalf

Jane E. Metcalf
Director of Public Information

JEM/mk

Letter 13

MIAMI VALLEY POWER COMPANY
P.O. BOX 1444
CLEMSON, IL 49305

(513) 264-4800

March 15, 19--

Ms. Kathryn J. Parsons
Howard Enterprises
P.O. Box 113
Clemson, IL 49301

Dear Ms. Parsons:

Thank you for inquiring about the heating system we would
recommend for use in homes designed according to the specifications outlined in our brochure "Insulating for Efficiency."

Since I cannot answer your specific questions, I have forwarded
your letter to Mr. Michael Stott, Engineering Assistant in our
development group. He should be able to answer the questions
you have raised.

Sincerely,

Jane E. Metcalf

Jane E. Metcalf
Director of Public Information

JEM/mk
cc: Michael Stott

Letter 14

them. *Insoluble* means "incapable of being dissolved." *Unsolvable* means "incapable of being solved."

EXAMPLES   The plastic is *insoluble* in most household solvents.
Until yesterday, the production problem seemed *unsolvable*.

instructions

When you explain how to perform a specific task—however simple or complex the task may be—you are giving instructions. Written instructions that are based on clear thinking and careful planning should enable a **reader** to carry out the task successfully.

To write instructions that are accurate and easily understandable, you first must thoroughly understand the task you are describing. Otherwise, your instructions could prove confusing or even dangerous. If you are unfamiliar with the task you are describing, you should watch someone who is familiar with the task go through each step. As you watch, ask questions about any step that is not clear to you. Direct observation should help you write instructions that are exact, complete, and easy to follow.

Keep in mind your reader's level of knowledge and experience. Is he skilled in the kind of task for which you are writing instructions? If you know that your reader has a good background, you might feel free to use fairly specialized technical words. If, on the other hand, he has little or no technical knowledge of the subject, you would more appropriately use simple, everyday language—and avoid specialized or technical terms as much as possible.

If the task requires any special tools or materials, say so at the beginning of the instructions. List all essential equipment at the beginning, in a section labeled "Tools Required" or "Materials Required."

The clearest, simplest instructions are written as commands, in the imperative.

CHANGE   The operator should raise the access lid. (indicative)
TO   Raise the access lid. (imperative)

Phrase instructions concisely but not as if they were telegrams. You can make **sentences** shorter by leaving out **articles** (*a, an, the*), some **pronouns** (*you, this, these*), and some **verbs,** but sometimes such sentences sound very telegraphic and are difficult to understand. The first version of the following instruction for cleaning a computer punch card assembly, for example, is confusing and difficult to understand.

CHANGE   Pass card through punch area for debris.

TO   Pass *a* card through *the* punch area *to clear away* any debris.

The meaning of the **phrase** *for debris* needed to be made clearer. The revised instruction is easily and quickly understood.

One good way to make instructions easy to follow is to divide them into short, simple steps. Be sure to order the steps in the proper sequence. Steps can be organized in one of two ways: with numbers or with words. You can label each step with a sequential number.

EXAMPLE   1. Connect each black cable wire to a brass terminal . . .

2. Attach one 4-inch green jumper wire to the back . . .

3. Connect both jumper wires to the bare cable wires . . .

Or you can use words that indicate time or sequence.

EXAMPLE   *First,* determine what the customer's problem is with the computer. *Next,* observe the system in operation. *At that time,* question the operator until you are sure that the problem has been explained completely.

Do not sacrifice clarity in an effort to be concise. Plan ahead for your reader. If the instructions in step 2 will affect a process in step 9, say so in step 2. Otherwise, your reader may reach step 9 before discovering that an important piece of information—that should have been given in advance—is missing.

Sometimes your instructions have to make clear that two operations must be performed at the same time. Either state this fact in an **introduction** to the specific instructions or include both operations in the appropriate step.

CHANGE   4. Hold the CONTROL key down.

5. Press the BELL key before releasing the CONTROL key.

TO   4. While holding the CONTROL key down, press the BELL key.

In any operation certain steps must be performed with more exactness than others. Alert your reader to those steps that require especially precise timing or measurement. Also warn of potentially hazardous steps or materials. If the instructions call for materials that are flammable or that give off noxious fumes, let the reader know before he or she reaches the step for which the material is needed.

Clear and well-planned **illustrations** can make even complex instructions easily understandable. Illustrations often simplify instructions by

INSTRUCTIONS

<u>Distribute the inoculum over the surface of
the agar in the following manner:</u>
(1) Beginning at one edge of the saucer, thin
the inoculum by streaking back and forth
over the same area several times, sweeping
across the agar surface until approximately
one quarter of the surface has been covered.
(2) Sterilize the loop in an open flame.
(3) Streak at right angles to the originally
inoculated area, carrying the inoculum out
from the streaked areas onto the sterile
surface with only the first stroke of the
wire. Cover half of the remaining sterile
agar surface.
(4) Sterilize the loop.
(5) Repeat as described in step (3), covering
the remaining sterile agar surface.

Figure 28

reducing the number of words necessary to explain a process or procedure. Appropriate **drawings** and diagrams will enable your reader to identify parts and the relationships between parts more easily than will long explanations. They will also free you, the writer, to focus on the steps making up the instructions rather than on the descriptions of parts. Not all instructions require illustrations, of course. Whether or not illustrations will be useful depends on your reader's needs as well as on the nature of the project. Instructions for inexperienced readers need to be more heavily illustrated than those for experienced readers.

To test the accuracy and **clarity** of your instructions, ask someone who is not familiar with the operation to follow your written directions. A novice will quickly spot missing steps or point out passages that should be worded more clearly.

The instructions in Figure 28 guide the reader through the steps of "streaking" a saucer-sized disk of material (called *agar*) used to grow bacteria colonies. The object is to thin out the original specimen (the inoculum) so that the bacteria will grow in small, isolated colonies. The streaking process makes certain that part of the saucer is inoculated heavily, while its remaining portions are inoculated progressively more lightly. The streaking is done by hand with a thin wire, looped at one end for holding a small sample of the inoculum.

insure/ensure/assure

Insure, ensure, and *assure* all mean "make secure or certain." *Assure* refers to persons, and it alone has the **connotation** of setting a person's mind at rest. *Ensure* and *insure* also mean "make secure from harm." Only *insure* is widely used in the sense of guaranteeing life or property against risk.

EXAMPLES   I *assure* you that the equipment will be available.
We need all the data to *ensure* the success of the project.
We should *insure* the contents of the building.

intensifiers

Intensifiers are **adverbs** that emphasize degree, such as *very, quite, rather, such,* and *too.* Although they serve a legitimate and necessary function, they can also seduce the unwary writer who is not on guard against overusing them. Too many intensifiers weaken your writing. When revising your draft, either eliminate intensifiers that do not make a definite contribution or replace them with specific details.

CHANGE   The team was *quite* happy to receive the *very* good news that it had been awarded a *rather* substantial monetary prize for its design.

TO   The team was happy to receive the good news that it had been awarded a $5,000 prize for its design.

The difference is not that the first example is wrong and the second one right, but that the intensifiers in the first example add nothing to the **sentence** and are therefore superfluous.

Some words (such as *unique, perfect, impossible, final, permanent, infinite,* and *complete*) do not logically permit intensification because they do not permit degrees of **comparison.** Although **usage** often ignores this logical restriction, the writer should be aware that to ignore it is, strictly speaking, to defy the basic meanings of these words. (For more detailed information, see **comparative degree.**)

CHANGE   It was *quite* impossible for the part to fit into its designated position.

TO   It was impossible for the part to fit into its designated position.

interface

An *interface* is the "surface providing a common boundary between two bodies or areas." The bodies or areas may be physical (the *interface* of piston and cylinder) or conceptual (the *interface* of mathematics and statistics). Do not use *interface* as a substitute for the **verbs** *cooperate, interact,* or even *work.*

CHANGE   The Water Resources Department will *interface* with the Department of Marine Biology on the proposed project.

TO   The Water Resources Department will *work* with the Department of Marine Biology on the proposed project.

interjections

An interjection is a word or **phrase** of exclamation that is used independently to express emotion or surprise or to summon attention. *Hey! Ouch!* and *Wow!* are strong interjections. *Oh, well,* and *indeed* are mild ones. An interjection functions much as do *yes* or *no,* in that it has no grammatical connection with the rest of the **sentence** in which it appears. When an interjection expresses a sudden or strong emotion, punctuate it with an **exclamation mark.**

EXAMPLE   His only reaction was a resounding *Wow!*

Because they get their main expressive force from sound, interjections are more common in speech than in writing. They are rarely appropriate to technical writing.

internal proposals

The purpose of an internal proposal is to suggest a change or an improvement within an organization. For example, one might be written to propose initiating a new management reporting system, expanding the cafeteria service, or changing the policy on parking privileges. An internal proposal is prepared by a person or a department, and it is sent to a higher-ranking person who has the authority to accept or reject the proposal.

In the **opening** of a **proposal,** you must establish that a problem exists which needs a solution. If the person to judge the proposal is not convinced that there is a problem, your proposal will have no chance of success. Notice how the following proposal opening states its problem directly.

> To: Harriet Sullivan, Office Supervisor
> From: Christina Lopez, Administrative Assistant
> Date: December 3, 19—
> Subject: Purchase of a Bently XL-100 System
>
> The number of direct mailings requiring an original copy for each recipient is increasing. The cost of calling in temporary help to get out such a mailing is very high, and the normal office work suffers in our effort to get out the direct mailing. In addition, the pressure causes much confusion and many frayed nerves.

The **body** of a proposal should offer a practical solution to the problem. In building a case for a solution, be as specific as possible. Where it is appropriate, include (1) a breakdown of costs; (2) information about equipment, material, and personnel requirements; and (3) a schedule for completing the task. Such information can help your **readers** to think about the proposal and thus may stimulate him or her to act. Consider the following example, which continues the proposal introduced above.

> The following details show the savings and efficiency of the Bently XL-100 System compared with our present method of handling the typing for our direct mailings. Based on these details, I suggest we consider purchasing this system.

COST SAVINGS

Purchase of the Bently XL-100 System would result in savings of over $2,500 in the first years, as shown below:

Purchase price of Bently XL-100	$1,495.00
Cost of one-year supply of XL-100 magnetic cards	238.00
	$1,733.00 TOTAL
Cost of temporary typists from Wilson Secretarial Agency last year	$3,700.00
Temporary typewriter rental from World Office Machines last year	560.00
	$4,260.00 TOTAL
Savings for the first year	$2,527.00

Since purchase of the Bently XL-100 would be a one-time investment, the savings would increase to over $3,000.00 a year after the first year—even considering the cost of magnetic cards and repairs.

EFFICIENCY

The Bently XL-100 System types without error and four times faster than a human typist, once programmed. Because it requires only one operator, the XL-100 would eliminate the need for temporary typists. It would also eliminate the confusion caused by our training these typists while attempting to meet our own deadlines. In short, we could carry on with our normal office routine. A normal routine would not only improve our employees' morale but also increase their efficiency.

The **conclusion** should be brief but should tie everything together. It is a good idea to restate recommendations here. Be careful, of course, to conclude in a spirit of cooperation, offering to set up a meeting, supplying further facts and figures, or providing any other assistance that might solve the problem, as is done in the following example.

On the basis of these details, I recommend that we consider purchasing a Bently XL-100 System to handle our direct mail program.

Enclosed is a brochure that describes the Bently XL-100 System in detail. I will be happy to provide any additional details about the system as I understand it, at your request.

Following is an example of a complete internal proposal.

To: Ronald Wepner, General Manager
From: Nevil Broadmoor, Data Center Manager
Date: June 14, 19—
Subject: Expansion of the Data Center

Problem

Because the data center is now operating at its maximum capacity, we have found ourselves having to decline new business. By expanding our equipment, staff, and facility we could attract new customers while offering our present customers additional and faster services. I have outlined a plan to expand the data center that would allow us to increase our business—and thus our net income about $300,000 a year.

REQUIREMENTS

Equipment. To provide the computer time required to process new applications, we need to reduce our present time requirements. By upgrading our processor to a V8565M, upgrading our Century software to VRX software, and adding four 658 disc units, we would reduce the time required for our present applications by as much as 50%. The addition of six online terminals and a

Solution

multiplexer would reduce our program maintenance time as much as 60%, which would allow our staff to convert applications to take advantage of the new software.

Staff. To convert present applications to VRX software and to develop and support new applications, we need to increase our programming staff by three people (one senior programmer and two junior programmers).

Facility. To house the additional staff, we need to expand our office area and to purchase additional office furniture for the six online terminals we would place in the programmers' offices. The V8565M requires less space than the present processor, however, and so we could rearrange the computer room to provide space for the multiplexer and additional disc units.

BENEFITS

On the basis of the business available in this area, we could expect to increase our net income by $300,000 a year. We would be able to attract new customers—as well as new business from our present customers' customers—by offering additional and faster services. The increase in our operating budget would actually be small because of the reduction in time requirements provided by the new equipment.

COSTS

Equipment. The purchase of new equipment—reduced by the trade-in allowance on our old processor—would be $383,000.

Cost
Breakdown

Processor	$250,000
Integrated Disc Controller	20,000
Disc Units (four @ $20,000 each)	80,000
Integrated Multiplexer	10,000
Online Terminals (six @ $3,000 each)	18,000
VRX Software	5,000

Budget. The annual increase in our operating budget would be $45,800.

Salaries	$34,000
Benefits	6,800
Office Supplies	1,000
Utilities	1,000
Equipment Maintenance	3,000

Facility. The cost of expanding our facility would be $33,000.

Office Area	$20,000
Computer Room	10,000
Furniture	3,000

SCHEDULE

To achieve the expected net income increase, we would need to meet the following schedule in the expansion.

July 1–Order the required equipment
September 1–Begin recruiting efforts.
September 15–Start expanding the office area.
October 1–Order the required office furniture.
November 1–Start soliciting new business.
November 5–Finish expanding office area.
November 5–Fill programming positions.
November 10–Start new applications design.
November 10–Start conversion effort.
December 1–Install new equipment.
December 15–Implement the first new customer application.

I feel that this expansion plan would enable us to meet the requirements of the businesses in this area. I have looked at the

Conclusion   alternative equipment and believe that the V8565M offers better performance than the other hardware available in this price range. The design of the V8565M would allow us to grow with a minimum of effort as the business in this area grows. I would like to meet with you Friday to discuss this plan and will be happy to explain any details or to gather any additional information that you may desire. Please let me know if we can get together then to discuss this proposed expansion of the data center.

interrogative adverbs

Interrogative adverbs ask questions. Common interrogative adverbs include *where, when, why,* and *how.*

EXAMPLES   *How* many hours did you work last week?
            Where are we going when the new project is finished?
            Why did it take so long to complete the job?

interrogative pronouns

Interrogative pronouns ask questions. The interrogative pronouns are *who, what,* and *which.*

EXAMPLES   *Who* said that?
            What is the trouble?
            Which of these is best?

Notice that interrogative pronouns normally lack expressed antecedents; the antecedent is actually in the expected answer.

interviewing

The **scope** of coverage that you have established for your writing project may require more information (or more current information) than is available in written form (in the library, in corporate specifications, in trade journals, etc.). When you reach this point, consider a personal interview with an expert in your subject.

A discussion of interviewing can be divided into three parts: (1) determining the proper person to interview, (2) preparing for the interview, and (3) conducting the interview.

DETERMINING THE PROPER PERSON TO INTERVIEW

Many times your subject, or your **objective** in writing about the subject, logically points to the proper person to interview. If you were writing about the use of a freeway onramp metering device on Highway 103, for example, you would want to interview the Director of the Traffic Engineering Department; if you were writing about the design of the device, you would want to interview the manufacturer's design engineer. You may interview a professor who specializes in your subject, in addition to getting leads from him as to others you might interview. Other sources available to help you determine the appropriate person to interview are (1) the city directory in the library, (2) professional societies, (3) the yellow pages in the local telephone directory, or (4) a local firm that is involved with all or some aspects of your subject.

PREPARING FOR THE INTERVIEW

After determining the name of the person you want to interview, you must request the interview. You can do this either by telephone or by letter, although a letter may be too slow to allow you to meet your deadline.

Learn as much as possible about the person you are going to interview, and learn as much as possible about the company or agency for which he works. When you make contact with your interviewee, whether by letter or by telephone, explain (1) who you are, (2) why you are contacting him, (3) why you chose him for the interview, (4) the subject of the interview, (5) that you would like to arrange an interview at his convenience, and (6) that you will allow him to review your draft.

Prepare a list of specific questions to ask your interviewee. The natural temptation for the untrained writer is to ask general questions rather than specific ones. "What are you doing about air pollution?" for example, may be too general to elicit specific and useful information. "Residents in the east end of town complain about the black smoke that pours from your east-end plant; are you doing anything to relieve the problem?" would be a more specific question. Analyze your questions to be certain that they are specific and to the point.

CONDUCTING THE INTERVIEW

When you arrive for the interview, be prepared to guide the discussion. The following list of points should help you to do so.

1. Be pleasant, but purposeful. You are there to get information, so don't be timid about asking leading questions on the subject.
2. Use the list of questions you have prepared.
3. Let your interviewee do the talking. Don't try to impress him with your own knowledge of the subject on which he is presumably an expert.
4. Be objective. Don't offer your opinions on the subject. You are there to get information, not to debate.
5. Some answers prompt additional questions; ask them. If you do not ask these questions as they arise, you may find later that you have forgotten to ask them at all.
6. When the interviewee gets off the subject, be ready with a specific and direct question to guide him back onto the track.
7. Take only memory-jogging notes that will help you recall the conversation later. Do not ask your interviewee to slow down so you can take detailed notes. To do so would not only be an undue imposition on his time but might also disturb or destroy his train of thought.
8. If you use a tape recorder, do not let it lure you into a relaxation of discipline so that you neglect to ask crucial questions.

Immediately after leaving the interview, use your memory-jogging notes to help you mentally review the interview and record your detailed notes. Do not postpone this step. Anyone, no matter how good his memory, will forget some important points if he does not do this at once.

intransitive verbs

An intransitive verb is a **verb** that does not need an **object** to complete its meaning. It is able, in its own right, to make a full assertion about its **subject.** Notice in the following examples that the same verb can often be transitive or intransitive, depending upon how it is used.

EXAMPLES The conveyor belt *runs* constantly. (intransitive)
This motor *runs* the conveyor belt. (transitive)

Linking verbs are inherently intransitive since they simply connect a subject to a **modifier** or a **complement.**

EXAMPLES The winch *is* rusted (*Rusted* is an **adjective** modifying *winch.*)
A calculator *remains* a useful tool. (*A useful tool* is a **subjective complement** renaming *calculator.*)

Some verbs, such as *lie* and *sit,* are inherently intransitive.

EXAMPLE   The patient should *lie* on his back during treatment. (*On his back* is
an **adverb** phrase modifying the verb *lie.*)

introductions

The purpose of an introduction is to provide your **readers** with any
general information they must have to understand the detailed information in the rest of your **report.** If you don't need to set the stage in this
way, turn to the entry on **openings** for a variety of interesting ways to
begin without writing an introduction. If you are writing a **house organ
article,** an **annual report,** a **proposal,** a **police report,** a **memorandum,** or something similar, you will probably find one of the various
types of openings more appropriate than an introduction. If you are
preparing a **formal report,** a **laboratory report,** a **technical manual,** a
specification, a **journal article,** a conference paper, or some similar
writing project, however, you will need to present your readers with
some general information before getting into the body of your writing.

You may find that you need to write one kind of introduction for a
report or **journal article** and a different kind for a **technical manual** or
specification.

WRITING INTRODUCTIONS FOR REPORTS AND JOURNAL ARTICLES

In writing an introduction for a report or journal article, you need to
state the subject, the purpose of your report or article, its scope, and the
way you plan to develop the topic.

Stating the Subject.   In stating the subject, it may be helpful to the reader
if you include a small amount of information on the history or theory of
the subject. In addition to stating the subject, you may need to define it if
it is one with which some of your readers may be unfamiliar. (See
definition method of development.)

Stating the Purpose.   The statement of purpose in your introduction
should function in your article or paper much as a **topic sentence**
functions in a **paragraph.** It should make your readers aware of your
goal as they read your supporting statements and examples. It should
also tell them why you are writing about the subject and whether your
material provides a new perspective or clarifies an existing perspective.

Stating the Scope.   Stating the **scope** of your article or report tells your
reader how much or how little detail to expect. Does your article or

report present a broad survey of your **topic,** or does it concentrate on one facet of the topic? Keep the statement of scope brief and simple, however; lack of restraint could lead you to write an **abstract** of your article of paper instead of a simple statement of scope in your introduction.

Stating the Development of the Subject. In a larger work, it may be helpful to your reader if you state how you plan to develop the subject. Is the work an analysis of the component parts of some whole? Is the material presented in chronological sequence? Or does it involve an inductive or deductive approach? Providing such information makes it easier for your readers to anticipate how the subject will be presented (see **methods of development**), and it gives them a basis for evaluating how you arrived at your **conclusions** or recommendations.

The following introduction is from a journal article.

The recent sharp increase in crude oil prices, coupled with the current drive toward energy independence in the United States, has suddenly made the development of processes for the large-scale production of synthetic fuels from domestic coal of crucial importance. Methanol, which is one such fuel that can be derived from coal, has projected manufacturing costs that, on an energy-equivalent basis, are roughly comparable to the other two likely conversion productions: synthetic gasoline and substitute natural gas. In addition to the possibility of producing methanol from coal, it has been suggested that methanol might be produced from the currently flared Middle East natural gas. These two potential sources could result in large quantities of methanol becoming available for fuel usage within the next few years.

Subject, its background and its value   One potential use for methanol is as a motor fuel, either in the "pure" form or as a gasoline blending component. The possible use of pure methanol is complicated by the lack of interchangeability between methanol and gasoline; the two fuels require very different carburetor settings. For vehicles used in captive fleet operations methanol has definite possibilities. However, this would only have a small impact on the overall consumption of motor fuel. Wide-spread use of pure methanol as a motor fuel is clearly a long-range proposition.

Scope of article   A detailed discussion of the use of pure methanol as a motor fuel is beyond the scope of this article; an evaluation of the use of methanol as an extender of gasoline is of greater interest because this use of methanol could have a greater impact near term. The

Tentative conclusions and their importance   advocates of this application for methanol point to better fuel economy, lower exhaust emissions, and better performance as compelling reasons for including methanol in motor gasoline at the earliest possible date. Fuel economy and lower emissions stand out as being particularly important factors for consideration since they are receiving a great deal of attention in today's climate of energy shortages and environmental awareness. While these claimed advantages for methanol use appear to be valid in certain cases, recent experiments conducted at this laboratory indicate that they are only significant for the older cars with richer carburetion which are rapidly disappearing from the roads. The main purpose of this article is to put fuel economy and emissions issues, as they relate to methanol-gasoline blends, into proper perspective. A secondary purpose is to draw attention to some product quality considerations that are associated with the use of methanol and have sometimes been overlooked.

Primary and secondary purposes of article

—E. E. Wigg, "Methanol as a Gasoline Extender: A Critique," *Science* 186 (November 1974), 785-790. Copyright © 1974 by the American Association for the Advancement of Science.

The following sample introduction comes from a report that presents a method for analyzing the efficiency with which cooling ponds can dissipate heat from nuclear power plants.

Subject and definition

The ultimate heat sink refers to the complex of water sources necessary to safely operate, shutdown, and cool a nuclear power plant. Cooling ponds, spray ponds, and mechanical draft cooling towers are examples of some types of ultimate heat sinks currently in use.

The U.S. Nuclear Regulatory Commission has set forth the following positions on the design of ultimate heat sinks. They must do the following:

1. be able to dissipate the heat of a design-basis accident (for example, a loss-of-coolant accident) of one unit plus the heat of a safe shutdown and cooldown of all other units it serves

2. provide a 30-day supply of cooling water at or below the design-basis temperature for all safety-related equipment

3. be capable of performing under the meteorologic conditions leading to the worst cooling performance and under the conditions leading to the highest water loss

Purpose and scope

This report identifies a procedure that may be used to select the most severe combinations of meteorologic parameters that control surface cooling pond heat transfer and evaporative water loss. In this procedure, a long weather record, usually available from the National Weather Service, is reviewed and used to predict the period for which either pond temperature or water loss would be maximized for a hydraulically simple cooling pond. The principle of linear superposition is assumed, which allows the peak ambient pond temperature to be superimposed on the peak "excess" temperature due to plant heat rejection. This procedure determines the timing within the weather record of the peak ambient pond temperature. The true peak can then be determined in a subsequent, more rigorous calculation.

Maximum evaporative water loss is determined by picking the 30-day continuous period of the record that has the highest evaporation losses and by assuming that all heat rejected by the plant results in the evaporation of pond water.

To be effective, the data-scanning procedure requires a data record on the order of tens of years in length. Since these data will usually come from somewhere other than the site (such as a nearby airport), methods to compare these data with the limited onsite data are developed so that the adequacy, or at least the conservatism, of the offsite data can be established. Conservative correction factors to be added to the final results are suggested.

(margin note) Method of development

WRITING INTRODUCTIONS FOR TECHNICAL MANUALS AND SPECIFICATIONS

In writing an introduction for a technical manual or specification, identify the topic and its primary purpose or function in the first sentence or two. Be specific, but do not get into details. Your introduction sets the stage for the entire document, and it is important that your reader get from the introduction a broad frame of reference. Only later, when your reader has some understanding of the overall topic and can appreciate its details in proper perspective, should you introduce technical details. Following is an example.

The System Constructor is a program that can be used to create operating systems for a specific range of microcomputer systems. The constructor selects requested operating software modules from an existing file of software modules and combines those modules with a previously compiled application program to create a functional operating system designed for a specific hardware configuration. . . .

The constructor selects the requested software modules, generates the necessary linkage between the modules, and generates the necessary control tables for the system according to parameters specified at run time. These parameters can either be input as online responses to sequentially presented display messages or prepared on punched cards and input to the constructor in a batch mode.

How technical your introduction should be depends on your readers: What is their technical background? What kind of information are they seeking in the manual or specification? A computer programmer, for example, has different interests in an application program or utility routine than does an operator—and the topic should be introduced accordingly. The assumptions you can make without providing explanations and the terminology you can use without providing definitions differ greatly from one reader to another.

Remember that this is an introduction to a topic with which your readers may be totally unfamiliar. Concepts and techniques that may be new to them need to be explained at a general level. Do not lose your readers by using terms and **acronyms and initialisms** that they cannot be expected to recognize. Some writers make a point of writing their introduction last, so that they are sure exactly which terms need defining.

You may encounter a dilemma that is very common in technical writing: Though you can't explain topic A until you have explained topic B, you can't explain topic B before explaining topic A. The solution is to explain both topics in broad, general terms in the introduction. Then, when you need to write a detailed explanation of topic A, you will be able to do so because your reader will know just enough about both topics to be able to understand your detailed explanation.

EXAMPLE   The NEAT/3 programming language, which treats all peripheral units as file storage units, allows your program to perform data input or output operations depending on the specific unit. Peripheral units from which your program can only input data are referred to as *source units*. Units to which your program can only output data are referred to as *destination units*. Units that permit both input and output may be used as source units, as destination units, or as combination *source-destination units*.

Thus, when the writer needs to explain any one of the three units in detail, readers will have at least a general knowledge of the topic.

TIPS ON WRITING INTRODUCTIONS

Not every subject needs an introduction. When your readers are already familiar with the major elements of your topic, you can save them time and hold their interest better by getting on to the details. If you do not need a full-scale introduction, turn to the entry on openings for a variety of interesting ways to begin without an introduction.

Introductions should be purposeful and concise. Wordy or rambling introductions usually mean that the subject has not been clearly focused. Adequate **preparation** will help solve this problem. The introduction should function like an abstract by giving readers a quick overview of the material, thus allowing them to decide whether they need to read the full text. A wordy introduction would defeat this purpose.

Make your **point of view** toward the subject obvious to the readers. Do not assume that readers knowledgeable about your topic will instinctively know your point of view toward it. Be brief, if that will suffice, but be explicit.

Finally, consider writing the introduction last. Many writers find it helpful to write the introduction last because they feel that only then do they have a full enough perspective on the writing to introduce it adequately.

inverted sentence order

An inverted **sentence** changes the normal **subject-verb-complement** sentence pattern. Inversion may occur for a variety of reasons.

Inverted sentence order is used in questions.

EXAMPLE   *Are* (verb) *you* (subject) *at the office* (object of preposition)?

It is used in exclamations.

EXAMPLE   What *a fool* (subjective complement) *he* (subject) *is* (verb)!

It is also used at times simply to achieve **emphasis** and variety.

EXAMPLES   A better *job* (direct object) *I* (subject) never *had* (verb).
A sorry *sight* (direct object) *we* (subject) *presented* (verb).

In general, however, most technical writing follows the subject-verb-complement word order. Before varying this order, be sure you have a good reason for doing so.

investigative reports

When a person investigates, or checks into, a particular topic, he or she might write up the results in an investigative report. For example, the report could summarize the findings of an opinion survey, of a product evaluation, or of a marketing study. It could review the published work on a particular topic or compare the different procedures used to perform the same operation. An investigative report gives a precise analysis of its topic and then offers the writer's conclusions and recommendations.

Open such a **report** with a statement of the purpose of the report. In the body of the report, first define the **scope** of your investigation. If the report is on a survey of opinions, for example, you might need to indicate the number of people and the geographical areas surveyed as well as the income categories, the occupations, the age groups, and possibly even the racial groups of those surveyed. Include any information that is pertinent in defining the extent of the investigation. Then report your findings and, if necessary, discuss their significance. End your report with your conclusions and, if appropriate, any recommendations.

Notice how the following sample investigative report opens by stating its purpose and ends with its recommendations.

MEMORANDUM

To: Noreen Rinaldo, Training Manager
From: Charles Lapinski, Senior Instructor
Date: February 14, 19--

Subject: The Addison Corporation's Basic
          English Program

      As you requested, I have investigated the subject program to determine whether we might also adopt such a program. The purpose of the Addison Basic English course is to teach foreign mechanics who do not speak or read English to understand repair manuals written in a special 800-word vocabulary called "Basic English," and thus eliminate the need for Addison to translate its manuals into a number of different languages. The Basic English Program does not attempt to teach the mechanics to be fluent in English but, rather, to recognize the 800 basic words that appear in the repair manuals.

      The course does not train mechanics. Students must know, in their own language, what a word like <u>torque</u> means; the course simply teaches them the English term for it. As prerequisites for the course, students must have a basic knowledge of their trade, must be able to identify a part in an illustrated parts book, must have served as a mechanic on Addison products for at least one year, and must be able to read and write in their own language.

Students are given the specially prepared
instruction manual, an illustrated book of parts
and their English names, and a pocket reference
containing all 800 words of the Basic English
vocabulary plus the English names of parts
(students can write the corresponding word in
their language beside the English words and then
use the pocket reference as a bilingual
dictionary). The course consists of thirty two-
hour lessons, each lesson introducing approximately 27 words. No effort is made to teach
pronunciation; the course teaches only recognition of the 800 words, which include 450 nouns,
70 verbs, 180 adjectives and adverbs, and 100
articles, prepositions, conjunctions, and
pronouns.

The 800-word vocabulary enables the writers
of the manuals to provide mechanics with any
information that might be required, because the
area of communication is strictly limited to
maintenance, inspection, troubleshooting, safety,
and the operation of Addison equipment. All
nonessential words (such as apple, father,
mountain, and so on) have been eliminated, as
have most synonyms (for example, under appears,
but beneath does not).

Conclusions

I see three possible ways in which we might be
able to use some or all of the elements of the
Basic English Program: (1) in the preparation
of all our student manuals, (2) in the preparation of student manuals for the international
students in our service school, or (3) as Addison
uses the program.

I think it would be unnecessary to use the
Basic English methods in the preparation of
student manuals for all our students. Most of
our students are English-speaking people to whom
an unrestricted vocabulary presents no problem.

In conjunction with the preparation of
student manuals for international students, the
program might have more appeal. Students would
take the Basic English course either before
coming to this country to attend school or after
arriving but before beginning their technical
training.

As for our initiating a Basic English
Program similar to Addison's, we could create
our own version of the Basic English vocabulary
and write our service manuals in it. Since our
product lines are much broader than Addison's,
however, we would need to create illustrated
parts books for each of the different product
lines.

irregardless/regardless

Irregardless is **nonstandard English** because it expresses a **double negative**. The word is a negative of *regardless*, which is itself already negative, meaning "unmindful." Always use *regardless*.

> CHANGE   *Irregardless* of the difficulties, we must increase the strength of the outer casing.
> TO   *Regardless* of the difficulties, we must increase the strength of the outer casing.

it

The **pronoun** *it* has a number of uses. First, it can refer to a preceding **noun** that names an object or idea or to a baby or animal whose sex is unknown or unimportant to the point. *It* should have a clear antecedent.

> EXAMPLES   Darwinism made an impact on nineteenth-century American thought. *It* even influenced economics.
> The Moodys' baby is generally healthy, but *it* has a cold at the moment.

It can also serve as an **expletive.**

> EXAMPLES   *It* is necessary to sand the hull before painting.
> *It* is a truth universally acknowledged that there is no such thing as a free lunch.

Be on guard against overusing expletives, however.

> CHANGE   *It* is seldom that we go.
> TO   We seldom go.

italics

Italics (indicated on the typewriter by underlining) is a style of type used to denote **emphasis** and to distinguish foreign expressions, book titles, and certain other elements. *This sentence is printed in italics.* You may need to italicize words that require special emphasis in a **sentence.**

> EXAMPLE   Contrary to projections, sales have *not* improved since we started the new procedure.

Do not overuse italics for emphasis, however.

> CHANGE   This will hurt *you* more than *me*.
> TO   This will hurt you more than me!

TITLES

Italicize the titles of books, periodicals, newspapers, movies, and paintings.

EXAMPLES   The book *Statistical Methods* was published in 1970.
The *Cincinnati Enquirer* is one of our oldest newspapers.
Chemical Engineering is a monthly journal.

Abbreviations of such titles are italicized if their spelled-out forms would be italicized.

EXAMPLE   The *Journal of the ABCA* is an informative publication.

Titles of chapters or articles within publications are placed in **quotation marks,** not italicized.

EXAMPLE   "Clarity, the Technical Writer's Tightrope" was an article in *Technical Communications.*

Titles of holy books and legislative documents are not italicized.

EXAMPLE   The Bible and the Magna Carta changed the history of Western civilization.

Titles of long poems and musical works are italicized, but titles of short poems and musical works and songs are enclosed in quotation marks.

EXAMPLES   Milton's *Paradise Lost* (long poem)
Handel's *Messiah* (long musical work)
T.S. Eliot's "The Love Song of J. Alfred Prufrock" (short poem)
Leonard Cohen's "Suzanne" (song)

PROPER NAMES

The names of ships, trains, and aircraft (but not the companies that own them) are italicized.

EXAMPLE   They sailed to Africa on the Onassis *Clipper* but flew back on the TWA *New Yorker.*

Craft that are known by model or serial designations are exceptions. These are not italicized.

EXAMPLES   DC-7, Boeing 747

WORDS, LETTERS, AND FIGURES

Words, letters, and figures discussed as such are italicized.

EXAMPLES   The word *inflammable* is often misinterpreted.
I should replace the *s* and the *6* keys on my old typewriter.

FOREIGN WORDS

Foreign words that have not been assimilated into the English language are italicized.

EXAMPLES   *sine que non, coup de grâce, in res, in camera*

Foreign words that have been fully assimilated into the language, however, need not be italicized.

EXAMPLES   cliché, etiquette, vis-à-vis, de facto, siesta

When in doubt about whether to italicize a foreign term, consult a current **dictionary**. (See also **foreign words in English**.)

SUBHEADS

Subheads in a report are sometimes italicized—on a typewriter, they are of course underlined. (See also **heads**.)

EXAMPLE   There was no publications department as such, and the writing groups were duplicated at each plant or location. Wellington, for example, had such a large number of publications groups that their publications effort can only be described as disorganized. Their duplication of effort must have been enormous.

Training Writers

We are certainly leading the way in developing first-line managers (or writing supervisors) who are not only technically competent but can also train the writers under their direction and be responsible for writing quality as well.

its/it's

Be careful never to confuse these two words: *Its* is a possessive **pronoun**, whereas *it's* is a **contraction** of *it is*.

EXAMPLE   *It's* important that the crew meet *its* schedule.

Although words normally form the possessive by the addition of an **apostrophe** and an *s*, the contraction of *it is (it's)* has already used that device; therefore, the possesive form of the pronoun *it* is formed by adding only the *s*.

J

jammed modifiers

Some writing is unclear or difficult to read because it contains jammed modifiers, or strings of **modifiers** preceding **nouns.**

CHANGE   Your *staffing level authorization reassessment* plan should result in a major improvement.

In this sentence the noun *plan* is preceded by four modifiers; this string of modifiers slows the **reader** down and makes the sentence awkward and clumsy. Jammed modifiers often result from an overuse of **jargon** or **vogue words.** Occasionally, they occur when writers mistakenly attempt to be concise by eliminating short **prepositions** or **connectives**—exactly the words that help to make sentences clear and readable. See how breaking up the jammed modifiers makes the previous example easier to read.

TO   Your plan for the reassessment of staffing-level authorizations should result in a major improvement.

OR   Your plan to reassess authorizations for staffing levels should result in a major improvement.

See also **conciseness, gobbledygook,** and **telegraphic style.**

jargon

Jargon is a highly specialized technical slang that is unique to an occupational group. If all your **readers** are members of a particular occupational group, jargon may provide a time-saving and efficient means of communicating with them. For example, finding and correcting the errors in a computer program is referred to by programmers as "debugging." If you doubt that your entire reading audience is a part of this group, however, avoid the use of jargon.

When jargon becomes so specialized that it applies only to one company or a subgroup of an occupation, it is referred to as "shop talk." Obviously, shop talk is appropriate only for those familiar with its special vocabulary and should be reserved for speech or informal **memorandums.**

When jargon becomes enmeshed in a tangle of **abstract words,** many of them pseudo-scientific or pseudo-legal, it becomes an **affectation** known as **gobbledygook.** Jargon, used to this extreme, is "language more technical than the ideas it serves to express."*

Jargon and shop talk are similar in many ways to "lingo" and "cant," which are terms used to describe language that is intended to exclude outsiders from a select circle of the initiated.

job descriptions

Most large companies and many small ones specify, in a formal job description, the duties of and requirements for many of the jobs in the firm. Job descriptions fill several important functions: They provide information on which equitable salary scales can be based; they help management determine whether all responsibilities within a company are adequately covered; and they let both prospective and current employees know exactly what is expected of them. Together, all of the job descriptions in a firm present a picture of the organization's structure.

Sometimes plant or office supervisors are given the task of writing the job descriptions of the employees assigned to them. In many organizations, though, an employee may draft his or her own job description, which the immediate superior then checks and approves.

A FORMAT FOR WRITING JOB DESCRIPTIONS

Although job description formats vary from organization to organization, they commonly contain the following sections.

Accountability. This section identifies, by title, the person or persons to whom the employee reports.

Scope of responsibilities. This section provides an overview of the primary and secondary functions of the job and states, if it is applicable, who reports to the employee.

Specific duties. This section gives a detailed account of the specific duties of the job, as concisely as possible.

Personal requirements. This section lists the education, training, experience, and licensing required or desired for the job.

TIPS FOR WRITING JOB DESCRIPTIONS

If you have been asked to prepare a job description for your position, the following guidelines should help.

* Suzanne K. Langer, *Mind: An Essay on Human Feeling* (Baltimore, Md.: The Johns Hopkins Press, 1967), p. 36.

1. Before attempting to write your job description, make a list of all the different tasks you do for a week or a month. Otherwise, you will almost certainly leave out some of your duties.
2. Focus on content. Remember that you are writing a description of your job, not of yourself.
3. List your duties in decreasing order of importance. Knowing how your various duties rank in importance makes it easier to set valid job qualifications.
4. Begin each statement of a duty with a **verb,** and be specific. Write "Answer and route incoming telephone calls" rather than "Handle telephone calls."
5. Review existing job descriptions that have been successful.

The following job description, which is a typical one, never mentions the person holding the job described; it focuses, instead, on the job and on the qualifications any person must possess to fill the position.

Manager, Technical Publications
Acme Electrical Corporation

ACCOUNTABILITY

Reports directly to the Vice President, Customer Service.

SCOPE OF RESPONSIBILITIES

The Manager of Technical Publications is expected to plan, coordinate, and supervise the design and development of technical publications and documentation required in the support of the sale, installation, and maintenance of Acme products. The manager is responsible for the administration and morale of the staff. The supervisor for instruction manuals and the supervisor for parts manuals report directly to the manager.

SPECIFIC DUTIES

Direct an organization presently composed of twenty people (including two supervisors), over 75 percent of whom are writing professionals and graphics specialists.
Screen, select, and hire qualified applicants for the department.
Prepare a formal program designed to orient writing trainees to the production of reproducible copy and graphic arts.
Evaluate the performance of departmental members and determine salary adjustments for all personnel in the department.
Plan documentation to support new and existing products.

Determine the need for subcontracted publications and act as a purchasing agent when they are needed.

Offer editorial advice to supervisors.

Develop and manage an annual budget for the Technical Publications Department.

Cooperate with the Engineering, Parts, and Service Departments to provide the necessary repair and spare parts manuals upon the introduction of new equipment.

Serve as a liaison between technical specialists, the publications staff, and field engineers.

Recommend new and appropriate uses for the department within the company.

Keep up with new techniques in printing processes, typesetting, computerized text editing, art, and graphics and use them to the advantage of Acme Electrical Corporation where applicable.

PERSONAL REQUIREMENTS

B.S.E.E. (or equivalent electrical background) desired.

Minimum of three years professional writing experience with a general knowledge of graphics and production.

Minimum of two years management experience with a knowledge of the general principles of management.

Must be conversant with the needs of support people, technical people, and customers.

For additional and detailed information on the use and creation of job descriptions, see: Joseph J. Famularo, *Organization Planning Manual,* New York: American Management Association, 1979.

job interviews

A job interview may last for thirty minutes, or it may take several hours; it may be conducted by one person or by several, either at one time or in a series of interviews. Since it is impossible to know exactly what to expect, it is very important to be as well prepared as possible.

BEFORE THE INTERVIEW

Learn as much as you can about the potential employer before the interview. What kind of business is it? Is the company locally owned? Is it a nonprofit organization? If the job is public employment, at what level of government is it? Does the business provide a service, and, if so, what kind? How large is the firm? Is the owner self-employed? Is it a subsidiary of a larger operation? Is it expanding? Where will you fit in? This kind of information can be obtained from current employees, from company publications, or from back issues of the local newspaper's business section (available at the public library). You should try to learn about the company's size, sales volume, products and new products, credit rating, and subsidiary companies from its **annual reports,** and from other business reference sources such as *Moody's Industrials, Dunn and Bradstreet, Standard and Poor's,* and *Thomas Register.* Such preparation will help you to speak knowledgeably about the firm or industry as well as to ask intelligent questions of the interviewer.

It is a good idea to try to anticipate the questions an interviewer might ask and to prepare your answers in advance. The following list includes some typical questions posed in job interviews.

What are your short-term and long-term occupational goals?
What are your major strengths and weaknesses?
Do you work better alone or with others?
Why do you want to work for this company?
How do you spend your free time?

DURING THE INTERVIEW

Promptness is very important. Be sure to arrive for an interview at the appointed time. It is usually a good idea, in fact, to arrive ahead of time, since you may be asked to fill out an application before meeting the interviewer. Take along your **résumé,** even if the company already has a copy. For one thing, the interviewer may want another copy; for another, the résumé contains most of the information required on an application.

Remember that the interview will actually begin before you are seated. What you wear and how you act will be closely observed. The way you dress matters: It is usually best to dress conservatively and to be especially well groomed.

Remain standing until you are offered a seat. Then sit up straight— good posture suggests self-assurance—and look at the interviewer, trying to appear relaxed and confident. It is a natural reaction to be nervous during an interview, but be careful to remain alert. Listen carefully and make an effort to remember especially important information. (See **listening.**) Do not attempt to take extensive notes during an interview, although it is acceptable to jot down a few facts and figures.

When answering questions, don't ramble or stray from the subject. Say only what you must in order to answer each question and then stop; however, avoid giving just yes and no answers, which usually don't permit the interviewer to learn enough about you. Some interviewers allow a silence to fall just to see how you will react. The burden of conducting the interview is the interviewer's, not yours—and he or she may interpret it as a sign of insecurity if you rush in to fill a void in the conversation. If such a silence would make you uncomfortable, be ready to ask an intelligent question about the company.

Highlight your qualifications for the job but admit obvious limitations as well. Remember also that the job, the company, and the location must be right for you. Ask about such factors as opportunity for advancement, fringe benefits (but don't create the impression that your primary interest is security), educational opportunity and assistance, and community recreational and cultural activities (if the job would require you to relocate).

If the interviewer overlooks important points, bring them up. But, if possible, let the interviewer mention salary first. If you are forced to bring up the subject, ask it as a straightforward question. Knowing the prevailing salaries in your field will make you better prepared to discuss salary. It is usually unwise to bargain, especially if you are a recent graduate. Many firms have inflexible starting salaries for beginners.

Interviewers look for self-confidence and an understanding on the part of the candidate of the field in which he or she is applying. Less is expected of a beginner, but even a newcomer must show some knowledge of the field. One way to impress your interviewer is to ask questions about the company that are related to your line of work. Interviewers respond favorably to people who can communicate easily and present themselves well. Jobs today require interactions of all kinds: person-to-person, department-to-department, division-to-division.

At the conclusion of the interview, thank your interviewer for his or her time. Indicate that you are interested in the job (if true), and try tactfully to get an idea of when you can expect to hear from the company.

AFTER THE INTERVIEW

When you leave, jot down pertinent information you learned during the interview. (This information will be especially helpful in comparing job offers.) A day or two later, send the interviewer a brief note of thanks, saying that you find the job attractive and feel you can fill it well. Letter 15 is typical.

See also **job search, résumés, application letters,** and **acceptance letters.**

```
                                  2647 Sitwell Road
                                  Charlotte, NC  28210
                                  March 17, 19--


Mr. F. E. Vallone
Personnel Manager
Calcutex Industries, Inc.
3275 Commercial Park Drive
Raleigh, NC  27609

Dear Mr. Vallone:

I want to express my appreciation for the informative and
pleasant interview we had last Wednesday.  Please extend my
thanks to Mr. Wilson of the Servocontrol Group as well.  I
came away from our meeting most favorably impressed with
Calcutex Industries.  I find the position to be an
attractive one and feel confident that my qualifications
would enable me to perform the duties to everyone's
advantage.

                              Sincerely yours,

                              Philip Ming
                              Philip Ming
```

Letter 15

job search

Before beginning to search for a job, you should always do some serious thinking about your future. Decide first what you would most like to be doing in the immediate future. Then think about the kind of work you'd like to be doing two years from now *and* five years from now. Once you have established your goals, you can begin your job hunt with greater confidence. You will have a better idea of what kind of job you want and what companies or organizations are most likely to have such a job.

There are a number of ways to find a job: classified ads, letters of inquiry, trade and professional journals, school placement services, employment agencies, and other sources.

CLASSIFIED ADS

Many employers advertise in the classified sections of newspapers. For the widest selection, look in the Sunday editions of local and big-city newspapers. Although reading want ads can be tedious, an item-by-item check is necessary if you are to do a thorough job search. The job you are looking for might be listed in the classified ads under any number of titles. A clinical medical technologist seeking a job, for example, might find the specialty listed under "Medical Technician," "Clinical Medical Technician," or "Laboratory Technician." Depending on a hospital's or a pathologist's needs, the listing could be more specific yet, such as "Blood Bank Technician," "Hematology Technician," or "Clinical Chemistry Technician." So play it safe—read *all* the ads. Occasionally, newspapers print special employment supplements that provide valuable information on many facets of the job market. Watch for these.

As you read the ads, take notes (see **note-taking**) on such factors as salary ranges, job locations, job duties and responsibilities, and even the terminology used in the ads. (A knowledge of the words and expressions that are generally used to describe a particular type of work can be helpful when you prepare your **résumé** and **application letter.**)

INQUIRY LETTERS

If you would like to work for a particular firm, write and ask if it has an opening for someone with your qualifications. Normally, you should send the letter either to the director of personnel or to the specific department head; for a small firm, however, you can write to the head of the firm. Your letter should present a general summary of your employment background or training. It should be brief and to the point, expressing interest in such a job but leaving everything else to your résumé, which you enclose with the letter.

TRADE AND PROFESSIONAL JOURNALS

Many occupations have associations that publish periodicals of general interest to members. Such periodicals often contain a listing of current job opportunities. If you were seeking a job in forestry, for example, you could check the job listings in the *Journal of Forestry*, published by the Society of American Foresters. To learn about the trade or professional associations for your occupation, consult the following **reference books.**

Encyclopedia of Associations
Encyclopedia of Business Information Sources
National Directory of Employment Services

SCHOOL PLACEMENT SERVICES

Check with the career counselors in your school's job-placement office. Government, business, and industry recruiters often visit job-placement offices to interview prospective employees; the recruiters also keep college placement offices aware of their company's current employment needs. While you are in the placement office, ask to see a current issue of the *College Placement Annual*. This publication lists the occupational requirements and addresses of over a thousand industry, business, and government employers.

STATE EMPLOYMENT AGENCIES

Most states operate free employment agencies that are in business specifically to match applicants and jobs. If your state has one, register with the local employment office. It may have just the job you want; if not, it will keep your résumé on file and call you if such a job comes along.

PRIVATE EMPLOYMENT AGENCIES

Private employment agencies are profit-making organizations that are in business to help people find jobs—for a fee. Choose a private employment agency carefully. Some are well established and quite reputable, but others have questionable reputations. Check with your local Better Business Bureau as well as with friends and acquaintances before signing an agreement with a private employment agency.

Reputable private employment agencies provide you with job leads and help you organize your campaign for the job you want. They may also provide useful information on the companies doing the hiring.

Who pays the fee if you are offered and accept a job through a private agency? Sometimes the employer will pay the agency's fee. Otherwise, you must pay either a set fee or a percentage of your first month's salary. Before signing a contract, be sure you understand who is paying the fee and, if *you* are, how much you are agreeing to pay. As with any written agreement, read the fine print carefully.

OTHER SOURCES

When searching for a job, consult with people you know whose judgment you respect. It is especially useful to speak to people who are already working in your chosen field.

Local, state, and federal government agencies offer many employment opportunities. Local government agencies are listed in the white pages of your telephone directory under the name of your city, county, or state. For information about jobs with the federal government, contact the U.S. Office of Personnel Management or the Federal Job Information Center. Both have branches in most major cities. The offices are listed in the white pages of the telephone book under "U.S. Government."

If you are a veteran, local and campus Veterans' Administration offices can provide material on special placement programs for veterans. Such agencies will supply you with the necessary information about the particular requirements or entrance tests for your occupation.

Once you determine where you might like to work, you will have to prepare an effective résumé and application letter. Then, with luck, you will also need to write a gracious **acceptance letter.** (See also **job interviews.**)

journal articles

From time to time in your career you may wish to write a journal article. Such an article may, in fact, improve your chances for professional advancement; at the very least, publication of a journal article may get your name known among leaders in your field. In addition, a journal article can provide favorable publicity for your firm.

A journal article is an article written on a specific subject for a professional periodical. These periodicals are often the official publications of professional societies. *Technical Communication,* for example, is the official voice of the Society for Technical Communication. Other professional publications include *Electrical Engineering Review, Chemical Engineering, Journal of Marketing,* and *Business Education Forum.* There are hundreds of periodicals, of course, ranging from scientific journals that publish articles on advanced research in highly specialized areas to industrial and business periodicals that publish material of more general interest.

Although the primary effort in writing a journal article is in the **research** phase, all the other steps of the writing process—**preparation, organization, writing the draft,** and **revision**—are crucial, and you should review the entries for these steps in this handbook.

Before you begin writing, learn something about the periodical to which you will be submitting the article. Remember that, although the article is aimed at specialists in your field, your audience may also include specialists in related fields who may have little in-depth knowledge of the basics of your discipline. Read back issues of the periodical for such things as the amount and kind of detail in the articles, the length of the articles, and the **style** of the writing. Some publications may indicate the desired **format** for a manuscript that is being submitted; some may even send a style sheet (a guide to manuscript format, **footnotes,** and **bibliography**) upon request.

After you have written the article, the last (but hardly least important) item you must consider is its **title.** The title is often critical to the researcher, who may decide whether to read your article on the basis of its title. The title, therefore, must be specific—naming the main topic in as few words as possible. In addition, many editors like a title that arouses interest in the article by suggesting that it will help the reader discover something new or valuable.

The following excerpt is from an article for a professional journal addressed mainly, but not exclusively, to chemical engineers.

Profile of a Technical Innovator*

Recent research has confirmed that a majority of technical innovations are due mainly to the work of a minority of professional employees—the innovators.

How does one recognize a technical innovator? What personality traits is he or she likely to have? Do the work habits of innovators differ from those of less-creative employees? And

*Keller, Robert T. "Profile of a Technical Innovator," *Chemical Engineering* 5 (1980), 155-158.

what kind of workplace atmosphere is likely to encourage the best performance from such people?

BEYOND CREATIVITY

Efforts to resolve these questions have lately focused on how engineers and scientists communicate and make use of information. Recent findings (see table) indicate that the role of innovators goes far beyond their creative activities.

Often, these key professionals will also serve as central nodes in a technical organization's communication network. Their presence is vital to the flow of scientific and technological information. . . .

RECOGNIZING THE INNOVATOR

The personality traits of the technical innovator have been examined in a number of R&D environments. These studies show that the innovator is predisposed toward innovation in that he or she tries to "do things differently." In contrast, non-innovators exhibit an adaptive orientation whereby they try to "do things better. . . ."

MANAGING THE INNOVATOR

Which kinds of tasks or projects is a technical innovator best equipped for? Which ones is he or she not suited for? What mix of innovators and others is desirable? These questions are crucial to the effective management of innovators in a technical organization, and determine these persons' contributions to the overall performance of the organization.

The innovator tends to do well with a task or project that is flexible, that affords the opportunity to make an impact on important matters, and that does not tend to stretch out its time demands.

PROBLEM AREAS

There are certain environments, however, in which an innovator does not tend to perform well. These generally are restrictive and inflexible in nature, and include projects or tasks that must be completed by well-established techniques or procedures. The innovator often does not do well because he or she is not inclined to seek a routine solution, nor is there a motivating challenge. Furthermore, other professionals often do not listen to the innovator in these circumstances. This spells frustration and rejection for the innovator. . . .

SECURITY VS. CHALLENGE

Do innovators as well as other technical professionals perform better when they feel secure? Or do they do better when faced

Recent Findings About the Innovator's Role

Study	Sample	Major findings
Allen [1]	Over 1,100 R&D professionals from 13 laboratories and 33 project teams	A small number of key professionals emerge as prime sources of information in the laboratory, via a "two-step flow" model. These key professionals are high performers and often have supervisory duties. They have high competence, work well with others, and are sought out for important projects.
Keller and Holland [2,3]	256 professionals from three applied R&D organizations	A majority of peer nominations for innovativeness are received by a minority of the R&D professionals. These innovators are also important sources of scientific and technological information for others and are high performers. They are oriented toward "doing things differently" and have a low need for clarity.
Pelz and Andrews [4]	Over 1,300 R&D professionals from 11 laboratories	High performers communicate more with others. Their creativity enhances the success of projects that foster free communication, but impairs the success of less-flexible projects. Self-reliance is a hallmark of high performers. Achievement is highest when creative tensions are induced by the presence of both security and challenge.
Tushman [5,6]	345 R&D professionals from one industrial organization, having 58 project teams	Key professionals, often supervisors, serve as central communication nodes. Project success requires a match between types of communicators and communication patterns, and the complexity and nature of the project tasks.

with a challenging situation? In fact, technical innovators as well as other professional employees perform better when both security and challenge are present. In effect, a kind of "creative tension" seems to produce the best results. . . .

With the direction of the economy moving more and more toward high-technology products and services, the importance of the technical innovator will inevitably grow. It therefore becomes incumbent on management to recognize the innovator's role and to establish an organizational climate conducive to his or her productivity.

REFERENCES

1. Allen, T. J., *Managing the Flow of Technology: Technology Transfer and the Dissemination of Technological Information Within the Research and Development Organization*, MIT Press, Cambridge, Mass., 1977.
2. Keller, R. T., and Holland, W. E., "Individual Characteristics of Innovativeness and Communication in Research and Development Organizations," *J. of Applied Psychol.*, Vol. 63, Dec. 1978, pp. 759-762.
3. Keller, R. T., and Holland, W. E., "Toward a Selection Battery for Research & Development Professional Employees," *IEEE Transactions on Engineering Management*, Vol. EM-26, Nov. 1979, pp. 90-93.
4. Pelz, D. C., and Andrews, Frank M., *Scientists in Organizations: Productive Climates for Research and Development*, Revised ed., Institute for Social Research, U. of Michigan, Ann Arbor, Mich., 1976.
5. Tushman, M. L., "Special Boundary Roles in the Innovation Process," *Administrative Science Quarterly*, Vol. 22, Dec. 1977, pp. 587-605.
6. Tushman, M. L., "Technical Communications in R&D Laboratories: The Impact of Project Work Characteristics," *Academy of Management Journal*, Vol. 21, Dec. 1978, pp. 624-645.

judicial/judicious

Judicial is a term that pertains only to law.

EXAMPLE   The *judicial* branch is one of the three branches of the United States government.

Judicious refers to careful or wise judgment.

EXAMPLE   We intend to convert to the new refining process by a series of *judicious* steps.

K

kind of/sort of

Kind of and *sort of* should be used in writing only to refer to a class or type of thing.

> EXAMPLE   They used a special *kind of* metal in the process.

Do not use *kind of* or *sort of* to mean "rather," "somewhat," or "somehow."

> CHANGE   It was *kind of* a bad year for the firm.
> TO   It was a bad year for the firm.

know-how

A colloquial term for "special competence or knowledge," *know-how* should be avoided in **formal writing.**

> CHANGE   He has great technical *know-how.*
> TO   He has great technical *skill.*

(See also **English, varieties of.**)

L

laboratory reports

A laboratory report communicates information acquired from a laboratory test or investigation. (Simpler tests use the less formal **test report** form.) Laboratory reports should state the reason the laboratory investigation was conducted, the equipment and procedures used, any problems encountered, any conclusions reached, and any recommendations based on the conclusions.

A laboratory report often places special **emphasis** on the equipment and the procedures used in the investigation because these two factors can be critical in determining the accuracy of the **data** obtained.

Present the results of the laboratory investigation clearly and concisely. Write complete **sentences,** however; **sentence fragments** are not usually sufficient for clear communication. (Use the Checklist of the Writing Process to write and revise your **report.**) Read the entries on **graphs** and **tables** if your **report** requires graphic or tabular presentation of data.

Each laboratory usually establishes the **organization** of its laboratory reports. The following is a typical example of a laboratory report.

PCB Exposure from Oil Combustion
Wayne County Professional Fire Fighters

Submitted to:
Mr. Philip Landowe
President, Wayne County Professional Fire-Fighters Association
Wandell, IN 45602

Submitted by:
Analytical Laboratories, Incorporated
Mr. Arnold Thomas
Certified Industrial Hygienist
Mr. Gary Seabolm
Laboratory Manager
Environmental Analytical Services
1220 Pfeiffer Parkway
Indianapolis, IN 46223

February 28, 19—

INTRODUCTION

Waste oil used to train fire fighters was suspected of containing polychlorinated biphenyls (PCB). According to information provided by Mr. Philip Landowe, President of the Wayne County Professional Fire-Fighters Association, it has been standard practice in training fire fighters to burn 20-100 gallons of oil in a diked area of approximately 25-50M^3. Fire fighters would then extinguish the fire at close range. Exposure would last several minutes, and the exercise would be repeated two or three times each day for one week.

Oil samples were collected from three holding tanks near the training area in Englewood Park on November 11, 19—. To determine potential fire-fighter exposure to PCB, bulk oil analyses were conducted on each of the samples. In addition, the oil was heated and burned to determine the degree to which PCB is volatized from the oil, thus increasing the potential for fire-fighter exposure via inhalation.

TESTING PROCEDURES

Bulk oil samples were diluted with hexane, put through a clean-up step, and analyzed in electron-capture gas chromatography. The oil from the underground tank that contained PCB was then exposed to temperatures of 100° C without ignition and 200° C with ignition. Air was passed over the enclosed sample during heating, and volatized PCB was trapped in an absorbing medium. The absorbing medium was then extracted and analyzed for PCB released from the sample.

RESULTS

Bulk oil analyses are presented in Table 1. Only the sample from the underground tank contained detectable amounts of PCB. Aroclor 1260, containing 60% chlorine, was found to be present in this sample at 18 μg/g. Concentrations of 50 μg/g PCB in oil are considered hazardous. Stringent storage and disposal techniques are required for oil with PCB concentrations at these levels.

Results for the PCB volatization study are presented in Table 2. At 100° C, 1 μg PCB from a total of 18 μg PCB (5.6%) was released to the air. Lower levels were released at 200° C and during ignition, probably as a result of decomposition. PCB is a mixture of chlorinated compounds varying in molecular weight; lightweight PCBs were released at all temperatures to greater degree than the high molecular weight fractions.

TABLE 1. Bulk Oil Analysis

Source	Sample #	PCB Content (μg/g)
Underground Tank (11' deep)	#6062	18*
Circle Tank (3' deep)	6063a 6063b	<1 <1
Square Pool (3' deep)	6064a 6064b	<1 <1

*Aroclor 1260 is the PCB type. This sample was taken for volatilization study.

TABLE 2. PCB Volatilization Study
for the 11-Foot Deep Underground Tank*

Outgassing Temp. (°C)	Outgasing Time (Min)	Sample Outgassed (g)	PCB Total (μg)	PCB Outgassed (μg)
100	30	1	18	1
200	30	1	18	0.6
200 with ignition	30	1	18	0.2

*Bulk analysis of 18 μg/g PCB

DISCUSSION AND CONCLUSIONS

At a concentration of 18 μg/g, 100 gallons of oil would contain approximately 5.5 grams of PCB. Of the 5.5 grams of PCB, about 0.3 grams would be released to the atmosphere under the worst conditions.

The American Conference of Governmental Industrial Hygienists has established a TLV* of 0.5 μg/M^3 air for a PCB containing 54% Cl as a time-weighted average over an 8-hour workshift and has stipulated that exposure over a 15-minute period should not exceed 1 mg/M^3. The 0.3 gram of released PCB would have to be diluted to 600 M^3 air to result in a concentration of 0.5 mg/M^3 or less. Since the combustion of oil

*The Threshold Limit Value (TLV) is the safe average concentration that most individuals can be exposed to in an 8-hour day.

lasted several minutes, a dilution to more than 600 M³ is likely; thus exposure would be less than 0.5 mg/M³. Since an important factor in determining exposure is time and the fire fighters were exposed only for several minutes at intermittent intervals, adverse effects from long-term exposure to low-level concentrations of PCB should not be expected.

It should be stressed, however, that the above conclusions are based solely on oil containing 18 μg/g PCB. If, on previous occasions, the PCB content of the oil was much higher, greater exposure could have occurred. PCB is a known liver toxin and has also been classified as a suspected carcinogen. Although the primary route of entry into the body is by inhalation, PCB can be absorbed through the skin. PCB can cause a skin condition known as chloracne which results from a clogging of the pores. This condition, which is often associated with a secondary infection, should not occur at the PCB concentration found in this oil. A clinical test exists for determining if PCB has been absorbed by the liver.

In summary, because exposure to this oil was limited and because PCB concentrations in the oil were low, it is unlikely that exposure from inhalation would be sufficient to cause adverse health effects. However, we cannot rule out the possibility that excessive exposure may have occurred under certain circumstances, based on factors such as excessive skin contact and the possibility that higher-level PCB concentrations in the oil could have been used earlier. The practice of using this oil should be terminated.

lay/lie

Lay is a **transitive verb** (a **verb** that requires a **direct object** to complete its meaning) and means "place" or "put." Its present **tense** form is *lay.*

EXAMPLE   We *lay* the foundation of the building one section at a time.

The past tense form of *lay* is *laid.*

EXAMPLE   We *laid* the first section of the foundation on the 27th of June.

The perfect tense form of *lay* is also *laid.*

EXAMPLE   Since June we *have laid* all but two sections of the foundation.

Lay is frequently confused with *lie,* which is an **intransitive verb** (a verb that does not require an **object** to complete its meaning) meaning "recline" or "remain." Its present tense form is *lie.*

> EXAMPLE   Injured employees should *lie* down and remain still until the doctor arrives.

The past tense form of *lie* is *lay.* (This form causes the confusion between *lie* and *lay.*)

> EXAMPLE   The injured employee *lay* still for approximately five minutes.

The perfect tense form of *lie* is *lain.*

> EXAMPLE   The injured employee *had lain* still for approximately five minutes when the doctor arrived.

leave/let

As a **verb,** *leave* should never be used in the sense of "allow" or "permit."

> CHANGE   *Leave* me do it my way.
> TO   *Let* me do it my way.

As a **noun,** however, *leave* can mean "permission granted."

> EXAMPLE   Employees are granted a *leave* of absence if they have a chronic illness.

lend/loan

Lend or *loan* may each be used as a **verb,** but *lend* is more common.

> EXAMPLES   You can *loan* them the money if you wish.
> You can *lend* them the money if you wish.

Unlike *lend, loan* can also be a **noun.**

> EXAMPLE   We made arrangements at the bank for a *loan.*

letters

See **correspondence.**

libel/liable/likely

The term *libel* refers to "anything circulated in writing or pictures that injures someone's good reputation." When someone's reputation is injured in speech, however, the term is *slander.*

EXAMPLES   When an editorial charged our board of directors with bribing a representative, the board sued the newspaper for *libel.*
If the mayor supports the bribery charge in his speech tonight, we will accuse him of *slander.*

The term *liable* means "legally subject to" or "responsible for."

EXAMPLE   Employers are held *liable* for their employees' decisions.

In technical writing, *liable* should retain its legal meaning. Where a condition of probability is intended, use *likely.*

CHANGE   Rita is *liable* to be promoted.
TO   Rita is *likely* to be promoted.

library classification systems

Libraries classify and arrange books and journals by call numbers. Call numbers are normally based on the Dewey Decimal System or the Library of Congress System. These systems can help you in any library research by leading you to the place in the library where your topic can be found. If you were looking for books about technology, for example, you could go directly to the 600-699 shelves or the T shelves, depending on which system the library used.

The Dewey Decimal System divides all books and journals by sets of numbers into the following ten general subject categories.

000–099 General Works
100–199 Philosophy
200–299 Religion
300–399 Social Sciences
400–499 Language
500–599 Pure Sciences
600–699 Technology (Applied Sciences)
700–799 Fine Arts
800–899 Literature
900–999 History

Each of these general categories is in turn divided into ten smaller categories. For example, the Technology classification (600-699) is broken down in the following way.

600–609 Technology
610–619 Medical Sciences
620–629 Engineering
630–639 Agriculture
640–649 Home Economics
650–659 Business and Business Methods
660–669 Chemical Technology and Industrial Chemistry
670–679 Manufacturing (metal, textile, paper, etc.)
680–689 Miscellaneous Manufacturers (hardware, furniture, etc.)
690–699 Building Construction

The Library of Congress System is divided by letters into the following major subject categories.

A	General Works
B	Philosophy and Religion
C	History and Auxiliary Sciences
D	Universal History and Topography
E/F	American History
G	Geography, Anthropology, Folklore, etc.
H	Social Sciences
J	Political Science
K	Law
L	Education
M	Music
N	Fine Arts
P	Language and Literature
Q	Science
R	Medicine
S	Agriculture
T	Technology
U	Military Science
V	Naval Science
Z	Library Science and Bibliography

Technology and related topics occur under the Technology category (T) as follows.

T	Technology (General)
TA	Engineering (General). Civil engineering (General)
TC	Hydraulic engineering
TD	Environmental technology. Sanitary engineering
TE	Highway engineering. Roads and pavements
TF	Railroad engineering and operation
TG	Bridge engineering
TH	Building construction
TJ	Mechanical engineering and machinery
TK	Electrical engineering. Electronics. Nuclear engineering
TL	Motor vehicles. Aeronautics. Astronautics
TN	Mining engineering. Metallurgy
TP	Chemical technology
TR	Photography
TS	Manufactures
TS	Packaging
TT	Handicrafts. Arts and crafts
TX	Home economics

library research

The key tools for doing library research include the card catalog, the periodical indexes, the bibliographies, the computer-search services, and the reference books. For help in using any of these as well as in using the library in general, ask a librarian.

THE CARD CATALOG

The card catalog contains a listing of each book a library owns. It is usually found in small cabinets near the reference or circulation section, though some larger libraries keep their card catalogs in bound books or on computer printouts. Go to the card catalog to determine whether the library has the book you need and where to find it.

Most books have three types of cards in the catalog—an *author card* and a *title card* (found in the Author/Title Index) and a *subject card* (found in the Subject Index). Cards are arranged alphabetically: Author cards are filed by the author's last name; title cards are filed by the first important word in the title (excluding *a, an,* and *the*); and subject cards are filed by the first word of the general subject. All three cards include the author; the title; the publisher; the date and place of publication; the number of pages; and the major **topics** covered. They indicate if a book has an **index,** a **bibliography,** or any **illustrations.** At the bottom are subject headings that lead you to more materials on the same topic.

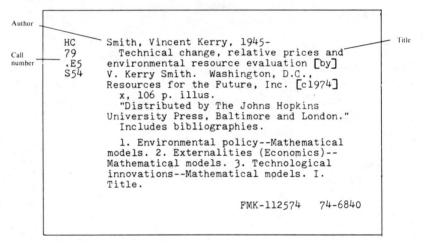

Author

Call
number

Title

```
          HC      Smith, Vincent Kerry, 1945-
          79          Technical change, relative prices and
          .E5     environmental resource evaluation [by]
          S54     V. Kerry Smith.  Washington, D.C.,
                  Resources for the Future, Inc. [c1974]
                      x, 106 p. illus.
                  "Distributed by The Johns Hopkins
                  University Press, Baltimore and London."
                  Includes bibliographies.

                  1. Environmental policy--Mathematical
                  models. 2. Externalities (Economics)--
                  Mathematical models. 3. Technological
                  innovations--Mathematical models. I.
                  Title.

                                  FMK-112574    74-6840
```

Author Card

Title

Number
of pages

```
                  Technical change, relative prices, and
                      environmental resource evaluation
          HC      Smith, Vincent Kerry, 1945-
          79          Technical change, relative prices, and
          .E5     environmental resource evaluation [by]
          S54     V. Kerry Smith. -- [Washington]
                  Resources for the Future; distributed by
                  Johns Hopkins University Press,
                  Baltimore [1974]
                      x, 106 p. illus. 24 cm.
                  Includes bibliographies.

                  1. Environmental policy--Mathematical
                  models. 2. Externalities (Economics)--
                  Mathematical models. 3. Technological
                  innovations--Mathematical models. I.
                  Title.

                  0163231  UM  C               74-6840
```

Title Card

Subject
categories

```
         ENVIRONMENTAL POLICY--MATHEMATICAL
            MODELS.
HC       Smith, Vincent Kerry, 1945-
79          Technical change, relative prices, and
.E5      environmental resource evaluation [by]
S54      V. Kerry Smith. -- [Washington]
         Resources for the Future; distributed by
         Johns Hopkins University Press,
         Baltimore [1974]
            x, 106 p. illus. 24 cm.
            Includes bibliographies.

            1. Environmental policy--Mathematical
         models. 2. Externalities (Economics)--
         Mathematical models. 3. Technological
         innovations--Mathematical models. I.
         Title.

         0163231  UM  C                    74-6840
```

Subjects
covered
and other
subjects in
catalog with
related
information

Subject Card

Each card also includes, in the upper left-hand corner, the *call number,* which indicates where the book is kept in the library. The call number is also written on the book's spine, which enables you to find the book on the shelf. Call numbers are actually the book's classification according to the Dewey Decimal System, the Library of Congress System, or the library's own system. (For details of the first two systems, see **library classification systems.**)

A card catalog also contains *see* cards and *see also* cards. *See* cards indicate that the heading you are looking under is not the one used for your topic and refer you instead to another heading. For example, if you looked up *instrumentation,* a *see* card might refer you to *control systems. See also* cards list additional headings related to the topic. Under *mining,* for example, a *see also* card might refer you to *metallurgy.*

BIBLIOGRAPHIES AND PERIODICAL INDEXES

Bibliographies and periodical indexes are lists of books and articles. Bibliographies list books, periodicals, and other research materials published in a particular subject area—for example, in business or engineering. Periodical indexes list journal, magazine, and newspaper articles. (*Periodicals* are publications that are issued at regular intervals— daily, weekly, monthly, and so on.) For a list of some useful bibliographies and periodical indexes, see **reference books.**

Instructions for using bibliographies and periodical indexes are found in the front of each volume (or the first volume of a series). There you will find a key to the **abbreviations** and **symbols** used as well as an explanation of the way that information is arranged in the book. You will also find a list of the specific topics covered and, for periodical indexes, a list of which newspapers, magazines, and journals are included. Following are some indexes that are useful for technical subjects.

Applied Science and Technology Index, 1913-- (monthly)
Monthly Catalog of U.S. Government Publications, 1895--
New York Times Index, 1851-- (bimonthly)

Libraries generally keep periodicals in a separate room or area. Current issues are often in one area, back issues (which are bound in volumes) in another. Back issues of some journals and most newspapers are generally in microform. If your library does not have the periodicals you need, you can get them through interlibrary loan, a service that permits you to borrow books and buy photocopied periodical articles from other libraries. Consult your librarian for specific details of the system.

COMPUTER-SEARCH AND INDEXES

Many academic, industrial, and medical libraries now offer computer-search services. Indexes are electronically stored as data bases in a computer. Librarians can then quickly find, or retrieve, information from the indexes. This computer-search service is an automated method of preparing a bibliography; it provides you with a computer printout of a list of citations, some with **abstracts,** appropriate to your subject. Data bases are available in medicine, business, psychology, biology, management, engineering, environmental studies, and many other subjects. Ordinarily there is a charge for this service, but the cost is relatively low when the speed and convenience of the search are taken into account.

Computer-search services are usually located in the reference section of the library. (Searches are almost always conducted by reference librarians skilled in searching computer data bases.) At some large universities, these services are located in satellite libraries. The reference staff can provide information about the range of subjects in the data base, the types and cost of search services, and procedures for using the service.

It is generally necessary to set up an appointment with a reference librarian to arrange for the search. Some libraries provide a form to fill out in advance that asks for a brief statement of the research topic and for any pertinent information necessary or helpful for conducting the search.

After completing the search, the librarian generally meets with the researcher to go over the printout. He or she can evaluate the citations for their relevance to your project and can also assist you in finding them. Some libraries provide with the printout a brief guide that indicates where books, periodicals, and microform sources on the citation list are located.

REFERENCE WORKS

Additional sources of helpful information include encyclopedias, specialized dictionaries, handbooks and manuals, statistical sources, and atlases.

Encyclopedias.   Alphabetically arranged collections of articles, encyclopedias are often illustrated and usually published in multivolume sets. Some encyclopedias cover a wide range of general subjects, while others specialize in a particular subject. General encyclopedias provide the kind of overview that can be helpful to someone new to a particular subject. The articles in general encyclopedias are useful sources of background information and can be especially helpful in defining the terminology essential for understanding the subject. Some articles contain bibliographies, which list additional sources. Three major general encyclopedias are the following.

Encyclopedia Americana (30 volumes)
Encyclopaedia Britannica (24 volumes)
New Columbia Encyclopedia (1 volume)

Specialized encyclopedias provide detailed information on a particular field of knowledge. Their treatment of a subject is rather thorough, so the researcher ought to have some background information on the subject in order to use the encyclopedia's information to full advantage. Subject encyclopedias contain comprehensive bibliographies. There are specialized encyclopedias on many subjects; the following two are especially useful to technical writers.

Encyclopedia of Chemical Technology
McGraw-Hill Encyclopedia of Science and Technology

Dictionaries.   Alphabetical arrangements of words with information about their forms, pronunciations, origins, meanings, and uses, dictionaries are essential to all writers. For a list of major abridged and unabridged dictionaries, look in the **dictionary** entry of this book. Specialized dictionaries define those terms used in a particular field—for example, technology, architecture, geology. Their definitions are generally more complete and are more likely to be current. Following are three specialized dictionaries that prove helpful to some technical writers:

Harris, C. M., ed. *Dictionary of Architecture and Construction*
McGraw-Hill Dictionary of Scientific and Technical Terms
Monkhouse, F. J., ed. *A Dictionary of Geography*

Handbooks and manuals. Compilations of frequently used information in a particular field of knowledge, handbooks and manuals are usually single volumes that generally provide such information as brief definitions of terms or concepts, explanations and details about particular organizations, numerical data in **graphs** and tables, **maps,** and the like. Handbooks and manuals are useful sources of fundamental information on a particular subject, although they are most valuable for someone with a basic knowledge of the topic, particularly in scientific or technical fields. Every field has its own handbook or manual; following are some useful ones.

Becker, Ester, and Evelyn Anders. *The Successful Secretary's Handbook*
CRC Handbook of Chemistry and Physics
Environmental Regulation Handbook
Merritt, F. S., ed. *Building Construction Handbook*
U.S. Government Manual

Statistical sources. Collections of numerical data, these are the best source for such information as the height of the Washington Monument; the population of Boise, Idaho; the cost of living in Aspen, Colorado; and the annual number of motorcycle fatalities in the United States. The answers to many statistical questions can be found in almanacs and encyclopedias. The answers to difficult or comprehensive questions, however, are most likely to be found in works devoted exclusively to statistical data, a selection of which follows.

American Statistics Index (1978-- with monthly, quarterly, and annual supplements. Lists and abstracts all statistical publications issued by agencies of the U.S. Government, including periodicals, reports, special surveys, and pamphlets.)

U.S. Bureau of the Census. *County and City Data Book* (1952--. issued every 5 years. Includes a variety of data from cities, counties, metropolitan areas, and the like. Arranged by geographic and political areas, it covers such topics as climate, dwellings, population characteristics, school districts, employment, and city finances.)

U.S. Bureau of the Census. *Statistical Abstract of the United States* (1979--. annual. Includes statistics on the U.S. social, political, and economic condition and covers broad topics such as population, education, public land, and vital statistics. Some state and regional data.)

Atlases. Collections of maps, atlases are classified into two categories based on the type of information their maps present—general atlases show physical and political boundaries and thematic atlases give special information, such as climate, population, natural resources, or agricultural products. Following are several general atlases.

Hammond Medallion World Atlas
National Geographic Atlas of the World
Rand McNally New Cosmopolitan World Atlas
The Times Atlas of the World
U.S. Geological Survey. *The National Atlas of the United States of America*

MICROFORMS

Some library source materials may require you to use microforms. Reduced photographic images of printed pages, microforms are used by many libraries for storing magazines, newspapers, and other materials. Because they are reduced, microforms must be magnified by machines called microreaders in order to be read. The most common kinds of microforms are microfilm and microfiche.

Microfilm. Rolls of 35-millimeter film, usually with four printed pages per frame, microfilm must be read on manually or electrically operated machines that advance the roll, frame by frame, for viewing.

Microfiche. A flat sheet of film, usually 4 × 6 inches, that can contain as many as 98 pages, microfiche is read on a microreader, where it can be moved horizontally or vertically from frame to frame for viewing.

If you are unsure about where to get microform materials or how to use microreaders, ask a librarian. Microform materials and microreaders are usually located in a special section of the library. Some microreaders are equipped with photocopying devices that for a fee permit you to make paper copies of microforms.

TIPS FOR DOING LIBRARY RESEARCH

Once you find sources with the information you need and actually get your hands on those sources, you must be very conscientious about keeping track of them. Besides taking careful notes of any information they contain, you should methodically note down certain information about the sources themselves. (See also **note-taking.**)

When you get the book or magazine or microform, there are certain shortcuts that can help you to decide quickly whether it has information that will be useful to your research. In a book it is best to start by looking over the table of contents and then by seeing if it has any informative diagrams, charts, or tables.

Does the book have an index or a bibliography? For a magazine article, it helps first to scan the **heads,** to get an idea of its major topics. Is the magazine an authoritative one, or is it brief and topical in its scope? With microforms, you can quickly flip through the text also looking for major topics. No matter what form your source takes, always consider its date: How current is the information? Timeliness is important in any research; you don't want to use out-of-date information.

Prepare a 3 × 5 inch note card for any source that you include in your research. The card should include the source's call number, author, and publication information (publisher, city, year). Then, when you compile your bibliography, you will have all necessary information on these cards. Following are two examples of bibliography cards.

HC
79
.E5554 Smith, Vincent Kerry
Technical Change, Relative Prices, and
Environmental Resource Evaluation
Baltimore : Johns Hopkins, 1974

Bibliography Card for a Book

Aurello, M.D.
"Facing Up to Consumerism"
USA Today, 108 (March 1980), 49-50.

Bibliography Card for a Periodical

-like

The **suffix** *-like* is sometimes added to **nouns** to make them **adjectives.** The resulting **compound word** is hyphenated only if it is unusual or if it might not be immediately clear.

EXAMPLES   childlike, lifelike, dictionary-like, computer-like
Her new assistant works with machine-*like* efficiency.

like/as

To avoid confusion between *like* and *as,* remember that *like* is a **preposition** and *as* is a **conjunction.** Use *like* with a **noun** or **pronoun** that is not followed by a **verb.**

EXAMPLE   The new supervisor behaves *like a novice.*

Use *as* before **clauses** (which contain verbs).

EXAMPLES   He acted *as though he owned the company.*
He responded *as we expected he would.*

Like may be used in elliptical constructions that omit the verb.

EXAMPLE   She took to architecture *like a bird to nest building.*

If the omitted portions of the elliptical construction were restored, however, *as* would be used.

EXAMPLE   She took to architecture *as a bird takes to nest building.*

See also **connectives.**

linking verbs

A **verb** that functions primarily to link the **subject** to a **noun** or **modifier** is called a linking verb. The most common linking verb is a form of the verb *be.*

EXAMPLE   Each group *is* important to the firm's overall performance.

However, many other verbs may also function as linking verbs, including *become, remain,* and *seem.*

EXAMPLE   He *became* an industrial engineer.

Verbs that convey the senses *(look, feel, sound, taste, smell)* often perform only a linking function.

EXAMPLE   The repair crew *looked* tired.

A few verbs may function either as linking verbs or as **transitive** or **intransitive verbs.** When *be* may be substituted, the verb is probably a linking verb.

EXAMPLES   He *grew* angry. (linking verb, equivalent in meaning to *became*)
He *grew* African violets in his office. (transitive verb)
The African violets *grew.* (intransitive verb)

See also **subjective complements.**

listening

You probably spend a great deal of your time listening to others. You may receive oral instructions, attend workshops sponsored by your employer, or take courses as a part of your continuing education. Whatever the activity may be, it is important to develop effective listening skills.

You must be motivated to listen—either by an interest in the subject or by a desire to succeed on the job. The following guidelines may help you improve your listening skills.

1. Develop a positive attitude toward the speaker. Good listeners assume, for example, that even though the subject may sound dull, the speaker is likely to say something that they can use.
2. Do not be distracted by a speaker's personality or speaking style; rather, respond thoughtfully to the speaker's words and avoid making judgments too quickly.
3. Be prepared. On the job, think and talk about the subject of a meeting or a workshop ahead of time. Preparation should enable you to understand the material better and to remember it more easily.
4. Analyze the speaker's words and ideas while you are listening, but don't become so engrossed with your own analysis that you miss important points. Try to spot inconsistencies between the facts and the ideas that you hear. Listen "between the lines." Is the speaker using emotionally charged words? If so, are they appropriate? Can you think of other points that support or reject what the speaker is advocating? Do your own attitudes match those of the speaker? Analyzing a speaker's words and ideas not only helps you probe the meaning of the subject but also helps you remember key points.
5. Take notes. Being a good listener sometimes means being a good note taker. You may wish to record your own thoughts and questions as you are listening, so that you will remember to verify or ask them after the speaker is finished. The best note takers record only key words and **phrases** while they are listening, so that they won't miss anything that is being said. (See also **note-taking.**)

You may not have time or need to follow all these guidelines; if you don't, use the technique or techniques that best help *you* absorb and remember. You may also find that some of these techniques work better when you are listening to formal oral presentations and others work better when you are listening to one or two people in conversations.

lists

Lists can save **readers** time by allowing them to see at a glance specific items, questions, or directions. A list also helps the reader by breaking up complex statements that include figures and by allowing key ideas to stand out.

EXAMPLE   Before we agree to hold the convention at the Brent Hotel, we should be sure the hotel facilities meet the following criteria:

1. At least eight meeting rooms that can accommodate 25 people each
2. Ballroom and dining facilities that can accommodate 250 people
3. Duplicating facilities that are adequate for the conference committee
4. Overhead projectors, flip charts, and screens that are sufficient for eight simultaneous sessions
5. Ground-floor exhibit area that can provide room for thirty 8 × 15 foot booths

To confirm that the Brent Hotel is our best choice, perhaps we should take a look at its rooms and facilities during our stay in Kansas City.

Notice that all the items in this example have **parallel structure.** In addition, all items are balanced—that is, all points are relatively equal in importance and are of the same general length. Consider, however, the following example and its revision:

CHANGE   I believe we should consider several important items at the meeting.

1. Our 1982 Production Schedule
2. The Five-Year Corporate Plan
3. We should also develop an agenda for the next meeting that includes a discussion of criteria for relocating the Westdale Division. Perhaps you can give me some tentative ideas at this meeting.
4. The draft of our Year-End Financial Report

TO   Please bring to the meeting the following items:

1. Our 1982 Production Schedule
2. The Five-Year Corporate Plan
3. The draft of our Year-End Financial Report

In addition, you might give me some tentative ideas about criteria for relocating the Westdale Division. Since we must formally discuss this problem soon, we should include such a discussion in the agenda for the committee's next meeting.

In the original version, item 3 is neither parallel in structure nor balanced in importance with the other items. In the revision, item 3 is discussed separately since it deserves a more developed treatment.

In an attempt to avoid writing **paragraphs,** some writers tend to overuse lists. A **memorandum** or **report** that consists almost entirely of lists, for example, can be difficult to understand, for the reader is forced to connect the separate items and mentally to provide **coherence.** Do not expect a reader to deal with unexplained lists of ideas.

To ensure that the reader understands how a list fits with the surrounding sentences, always provide adequate **transitions** before and after any lists. If you do not wish to indicate rank or sequence, which numbered lists suggest, you can use bullets, as shown in the list of tips that follows. In typography a bullet is a small *o* that is filled in with ink.

TIPS FOR USING LISTS

• List only comparable items.
• Use parallel structure throughout.
• Use only words, **phrases,** or short **sentences.**
• Provide adequate transitions before and after lists.
• Use bullets when rank or sequence is not important.
• Do not overuse lists.

literature

In technical contexts, the word *literature* applies to a body of writing pertaining to a specific field, such as finance, insurance, or computers. We commonly speak, for example, of medical literature or campaign literature.

EXAMPLE   Please send me any available *literature* on computerized translations of foreign languages.

literature reviews

A literature review is a summary **report** on the **literature,** or printed material, that is available on a particular subject in a specific period of time. For example, a literature review might describe all material published within the past five years on a special technique for debugging computers, or it might describe all reports written in the past fifteen years on efforts to improve quality control at a particular company. A literature review tells the **reader** what is available on a particular subject and gives him or her an idea of what should be read in full. The review may be self-contained or it may be part of a larger work.

Industrial research departments often prepare literature reviews, which then serve as starting points for detailed **research.** Managers in business and industry also rely on literature reviews to keep specialists informed on the latest developments and trends in their fields. Some **journal articles** or theses begin with a brief literature review to bring the reader up to date on current research in the area. The author can then use the review as background for his or her own discussion of the subject.

To prepare a literature review, you must begin with **library research** or even a computer search of published material on your **topic.** Since your reader may begin research on the basis of your literature review, be especially careful to cite all bibliographic information accurately. As you review each source, note the scope of the book or article and judge its value to the reader. Save all printouts of computer-assisted searches— you might wish to attach them to the final version of your literature review.

You might arrange your discussion chronologically, beginning with a description of the earliest relevant literature and progressing to the most recent. You can also subdivide the topic, discussing works on various subcategories of the topic.

Begin a literature review by defining the area to be covered and the types of works to be reviewed. For example, a literature review might be limited to articles and reports, and not include any books. Remember also to make the review as concise as possible (much like an **abstract** or an **executive summary**). Mention material directly pertinent to your topic, but do not go into any details. (Your readers can go to the original literature if they want detail.) If you cite many works, number them, and provide a list of references at the end.

The following passage forms the introduction and first two sections of a literature review concerning the use of activated charcoal at nuclear power plants to remove radioactivity from the noble gases krypton and xenon.

Introduction

Among the most difficult components in the safety systems of a nuclear reactor to quantify the performance of are the activated charcoal beds used to remove radioactive noble gases. Knowledge of the basic parameters influencing adsorption behavior has often not been sufficient to permit the calculation of adsorption coefficients under the full range of necessary conditions. Further complicating this situation is that samples of charcoal taken from sequential lots, or even from the same lot, can be significantly different in their adsorption effectiveness.

In spite of this situation, existing uncertainties in analyzing the performance of fission gas adsorption systems can be reduced. These uncertainties can be reduced through better interpretation of data relating to the various parameters affecting the performance of these systems. The resulting information should provide not only a better understanding of the variables associated with adsorption system performance, but also a means of predicting their behavior under conditions that differ substantially from those described for specific nuclear facilities. The purpose of this report is to review and tabulate the published information of the adsorption of krypton and xenon to promote these goals.

Correlations with Surface Area

If the adsorption coefficients for krypton and xenon can be correlated successfully with commonly measured physical parameters, such as pore structure or the surface area of the adsorbent, then these parameters can serve as secondary standards in evaluating charcoals. However, current correlations between surface areas and adsorption coefficients of activated charcoals are not promising. As the data in Table 1.1 indicate, First, and others (Ref.1), were unable to observe any apparent correlation between the adsorption coefficients for krypton and the surface areas of several samples of charcoal tested. Table 1.2, from Kitani, and others (Ref. 2), and Table 1.3, from Nakhutin, and others (Ref. 3), also show no apparent relationship between either the pore volume or the surface area and the adsorption coefficient for krypton. In experiments with radon, Strong and Levins (Ref.4) observed a linear relationship between surface area and the adsorption coefficient. Why a similar relationship has not been observed for krypton and xenon is not known.

Correlations with Density

Nakhutin, and others (Ref.3), showed that there appears to be a positive correlation between the apparent density of a charcoal sample and its adsorption coefficients for krypton and xenon. This result may be related to the observation of Kovach and Etheridge (Ref. 5) that the adsorption coefficient for krypton was a maximum if the charcoal was activated to give a surface area of 900 square meters per gram (Figure 1.1). Schroeter, and others (Ref. 6), obtained a patent for the use of dense charcoals for the adsorption of krypton and xenon. The data developed by these authors (Table 1.4) also show clearly that the adsorption coefficients for krypton and xenon are not proportional to the surface area of the adsorbent, and that high surface area and/or pore volume are not necessarily indicators of high adsorption capacity for krypton and xenon. These results are also shown to Kovach and Etheridge in Figure 1.1 (Ref. 5).

—D. W. Underhill and D. W. Moeller, *The Effects of Temperature, Moisture, Concentration, Pressure and Mass Transfer on the Adsorption of Krypton and Xenon on Activated Carbon* (Washington, D.C.: U.S. Nuclear Regulatory Commission, 1980), pp. 1-1 - 1-2.

logic

Logic, or correct reasoning, is essential to convincing your **reader** that your conclusion is valid. Errors in logic can quickly destroy your credibility with your reader. Although a detailed discussion of logic is beyond the scope of a handbook (see **reference books** for books on logic), the following discussion points out some common errors in logic that you should watch out for in your writing.

LACK OF REASON

When a statement violates the reader's common sense, that statement is not reasonable. If, for example, you stated "New York City is a small town," your reader might immediately question your logic. Common sense would suggest that a city of over eight million people is not a small town. If, however, you stated "Although New York's population is over eight million, it is a city composed of small towns," your reader could probably accept the statement as reasonable—that is, if you then demonstrated the truth of the statement with reasonable examples. Always be sure that your statements are sensible and that they are supported by sound examples.

SWEEPING GENERALIZATION

Sweeping generalizations are statements that are too large to be supportable; they generally enlarge an observation about a small group to a generalization about an entire population.

EXAMPLES   Management is never concerned about employees.
            Computer instructions are always confusing.

These statements are too large; they ignore any possibility that some companies show great concern for employee welfare or that some computer instructions are clearly written. Such generalizations are simply not true; using them will weaken your credibility. No matter how certain you are of the general applicability of an opinion, exercise great caution when using such all-inclusive expressions as *anyone, everyone, no one, all,* and *in all cases.* Otherwise, your statement is likely to sound illogical and will not be taken seriously. Certain stereotypical **phrases** are based on sweeping generalizations and should also be avoided: "typical business person," "bureaucratic mentality," and "absent-minded professor."

NON SEQUITUR

A statement that does not logically follow from a previous statement or has little bearing on the previous statement is called a *non sequitur.*

> EXAMPLE   I arrived at work early today, so the weather is calm.

In this example, common sense tells us that arriving early for work does not produce calm weather, or any other kind. Thus, the second half of the sentence does not follow logically from the first. Notice how much more logical the following example is: It makes sense because each idea is logically linked to the next.

> EXAMPLE   The snow slowed traffic in the city, so although I left for work at the usual time, I arrived a bit late.

Of course, most non sequiturs are not so obvious as the first example. They often occur when a writer neglects to express adequately the logical links in a chain of thought.

> EXAMPLE   Last year our laboratory increased its efficiency by 50% by using part-time help. I suggest that we hire two full-time laboratory assistants.

Although the statements are both about the same subject—additional laboratory staff—a logical connection is missing. The reader is forced to guess the meaning. Does the writer mean that full-time assistants will be more efficient? Are part-time employees unavailable? Such non sequiturs can cause the reader to misunderstand what he is reading or to assume that it is illogical. In your own writing be careful that all points stand logically connected; non sequiturs cause difficulties for reader and writer alike.

POST HOC, ERGO PROPTER HOC

This term means literally "after this, therefore because of this," and it refers to the logical fallacy that because one event happened after another event the first somehow caused the second.

EXAMPLE I didn't bring my umbrella today. No wonder it is now raining.

Many superstitions are based on this error in logic. In on-the-job writing, this error in reasoning usually results from a hasty conclusion that events are related without examining the logical connection between the two events.

EXAMPLE We issued the new police uniforms on September 2nd. Arrests then increased 40%.

In this example, it is not logical to assume that new uniforms would produce such a dramatic increase in arrests. To demonstrate such a claim logically would require an explanation of the special circumstances that produced the result. For example, if the new uniforms resulted in a union settlement with police officers, which in turn caused the officers to end a work slowdown, then the conclusion would be reasonable. Even in that case, however, the new uniforms *per se* did not increase arrests; rather, a complete chain of events produced the result.

BIASED OR SUPPRESSED EVIDENCE

A **conclusion** reached as a result of self-serving data, questionable sources, or purposely incomplete facts is illogical—and probably dishonest. If you were asked to prepare a **report** on the acceptance of a new policy among employees and you distributed **questionnaires** only to those who thought the policy was effective, the resulting evidence would be biased. If you purposely ignored employees who did not believe the policy was effective, you would also be suppressing evidence. Any writer must be concerned about the fair presentation of evidence—especially in reports that recommend or justify actions.

FACT VS. OPINION

It is important in writing to distinguish between fact and opinion. Facts include verifiable data or statements, whereas opinions are personal conclusions that may or may not be based on facts. For example, it is a verifiable fact that distilled water boils at 100°C; that distilled water tastes better than tap water is an opinion. In order to prove a point, some writers mingle facts with opinions, thereby obscuring the differences between them. This is of course unfair to the reader. Be sure always to distinguish your facts from your opinions in your writing so that your reader can clearly understand and judge your conclusions.

EXAMPLES The new milling machines produce parts that are within 2% of specification. (This sentence is stated as a fact and can be verified by measurement.)

The milling machine operators believe the new models are safer than the old ones. (The word *believe* identifies the statement as an opinion—later statistics on the accident rates may or may not verify the opinion as a fact.)

The opinion of experts in their area of expertise is often accepted as valid evidence in court. In writing, too, such testimony can help you to convince your reader of the logic of your conclusion. Be on guard against flawed testimonials, however. For example, a chemist may well make a statement about the safety of a product for use on humans, but the chemist's opinion would not be as valid as that of a medical researcher who has actually studied the effects of that product on humans. When you quote the opinion of an authority, be sure that his or her expertise is appropriate to your point. Make sure also that the opinion is a current one and that the expert is indeed highly respected.

LOADING

When you embed an opinion in a statement and then reach conclusions that are based on that statement, you are loading the argument. Consider the following opening of a **memorandum.**

EXAMPLE I have several suggestions to improve the poorly written policy manual. First, we should change. . . .

By opening with the assumption that the manual is poorly written, the writer has loaded the statement to get readers to accept his arguments and conclusions. Yet, he has said nothing to establish *how* the manual is poorly written. In fact, it may be that only parts of the manual are poorly written. And if so, what are the exact problems? Be careful not to load arguments in your writing; conclusions reached with loaded statements are weak and ultimately unconvincing.

In your writing you should always take pains to express your ideas logically. Logic not only gives writing **coherence** but also keeps it fair and honest.

long variants

Guard against inflating plain words beyond their normal value by adding extra **prefixes** or **suffixes,** a practice that creates long variants. Here is a list of some normal words followed by their inflated counterparts.

analysis/analyzation
certified/certificated
commercial/commercialistic
connect/interconnect
finish/finalize
orient/orientate
priority/prioritization (or prioritize)
use/utilize (see also **utilize**)
visit/visitation

(See also **gobbledygook** and **word choice.**)

loose/lose

Loose is an **adjective** meaning ''not fastened'' or ''unrestrained.''

    EXAMPLE   He found a *loose* wire.

Lose is a **verb** meaning ''be deprived of'' or ''fail to win.''

    EXAMPLE   Did you *lose* the operating instructions?

lower-case and upper-case letters

Lower-case letters are small letters, as distinguished from **capital letters** (known as *upper-case* letters). The terms were coined in the early history of printing when the printer kept the small letters in a lower ''case,'' or tray, below the tray where the capital letters were kept. (See also **capitalization.**)

M

malapropisms

A malapropism is a word that sounds similar to one intended but that is ludicrously wrong in the context.

CHANGE The repairman cleaned the typewriter's *plankton.*
TO The repairman cleaned the typewriter's *platen.*

Intentional malapropisms are sometimes used in humorous writing; unintentional malapropisms can embarrass a writer. (See also **antonyms** and **synonyms.**)

male

The term *male* is usually restricted to scientific, legal, or medical contexts (a *male* patient or suspect). Keep in mind that this term sounds cold and impersonal. The terms *boy, man,* and *gentleman* are acceptable substitutes in other contexts; however, be aware that these substitute words also have **connotations** involving age, dignity, and social position. (See also **female.**)

maps

Maps can be used to show specific geographic features (roads, mountains, rivers, etc.) or to show information according to geographic distribution (population, housing, manufacturing centers, etc.). Bear these points in mind in creating and using maps:

1. Label the map clearly.
2. Assign the map a figure number if you are using enough **illustrations** to justify use of figure numbers. (See Figure 29.)
3. Make sure all boundaries within the map are clearly identified. Eliminate unnecessary boundaries.
4. Eliminate unnecessary information from your map (if population is important, do not include mountains, roads, rivers, etc.).
5. Include a scale of miles or feet to give your **reader** an indication of the map's proportions.
6. Indicate which direction is north.

7. Show the features you want emphasized by shading, dots, cross-hatching, or by using appropriate symbols when color reproduction cannot be used.
8. If you use only one color, remember that only three shades of a single color will show up satisfactorily.
9. Include a key telling what the different colors, shadings, or symbols represent.
10. Place maps as close as possible to the portion of the text that refers to them.

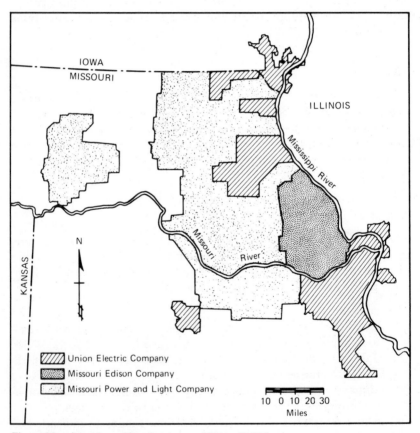

Figure 29.   Location of Service Areas of Three Utilities
Source: The U.S. Nuclear Regulatory Commission

mass nouns

Mass nouns, as opposed to **count nouns,** identify those materials that comprise a mass rather than individual units—a substance or property that cannot be separated and counted. Such words as *electricity, gold, silver, oil, water, beer, aluminum, wheat,* and *cement* are mass nouns. (For the **case** and **number** of mass nouns, see **nouns.** See also **fewer/less.**)

mathematical equations

Material with mathematical equations is easily prepared and easily read by following a few standard guidelines on displaying equations.

Short and simple equations, such as $x(y) = y^2 + 3y + 2$, should be set as part of the running text rather than being displayed, or set on a separate line. If a document contains many equations that are referred to in the text, however, short equations should be displayed and identified with a number, as the following example shows.

$$x(y) = y^2 + 3y + 2 \hspace{4em} (1)$$

Equations are usually numbered consecutively throughout the work. Place the equation number, in **parentheses,** at the right margin, on the same line as the equation (or on the first line if the equation runs longer than one line). Leave at least four spaces between the equation number and the equation. Number displayed equations consecutively, and refer to them in the text as "Equation 1" or "Eq. 1."

POSITIONING DISPLAYED EQUATIONS

Equations that are set off from the text need to be surrounded by space. Triple-space between displayed equations and the normal text. Double-space between one equation and another and between the lines of multiline equations. Count space above the equation from the uppermost character in the equation; count space below from the lowermost character.

Type displayed equations either at the left margin or indented five spaces from the left margin, depending on their length. When a series of short equations is displayed in sequence, however, align them on their equal signs.

$$p(x,y) = \sin(x + y), \qquad (2)$$
$$p(x_0,y_0) = 2 \sin x_0 \cos x_0,$$
$$q(x,y) = \cos(x + y),$$
$$q(x_0,y_0) = \cos^2 x_0 - \sin^2 x_0.$$

Break an equation requiring two lines at the equal sign, carrying the equal sign over to the second portion of the equation.

$$_0\int^1 (f_n - \tfrac{n}{r} f_n)^2 \, r \, dr + 2n \, _0\int^1 f_n f_n \, dr \qquad (3)$$
$$= {_0\int^1} (f_n - \tfrac{n}{r} f_n)^2 \, r \, dr + n f_n^2 \, (1).$$

If you cannot break it at the equal sign, break it at a plus sign or minus sign that is not within parentheses or **brackets.** Bring the plus sign or minus sign to the next line of the equation, which should be positioned to end near the right margin.

$$\phi(x,y,z) = (x^2 + y^2 + z^2)^{\frac{1}{2}} \quad (x - y + z)(x + y - z)^2 \qquad (4)$$
$$- [f(x,y,z) - 3x^2].$$

The next best place to break an equation is between parentheses or brackets that indicate multiplication of two major elements.

For equations requiring more than two lines, start the first line at the left margin, end the last line at the right margin, and center intermediate lines inside the margins. Whenever possible, break equations at operational signs, parentheses, or brackets.

EXPRESSING MATHEMATICAL EQUATIONS

Type mathematical equations rather than writing them out. If your typewriter does not have mathematical **symbols** or Greek letters, it is most professional to use commercially produced press-on letters and symbols. If press-on material is unavailable, handprint letters and symbols in ink as neatly as possible.

maybe/may be

Maybe (one word) is an **adverb** meaning "perhaps."

EXAMPLE   *Maybe* the legal staff can resolve this issue.

May be (two words) is a **verb phrase.**

EXAMPLE   It *may be* necessary to ask for an outside specialist.

media/medium

Media is the plural of *medium* and should always be used with a plural verb.

EXAMPLES   The *media are* a powerful influence in presidential elections.
The most influential *medium is* television.

meetings

When you chair a committee or run a meeting, you should take a number of steps to ensure a smooth discussion.

Before the meeting prepare an agenda and learn as much as you can about the topics that will be discussed. An agenda may be flexible (if, for example, the committee is simply studying possibilities) or inflexible (if the committee is implementing specific plans). Be sure to distribute the agenda well ahead of the meeting so committee members have sufficient time to study it and prepare. If the agenda is flexible, prepare a few questions or comments to start the discussion, such as the reason for the committee's existence or why the topic needs attention. Finally, make sure there are adequate facilities and equipment for the meeting, such as a conference room, chairs, a chalkboard, and so on.

During the meeting it is your duty as chairperson to keep the meeting moving in the appropriate direction. Encourage all the members to participate—do not allow a few to dominate the entire discussion. As chairperson you should build on the contributions of others—do not automatically reject a committee member's idea even though it may seem inappropriate. It is useful, occasionally, to summarize the committee's thinking up to a certain point. Finally, remember that the members have other jobs and responsibilities, so start the meeting on time and close on time. (See also **minutes of meetings.**)

PANEL DISCUSSIONS

Often professional societies and organizations use panel discussions, as well as speakers, during meetings and conferences. If you are asked to lead or participate in a panel discussion, the following advice may be useful.

If you are the leader, or moderator, of a panel discussion, decide ahead of time on the format that is best for the circumstances. Will there, for example, be spontaneous questions from the audience (making the discussion a "forum")? One successful format is to allow each panelist five or ten minutes to answer a specific question; then devote some time for interchange, or discussion, among the panelists. Afterwards, accept questions from the floor—for a definite period of time. Always make sure that panel members have the questions (or topics for discussion) and the format well in advance of the meeting date, so they can be fully prepared. Also, check the physical arrangements for the discussion well in advance. (If amplification is needed, for example, is it set up and adequate?)

During the discussion the leader should inform the audience of the ground rules, introduce the participants, accept and direct questions from the floor, and, if appropriate, sum up the presentation. Obviously, the leader must be as well prepared as the panelists.

memorandums

Communication between members of the same organization is written in the form of a memorandum, called a memo for short. Memos are routinely used for internal communications of all kinds—from short notes to small **reports.** In addition to transmitting information, the

```
                            MEMORANDUM


        To:  Ben Johnson, Systems Design
        From: Linda Coleman, Sales and Service
        Date:  February 12, 19--

        Subject:  Meeting with Ackley Brothers & Co.

        I thought you'd like to know that the meeting with Ackley went very
        well.  In fact, I think they'll accept our proposal.

        Keep your fingers crossed.
```

Figure 30

INTEROFFICE CORRESPONDENCE
INDUSTRIAL PUBLISHING CORPORATION

To: Hazel Smith, Publications Manager

From: Herbert Kaufman, Vice President

Date: April 14, 19--

Subject: Time Estimate for Acme Electronics Brochure

Acme Electronics has asked us to prepare a comprehensive brochure for
their Milwaukee office by August 1 of this year. We have worked with
electronic firms in the past, so this job should be relatively easy to
prepare. My guess is that the job will take nearly two months. Ted Harris
has requested time and cost estimates for the project. Fred Moore in
accounting will prepare cost estimates, and I would like you to prepare a
schedule for the estimated time.

Schedule
In preparing the schedule, check production schedules for all staff
writers, compile a list of available free-lance writers we can depend on,
and contact local graphic designers for art work, since our art department
is heavily scheduled through July. Also, don't forget to take staff
vacation time into account. Let me emphasize that we must finish by our
estimated date (most important), that the overall design must have a
uniform format, and that the art work must blend properly with the copy.

Time Estimates
Please give me time estimates by April 19. A successful job done on
time will give us a good chance for the contract to do Acme's publications
for their annual stockholder's meeting this fall.

I know your staff can do the job.

cc: Ted Harris

Figure 31

memo serves as a permanent record of decisions made and of actions taken.

In writing a memo, follow the rules for good writing that you would use for reports and longer works. An **outline,** even if only a number of ideas jotted down and arranged in a logical sequence, is helpful.

A memo should announce its subject in a subject line. To focus the **reader's** attention on particular topics in a long memo, you can insert appropriate **heads.** Headings divide the material into manageable segments, call attention to the main topics, and signal changes of topic—and they should aid you in organizing your material. If your memo presents a series of items for consideration, you may list them in a numbered sequence.

The **tone** of a memo can range from formal to informal, depending on your reader. A memo to your general manager would be more formal than a memo to an employee of equal rank in another department who is also a friend.

If your **reader** is not familiar with your subject, or with the background of the problem being dealt with, provide a brief introductory **paragraph.** A brief **introduction** will also provide necessary information months (or even years) later. Generally, longer memos or those dealing with complex subjects benefit most from such introductory information.

Job titles or department names are occasionally omitted if the sender and receiver know each other well. Since memos are often used as records of information, however, it is usually best to include them.

Figure 30 shows a short memo; Figure 31 shows a longer one.

metaphor

Metaphor is a figure of speech that points out similarities between two things by treating them as being the same thing. Metaphor states that the thing being described *is* the thing to which it is being compared.

EXAMPLE   *The building site was a beehive,* with iron workers still at work on the tenth floor, plumbers installing fixtures on the floors just beneath them, electricians busy on the middle floors, and carpenters putting the finishing touches on the first floor.

The use of metaphor often helps clarify complex theories or objects. For example, the life-sustaining tube connecting a space-walking astronaut to his oxygen supply in the spacecraft is called an "umbilical cord." The person who first drew this **comparison** knew that such a life support tube was new and unfamiliar, so to make its function immediately apparent he used a metaphor that compared it to something with which people were already familiar. A similar purpose was served when the term "window" was used to describe the abstract concept of the point at which a spacecraft reenters the earth's atmosphere. Here the strange and far-away was made near and familiar by the metaphorical use of a term familiar to everyone. Both of these expressions have gained widespread use because they helped clarify new and unfamiliar concepts.

A word of caution: Beginning writers sometimes mix metaphors, which usually results in an illogical statement.

EXAMPLE   Billingham's proposal *backfired* and now she is *in hot water.*

This example makes an incongruous statement because a backfiring automobile does not land someone in hot water. (See also **figures of speech.**)

methods of development

One of the most important elements in any writing is its method of development. After you have completed your **research,** but before beginning your **outline,** ask yourself how you can most effectively "unfold" your **topic** for your **reader.** An appropriate method of development will make it easy for your reader to understand your topic and will move the topic smoothly and logically from an **introduction** (or **opening**) to a **conclusion.** There are several common methods of development, each best suited to particular purposes.

If you are writing a set of **instructions,** for example, you know that your readers need the instructions in the order that will enable them to perform some task. Therefore, you should use a **sequential method of development** to write the instructions. If you wished to emphasize the time element of a sequence, however, you could follow a **chronological method of development.**

If writing about a new topic that is in many ways similar to another, more familiar topic, it is sometimes useful to develop the new topic by comparing it to the old one—thereby enabling your readers to make certain broad assumptions about the new topic, based on their understanding of the familiar topic. By doing so, you are using a **comparison method of development.**

When describing a mechanical device, you may divide it into its component parts and explain each part's function as well as how all the parts work together. In this case, you are using a **division-and-classification method of development.** Or you may use a **spatial method of development** to describe the physical appearance of the device from top to bottom, from inside to outside, from front to back, and so on.

If you are writing a report for a government agency explaining an airplane crash, you might begin with the crash and trace backwards to its cause; or you could begin with the cause of the crash (for example, a structural defect) and show the sequence of events that led to the crash. Either way is the **cause-and-effect method of development.** You can also use this approach to develop a report dealing with the solution to a problem—you could begin with the problem and move on to the solution, or vice versa.

If you are writing about the software for a new computer system, you might begin with a general statement of the function of the total software package, then explain the functions of the larger routines within the software package, and finally deal with the functions of the various subroutines within the larger routines. You would be using a **general-to-specific method of development.** (In another situation, you might use the **specific-to-general method of development.**)

To explain the functions of the departments in a company, you could present them in a sequence that reflects their importance within the company: the executive department first and the custodial department last, with all other departments (sales, engineering, accounting, etc.) arranged in the relative order of importance they are given in that company. You would be using the **decreasing-order-of-importance method of development.** (In another situation, you might use the **increasing-order-of-importance method of development.**)

Methods of development often overlap, of course. Rarely does a writer rely on only one method of development in a written work. The important thing is to select one primary method of development and then to base your outline on it. For example, in describing the organization of a company, you would actually use elements from three methods of development: you would *divide* the larger topic (the company) into departments; you would present the departments *sequentially;* and you would arrange the departments by their *order of importance* within the company.

To consider one example, assume that you are writing an article about a particular company. You could approach such a topic any number of ways. You could *compare* the company to its major competitor, touching upon such things as each one's historical development, market share, number of employees, revenue produced per employee, gross annual sales, net income, return on investment, research and development expenditures, debt as a percentage of assets, and the relative financial condition. Or you could *divide,* or break the company down, by departments and discuss each in turn by decreasing order of importance: the executive department, engineering, sales, production, accounting, data processing, inventory control, advertising, order processing. Or, in order to discuss and illustrate the company's operations, you could follow a purchase order through its entire life cycle, from the time an advertisement is placed in a trade journal, to the actual product sale, to the receipt of the order at the factory, to the product being pulled from stock and shipped to the customer, to the invoice being received by the accounting department and the bill being sent to the customer, to the payment being received from the customer, and finally to the account being closed. The method of development that you follow should be determined by your reader as well as by your **objective.**

minutes of meetings

Organizations and committees keep official records of their **meetings;** such records are known as *minutes.* If you attend many business-related meetings, you may be asked to serve as recording secretary and write and distribute the minutes of a meeting. At each meeting, the minutes of the previous meeting are usually read aloud if printed copies of the minutes were not distributed to the members beforehand; the group then votes to accept the minutes as prepared or to revise or clarify specific items. The minutes of the meetings should include the following:

1. Name of the group or committee holding the meeting
2. Place, time, and date of the meeting
3. Kind of meeting (a regular meeting or a special meeting called to discuss a specific subject or problem)
4. Number of members present and, for committees or boards of ten or fewer members, their names
5. A statement that the chairperson and the secretary were present, or the names of any substitutes
6. A statement that the minutes of the previous meeting were approved, revised, or not read
7. A list of any reports that were read and approved

8. All the main motions that were made, with statements as to whether they were carried, defeated, or tabled (vote postponed), and the names of those who made and seconded the motions (Motions that were withdrawn are not mentioned.)
9. A full description of resolutions that were adopted and a simple statement of any that were rejected
10. A record of all ballots with the number of votes cast for and against
11. The time that the meeting was adjourned (officially ended) and the place, time, and date of the next meeting
12. The recording secretary's signature and typed name and, if desired, the signature of the chairperson

Since minutes are often used to settle disputes, they must be accurate, complete, and clear. When approved, minutes of meetings are official and can be used as evidence in legal proceedings.

Keep your minutes brief and to the point. Give complete information on each topic, but do not ramble—conclude the topic and go on to the next one. Following a set **format** will help you keep the minutes concise. You might, for example, use the **heading** *TOPIC*, followed by the subheadings *Discussion* and *Action Taken*, for each major point discussed.

Avoid abstractions and generalities; always be specific. If you are referring to a nursing station on the second floor of a hospital write "the nursing station on the second floor" or "the second-floor nursing station," not just "the second floor."

Be especially specific when referring to people. Avoid using titles (the chief of the Word Processing Unit) in favor of names and titles (Ms. Florence Johnson, head of the Word Processing Unit). And be consistent in the way you refer to people. Do not call one person *Mr.* Jarrel and another *Janet* Wilson. It may be unintentional, but a lack of consistency in titles or names may reveal a deference to one person at the expense of another. Avoid **adjectives** and **adverbs** that suggest either good or bad qualities, as in "Mr. Sturgess's *capable* assistant read the *extremely comprehensive* report of the subcommittee." Minutes should always be objective and impartial.

If a member of the committee is to follow up on something and report back to the committee at its next meeting, state clearly the person's name and the responsibility he or she has accepted.

When assigned to take the minutes at a meeting, be prepared. Bring more than one pen and plenty of paper. If convenient, you may bring a tape recorder as backup to your notes. Have ready the minutes of the previous meeting and any other material that you may need. If you do not know shorthand, take memory-jogging notes during the meeting and then expand them with the appropriate details immediately after the meeting. Remember that minutes are primarily a record of specific actions taken. (See also **note-taking**.)

Following is a sample set of minutes.

```
Minutes of the Regular Meeting of the Credentials
                      Committee

DATE:  April 18, 19--

PRESENT:  M. Valden (Chairperson), R. Baron,
          M. Frank, J. Guern, L. Kingson,
          L. Kinslow (Secretary), S. Perry,
          B. Roman, J. Sorder, F. Sugihana


Dr. Mary Valden called the meeting to order at
8:40 p.m. The minutes of the previous meeting were
unanimously approved, with the following correc-
tion:  the name of the secretary of the Department
of Medicine is to be changed from Dr. Juanita
Alvarez to Dr. Barbara Golden.


Old Business

None.


New Business

The request by Dr. Henry Russell for staff
privileges in the Department of Medicine was dis-
cussed.  Dr. James Guern made a motion that
Dr. Russell be granted staff privileges.
Dr. Martin Frank seconded the motion, which passed
unanimously.
     Similar requests by Dr. Ernest Hiram and
Dr. Helen Redlands were discussed.  Dr. Fred
Sugihana made a motion that both physicians be
granted all staff privileges except respiratory-
care privileges, because the two doctors had not
had a sufficient number of respiratory cases.
Dr. Steven Perry seconded the motion, which was
passed unanimously.
     Dr. John Sorder and Dr. Barry Roman asked
for a clarification of general duties for active
staff members with respiratory-care privileges.
```

Dr. Richard Baron stated that he would present a clarification at the next scheduled staff meeting, on May 15.

Dr. Baron asked for a volunteer to fill the existing vacancy for Emergency Room duty. Dr. Guern volunteered. He and Dr. Baron will arrange a duty schedule.

There being no further business, the meeting was adjourned at 9:15 p.m. The next regular meeting is scheduled for May 15, at 8:40 p.m.

Respectfully submitted,

Leslie Kinslow

Leslie Kinslow
Medical Staff Secretary

Mary Valden, MD

Mary Valden, M.D.
Chairperson

misplaced modifiers

A **modifier** is misplaced when it modifies, or appears to modify, the wrong word or phrase. It differs from a **dangling modifier** in that a dangling modifier cannot *logically* modify any word in the **sentence** because its intended referent is missing. The best general rule for avoiding misplaced modifiers is to place modifiers as close as possible to the words they are intended to modify. A misplaced modifier can be a word, a **phrase,** or a **clause.**

MISPLACED WORDS

Adverbs are especially likely to be misplaced because they can appear in several positions within a sentence.

EXAMPLES   We *almost* lost all of the parts.
We lost *almost* all of the parts

The first sentence means that all of the parts were *almost* lost (but they were not), while the second sentence means that a majority of the parts (*almost* all) were in fact lost. Possible confusion in sentences of this type can be avoided by placing the adverb immediately before the word it is intended to modify.

CHANGE   All navigators are *not* talented in mathematics. (The implication is that *no* navigator is talented in mathematics.)

TO   *Not* all navigators are talented in mathematics.

MISPLACED PHRASES

To avoid confusion, place phrases near the words they modify. Note the two meanings possible when the phrase is shifted in the following sentences.

EXAMPLES   The equipment *without the accessories* sold the best. (Different types of equipment were available, some with and some without accessories.)
The equipment sold the best *without the accessories*. (One *type* of equipment was available, and the accessories were optional.)

MISPLACED CLAUSES

To avoid confusion, clauses should be placed as close as possible to the words they modify.

CHANGE   We sent the brochure to four local firms *that had three-color art*.
TO   We sent the brochure *that had three-color art* to four local firms.

SQUINTING MODIFIERS

A modifier "squints" when it can be interpreted as modifying either of two sentence elements simultaneously, so that the **reader** is confused about which is intended.

EXAMPLE   We agreed *on the next day* to make the adjustments.

The meaning of the preceding example is unclear; it could have either of the following two senses.

EXAMPLES   We agreed to *make the adjustments on the next day*.
On the next day we agreed to make the adjustments.

A squinting modifier can sometimes be corrected simply by changing its position, but often it is better to recast the sentence:

EXAMPLES   We agreed that *on the next day* we would make the adjustments. (The adjustments were to be made on the next day.)
On the next day we agreed that we would make the adjustments. (The agreement was made on the next day.)

mixed constructions

A mixed construction occurs when a **sentence** contains grammatical forms that are improperly combined. These constructions are improper because the grammatical forms are inconsistent with one another. The most common types of mixed constructions result from the following causes.

TENSE

CHANGE　The pilot *lowered* the landing gear and *is approaching* the runway. (shift from past to present tense)

TO　The pilot *lowered* the landing gear and *approached* the runway.

OR　The pilot *has lowered* the landing gear and *is approaching* the runway.

PERSON

CHANGE　The *technician* should take care in choosing *your* equipment. (shift from third to second person)

TO　The *technician* should take care in choosing *his* equipment.

NUMBER

CHANGE　My *car,* though not as fast as the others, *operate* on regular gasoline. (singular subject with plural verb form)

TO　My *car,* though not as fast as the others, *operates* on regular gasoline.

VOICE

CHANGE　*I will check* your report, and then *it will be returned* to you. (shift from active to passive voice)

TO　*I will check* your report, and then *I will return* it to you.

See also **parallel structure** and **agreement**.

modifiers

Modifiers are words, **phrases,** or **clauses** that expand, limit, or make more exact the meaning of other elements in a **sentence.** Although we could create sentences without modifiers, we often need the detail and clarification they provide.

EXAMPLES　Production decreased. (without modifiers)
　　　　　　Automobile production decreased *rapidly.* (with modifiers)

Most modifiers function as **adjectives** or **adverbs.** An adjective makes the meaning of a **noun** or **pronoun** more exact by pointing out one of its qualities or by imposing boundaries upon it.

EXAMPLES   *ten* automobiles   *this* crane
                an *educated* person   *loud* machinery

An adverb modifies an adjective, another adverb, a **verb,** or an entire clause.

EXAMPLES   Under test conditions, the brake pad showed *much* less wear than it did under actual conditions. (modifying the adjective *less*)
                The wear was *very* much less than under actual conditions. (modifying another adverb, *much*)
                The recording head hit the surface of the disc *hard*. (modifying the verb *hit*)
                Surprisingly, the machine failed even after all the tests that it had passed. (modifying a clause)

Adverbs become **intensifiers** when they increase the impact of adjectives (*very* fine, *too* high) or adverbs (*rather* quickly, *very* slowly). As a rule, be cautious in using intensifiers; their overuse can lead to exaggeration, and hence to inaccuracies.

DANGLING MODIFIERS

A modifying word or phrase ''dangles''when it has no clear word or subject to modify. Most dangling modifiers are **verbal** phrases **(gerund, participle, infinitive)** that should modify a noun or pronoun. Dangling modifiers usually occur because the writer fails to provide the **subject** of a clause. (See also **dangling modifiers.**)

MISPLACED MODIFIERS

A modifier is misplaced when it modifies, or appears to modify, the wrong word or phrase. It differs from a dangling modifier, in that a dangling modifier cannot *logically* modify any word in the sentence since its intended referent is missing. Place modifiers as close as possible to the words they are intended to modify to avoid misplacing them. (See also **misplaced modifiers.**)

mood

The grammatical term *mood* refers to the **verb** functions—and sometimes form changes—that indicate whether the verb is intended to (1) make a statement or ask a question (indicative mood), (2) give a command (imperative mood), or (3) express a hypothetical possibility (subjunctive mood).

The indicative mood refers to an action or a state that is conceived as fact.

EXAMPLES   *Is* the setting correct?
The setting *is* correct.

The imperative mood expresses a command, suggestion, request, or entreaty. In the imperative mood, the implied subject "you" is not expressed.

EXAMPLES   *Install* the wiring today.
Please *let* me know if I can help.

The subjunctive mood expresses something that is contrary to fact, that is conditional, hypothetical, or purely imaginative; it can also express a wish, a doubt, or a possibility. The subjunctive mood may change the form of the verb, but the verb *be* is the only one in English that preserves many such distinctions.

EXAMPLES   The senior partner insisted that he (I, you, we, they) *be* in charge of the project.
If we *were* to close the sale today, we would meet our monthly quota.
If I *were* you, I would postpone the trip.

Form change for the subjunctive is rare in most verbs other than *be*. Instead, we use **helping verbs** to show the subjunctive function. Be careful, however, not to use unnecessary verbs.

CHANGE   If I *would have known* that you were here, I would have come earlier.
TO   If I *had known* that you were here, I would have come earlier.
OR   *Had I known* that you were here, I would have come earlier.

The advantage of the subjunctive mood is that it enables us to express clearly whether or not we consider a condition contrary to fact. If so, we use the subjunctive; if not, we use the indicative.

EXAMPLES   If I *were* president of the firm, I would change several personnel policies. (subjunctive mood)
Although I *am* president of the firm, I don't feel that I control every aspect of its policies. (indicative mood)

Be careful not to shift from one mood to another within a **sentence;** to do so makes the sentence not only ungrammatical but unbalanced as well.

CHANGE   *Put* the clutch in first (imperative); then you *can* put the truck in
gear. (indicative)

TO   *Put* the clutch in first (imperative); then *put* the truck in gear
(imperative).

Ms./Miss/Mrs.

Ms. is a convenient form of addressing a woman regardless of her marital
status, and it is now almost universally accepted. *Miss* is used to refer to
an unmarried woman, and *Mrs.* is used to refer to a married woman.
Some women indicate a preference for *Miss* or *Mrs.*, and such a preference should be honored. An academic or professional title *(Dr., Prof.,
Capt.)* should take preference over *Ms., Miss,* or *Mrs.*

MS/MSS

MS (or ms) is the **abbreviation** for *manuscript.* The plural form is MSS
(or mss). Do not use this abbreviation (or any abbreviation) unless you
are certain that your **reader** is familiar with it.

mutual/in common

When two or more persons (or things) have something *in common,* they
share it or possess it jointly.

EXAMPLES   What we have *in common* is our desire to make the company
profitable.

The fore and aft guidance assemblies have a *common* power
source.

Mutual may also mean "shared" (as in *mutual friends*), but it usually
implies something given and received reciprocally, and it is used with
reference to only two persons or parties.

EXAMPLES   Smith mistrusts Jones, and I am afraid the mistrust is *mutual.*
(Jones also mistrusts Smith.)

Both business and government consider the problem to be of
mutual concern.

N

narration

Narration is the presentation of a series of events; that is, it tells a story. Narration avoids any unnecessary interruptions in movement from the beginning event to the concluding event. It uses a combination of the **sequential** and the **chronological methods of development,** presenting events as they occur both in time and in order—from start to finish sequentially and from beginning to end chronologically. Like good **description,** good narration makes effective use of details. Following is a good example of narration.

> One of the foundations engaged me to edit the manuscript of a socio-economic research report designed for the thoughtful citizen as well as for the specialist. My expectations were not high— no deathless prose, merely a sturdy, no-nonsense report of explorers into the wilderness of statistics and half-known fact. . . . Although I did not expect fine writing from a trained professional researcher, I did assume that a careful fact-finder would write carefully.
>
> And so, anticipating no literary treat, I plunged into the forest of words of my first manuscript. My weapons were a sturdy eraser and several batteries of sharpened pencils. My armor was a thesaurus. And if I should become lost, a near-by public library was a landmark, and the Encyclopedia of the Social Sciences on its reference shelves was an ever-ready guide.
>
> Instead of big trees, I found underbrush. Cutting through lumbering sentences was bad enough, but the real chore was removal of the burdocks of excess verbiage which clung to the manuscript. Nothing was big or large; in my author's lexicon, it was "substantial." When he meant "much," he wrote "to a substantially high degree." If some event took place in the early 1920's, he put it in "the early part of the decade of the twenties." And instead of "that depends," my author wrote, "any answer to this question must bear in mind certain peculiarities characteristic of the industry."
>
> So it went for 30,000 words. The pile of verbal burdocks grew—sometimes twelve words from a twenty-word sentence. The shortened version of 20,000 words was perhaps no more

thrilling than the original report; but it was terser and crisper. It took less time to read and it could be understood quicker. That was all I could do.

—Samuel T. Williamson, "How to Write Like a Social Scientist," *Saturday Review,* October 4, 1947.

Although most often associated with fiction, narration may just as effectively be used to follow the development of a scientific theory or to explain a process or procedure. (See also **explaining a process** and **forms of discourse.**)

nature

Nature, used to mean "kind" or "sort," can—like those words—often be vague. Avoid the word in your writing. Say exactly what you mean.

CHANGE The *nature* of the engine caused the problem.
TO The compression ratio of the engine caused the problem.

needless to say

Although the **phrase** *needless to say* often occurs in speech and writing, it is redundant because it always precedes a remark that is stated despite the inference that nothing further need be said.

CHANGE *Needless to say,* departmental cutbacks have meant decreased efficiency.
TO Departmental cutbacks have meant decreased efficiency.
OR Understandably, departmental cutbacks have meant decreased efficiency.

neo-

The **prefix** *neo-* is derived from a Greek word meaning "new." It is hyphenated when used with a **proper noun,** and it may optionally be hyphenated when used with a **noun** beginning with the vowel *o.*

EXAMPLES neo-Fascism, neo-Darwinism, neo-orthodoxy
neologism, neonatal, neoplasm

new words

New words continually find their way into the language from a variety of sources. Some come from other languages.

EXAMPLES   discotheque (French), skiing (Norwegian), whiskey (Gaelic)

Some come from technology.

EXAMPLES   software, vinyl, nylon

Some come from scientific research.

EXAMPLES   berkelium, transistor

Some come from brand or trade names.

EXAMPLES   Jello, Teflon, Kleenex

Some, called **blend words,** are formed by combining existing words.

EXAMPLE   smoke + fog = smog

Some come from **acronyms.**

EXAMPLES   scuba, radar, laser

Business and technology are responsible for many new words. Some are necessary and unavoidable; however, it is best to avoid creating a new word if an existing word will do. If you do use what you believe to be a new word or expression, be sure to define it the first time you use it; otherwise, your **reader** will be confused. (See also **long variants.**)

news releases

Companies and organizations write news releases, or press releases, to announce to the public new products or services, special events (such as plant openings, company anniversaries, or union contracts), management changes, or sponsorship of social-action or cultural programs. The purpose of the news release is both to inform the public about the company and its products and to create a favorable company image. Even the announcement of an unfavorable event, such as a strike, should try to present the company in a good light. Large corporations and institutions usually have their own public relations staffs or use outside agencies. However, a small company without a public relations staff or an agency will often want to give information to the public and will need

Bentley Plastics
3535 Michigan Avenue
Chicago, IL 60653

FOR IMMEDIATE RELEASE

For More Information Call:
Marjorie Kohls
(312) 712-1946

CHICAGO, ILLINOIS, MAY 20, 19-- . . . Mark Williams has
joined Bentley Plastics as vice president-marketing. He
will direct the Illinois and Indiana district sales offices
and coordinate overseas distribution through the company's
Singapore office. Williams will travel extensively to the
Far East while developing marketing channels for Bentley.

Formerly, Williams was director of services with
International Marketing Associates, a consulting group in New
York. While at IMA, he developed a computer-based marketing
center that linked textile firms in the United States,
Finland, and Great Britain. A graduate of the Columbia
University Graduate School of Business, Williams is the
author of <u>Marketing Dynamics</u>, an informal examination of
psychological appeals to the buying public. The book has
been used in marketing classrooms at several universities.

Bentley Plastics, headquartered in Chicago, produces
plastic tubing and coils. The company's main plants are in
Skokie, Illinois, and Gary, Indiana. Since 1978 Bentley
has also been producing tubing for French distribution
through the firm of Jourdan and Sons, Paris.

-30-

Figure 32

someone to write news releases. In case you are called upon to write one, the following guidelines may prove useful.

The news release must be clear and concise and should be written with particular attention to the five *W's: who, what, where, when,* and *why.* Write the release so that all critical information is in the first **paragraph,** information of the next level of importance is in the second paragraph, and so on. Editors must make your release fit the space they have available; if they must cut your release, they will delete the last paragraph first, and then the next to last, and so on. Be sure about the accuracy of your facts, and be careful to define any technical terms. Type the release, double-spaced, on company letterhead. Leave the top third of the page blank for editors to mark any necessary heads on the copy. Give the name and phone number of someone who can give further information. Begin with the place and date of the announcement, as is shown in the following sample. If the release is longer than one page, type *More* at the bottom of all pages (except the last). Type *-30-* or *End* at the bottom of the last page. Releases are usually sent to local newspapers, television and radio stations, and other special groups, such as trade publications or professional associations. Send the release to a specific person whenever possible. Otherwise, try to address the news release to a particular editor—for example, to the business editor or the education editor. Figure 32 is a typical news release.

no doubt but

In the **phrase** *no doubt but,* the word *but* is redundant.

> CHANGE   There is *no doubt but* that he will be promoted.
> TO   There is *no doubt* that he will be promoted.

none

None may be considered either a singular or a plural **pronoun,** depending on the context.

> EXAMPLES   *None* of the material *has* been ordered. (always singular with a singular noun—in this case, "material")
> *None* of the clients *has* been called yet. (singular even with reference to a plural noun ["clients"] if the intended emphasis is on the idea of *not one*)
> *None* of the clients *have* been called yet. (plural with a plural noun)

For **emphasis,** substitute *no one* or *not one* for *none* and use a singular **verb.**

> EXAMPLE I paid the full retail price for three of your firm's machines, *no one* of which was worth the money.

See also **agreement.**

nonstandard English

Nonstandard English is written language that does not conform to the conventions or regular grammatical features of **standard English.** Nonstandard English is therefore characterized by misspellings, unconventional **grammar** and **punctuation,** and faulty **word choice.** (See also **English, varieties of.**)

nor/or

Nor always follows *neither* in **sentences** with continuing negation.

> EXAMPLE They will *neither* support *nor* approve the plan.

Likewise, *or* follows *either* in sentences.

> EXAMPLE The firm will accept *either* a short-term *or* a long-term loan.

Two or more singular **subjects** joined by *or* or *nor* usually take a singular **verb.** But when one subject is singular and one is plural, the verb agrees with the subject nearer to it.

> EXAMPLES *Neither* the manager *nor* the secretary *was* happy with the new filing system. (singular)
> *Neither* the manager *nor* the secretaries *were* happy with the new filing system. (plural)
> *Neither* the secretaries *nor* the manager *was* happy with the new filing system. (singular)

See also **correlative conjunctions.**

notable/noticeable

Notable, meaning "worthy of notice," is sometimes confused with *noticeable,* meaning "readily observed."

> EXAMPLES His accomplishments are *notable.*
> The construction crew is making *noticeable* progress on the new building.

note-taking

The purpose of note-taking is to summarize and record the information that you extract from your **research** material. The great challenge in taking notes is to condense another's thoughts in your own words without distorting the original thinking. Meeting this challenge involves careful reading on your part, because to compress someone else's ideas accurately, you must first understand them.

When taking notes on abstract ideas, as opposed to factual data, be careful not to sacrifice **clarity** for brevity. You can be brief—if you are accurate—with statistics, but notes expressing concepts can lose their meaning if they are too brief. The critical test of a note is whether, after a week has passed, you will still know what the note means and from it be able to recall the significant ideas of the passage. If you are in doubt about whether or not to take a note, take it—it is much easier to discard a note you don't need than to find the source again if it is needed.

As you extract information, be guided by the **objective** of your writing and by what you know about the **readers.** (Who are they? How much do they know about your subject?)

Resist the temptation to copy your source word-for-word as you take notes. On occasion, when your source concisely sums up a great deal of information or points to a trend or development important to your subject, you are justified in quoting it verbatim and then incorporating that **quotation** into your paper. As a general rule, you will rarely need to quote anything longer than a **paragraph.**

If a note is copied word-for-word from your source, be certain to enclose it in **quotation marks** so you will know later that it is a direct quotation. In your finished paper, be certain to give the source of your quotation in a **footnote;** otherwise, you will be guilty of **plagiarism.**

The mechanics of note-taking are simple and if followed conscientiously can save you much unnecessary work. Consider the information contained in the following paragraph.

> Long before the existence of bacteria was suspected, techniques were in use for combatting their influence in, for instance, the decomposition of meat. Salt and heat were known to be effective and these do in fact kill bacteria or prevent them from multiplying. Salt acts by the osmotic effect of extracting water from the bacterial cell fluid. Bacteria are less easily destroyed by osmotic action than are animal cells because their cell walls are con-

structed in a totally different way, which makes them very much less permeable.

—Roger James, *Understanding Medicine* (London: Penguin Books, 1970), p. 103.

The paragraph says essentially three things:

1. Before the discovery of bacteria, salt and heat were used in combatting bacteria.
2. Salt kills bacteria by extracting water from their cells by osmosis, hence its use in curing meat.
3. Bacteria are less affected by the osmotic effect of salt than are animal cells because bacterial cell walls are less permeable.

If your **topic** involved tracing the origin of the bacterial theory of disease, you might want to note that salt was traditionally used to kill bacteria long before people realized what caused meat to spoil, though it might not be necessary to your topic to say anything about the relative permeability of bacterial cell walls. Again, your best note-taking guides are your objective, reader, and **scope**.

As you read the following paragraph, which contains only factual data, determine which facts you should record or summarize as notes in order to be complete and accurate.

> The arithmetic of searching for oil is stark. For all his scientific methods of detection, the only way the oilman can actually know for sure that there is oil in the ground is to drill a well. The average cost of drilling an oil well is over $300,000, and drilling a single well may cost over $8,000,000! And once the well is drilled, the odds against its containing enough oil to be commercially profitable are 49 to 1. The odds against its containing any oil at all are 8 to 1! Even after a field has been discovered, one out of every four holes drilled in developing the field is a dry hole because of the uncertainty of defining the limits of the producing formation. The oilman can never know what Mark Twain once called "the calm confidence of a Christian with four aces in his hand."

Now read the following list of notes and see whether they are the same as those you would have taken.

1. Many scientific methods of detection
2. Only sure way to know is to drill a well

3. Average cost of driling a well over $300,000; actual cost can go over
   $8,000,000
4. Odds against profitable well: 49 to 1
5. Odds against any oil: 8 to 1
6. If existence of oil is known, 1 of 4 wells still dry because limits of field
   unknown.

The following paragraph is less factual. As you read it, determine the
opinion or ideas you would record as notes.

> Energy does far more than simply make our daily lives more
> comfortable and convenient. Suppose you wanted to stop—and
> reverse—the economic progress of this nation. What would be
> the surest and quickest way to do it? Find a way to cut off the
> nation's oil resources! Industrial plants would shut down; public
> utilities would stand idle; all forms of transportation would halt.
> The country would be paralyzed! And our economy would
> plummet into the abyss of national economic ruin. Our
> economy, in short, is energy-based.
>
> —*The Baker World* (Los Angeles: Baker Oil Tools, 1964), p. 5.

Now read the following list of notes and see whether they are approximately the same as those you would have taken.

1. Best way to stop economic progress of nation—cut off oil resources
2. Plants would close, public utilities be idled, transportation stopped
3. Economy ruined because it is based on energy

These three notes sum up all the significant ideas contained in the
paragraph. A single note, such as "Our economy is energy-based,"
might tempt the writer because it is short and easy; however, although it
sums up the ultimate significance of the paragraph, it does *not* contain all
the important ideas of the paragraph. Remember that the critical test of a
note is whether, after a week has passed, you will still be able to recall all
the significant ideas of the passage from which it was taken.

Whether you record your notes on 3 × 5 cards or in a note pad, be sure
to include the following with the first note taken from a book: author,
title, publisher, place and date of publication, and page number. (On
subsequent notes from the same book, you will need to include only the
author and page number.) Without this bibliographic information, you
cannot give proper credit when you incorporate your notes into your
paper or **report.**

If you are rushed for time in the library, you may want to check out a book or books to review at home. If you need to take more information from periodical articles (which cannot be checked out) than can be conveniently committed to notes, you may want to make a photographic copy at the library to take home for fuller evaluation. (See also **library research.**)

nouns

A *noun* names a person, place, thing, concept, action, or quality. The two basic types of nouns are **proper nouns** and **common nouns.**

PROPER NOUNS

Proper nouns name specific persons, places, things, concepts, actions, or qualities. They are usually capitalized.

EXAMPLES   Abraham Lincoln, New York, U.S. Army, Nobel Prize, Montana, Independence Day, Amazon River, Butler County, Magna Carta, June, Colby College

COMMON NOUNS

Common nouns name general classes or categories of persons, places, things, concepts, actions, or qualities. Common nouns include all types of nouns except proper nouns. Some nouns (turkey/Turkey) may be both common and proper.

EXAMPLES   human, college, knife, bolt, string, faith, copper

Abstract nouns are common nouns that refer to something intangible that cannot be discerned by the five senses.

EXAMPLES   love, loyalty, price, valor, peace, devotion, harmony

Collective nouns are common nouns that indicate a group or collection of persons, places, things, concepts, actions, or qualities. They are plural in meaning but singular in form.

EXAMPLES   audience, jury, brigade, staff, committee

Concrete nouns are common nouns used to identify those things that can be discerned by the five senses.

EXAMPLES   house, carrot, ice, tar, straw, grease

A **count noun** is a type of concrete noun that identifies things that can be separated into countable units.

EXAMPLES   desks, chisels, envelopes, engines, pencils

A **mass noun** is a type of concrete noun that identifies things that comprise a mass and cannot be separated into countable units.

EXAMPLES   electricity, water, sand, wood, air, uranium

NOUN FUNCTION

Nouns may function as subjects of **verbs,** as objects of verbs and **prepositions,** as **complements,** or as **appositives.**

EXAMPLES   The *metal* bent as *pressure* was applied to it. (subjects)
The bricklayer cemented the *blocks* efficiently. (direct object)
The company awarded our *department* a plaque for safety. (indirect object)
The event occurred within the *year.* (object of a preposition)
An equestrian is a *horseman.* (subjective complement)
We elected the sales manager chairman. (objective complement)
George Thomas, the *treasurer,* gave his report last. (appositive)

Words usually used as nouns may also be used as **adjectives** and **adverbs.**

EXAMPLES   It is *company* policy. (adjective)
He went *home.* (adverb)

USING NOUNS

Forming the Possessive.   Nouns form the possessive most often by adding *'s* to the names of living things and by adding an *of* **phrase** to the names of inanimate objects.

EXAMPLES   The *chairman's* statement was forceful.
The keys *of the typewriter* were sticking.

However, either form may be used.

EXAMPLES   The *table's* mahogany finish was scratched.
Two personal friends *of the chairman* were on the committee.

Plural nouns ending in *s* need only to add an **apostrophe** to form the possessive case.

EXAMPLE   The *architects'* design manual contains many illustrations.

Plural nouns that do not end in *s* require both the apostrophe and the *s*.

EXAMPLE   The installation of the plumbing is finished except in the *men's* room.

With group words and compound nouns, add the *'s* to the last noun.

EXAMPLES   The *Chairman of the Board's* report was distributed.
My *son-in-law's* address was on the envelope.

To show individual possession with coordinate nouns, use the possessive with both.

EXAMPLES   Both the *Senate's* and *House's* galleries were packed for the hearings.
Mary's and *John's* presentations were the most effective.

To show joint possession with coordinate nouns, use the possessive with only the last.

EXAMPLES   The *Senate and House's* joint committee worked out a satisfactory compromise.
Mary and John's presentation was the most effective.

Forming the Plural.   Most nouns form the plural by adding *s*.

EXAMPLE   *Dolphins* are capable of communication with man.

Those ending in *s, z, x, ch,* and *sh* form the plural by adding *es*.

EXAMPLES   How many size *sixes* did we produce last month?
The letter was sent to all the *churches*.
Technology should not inhibit our individuality; it should fulfill our *wishes*.

Those ending in a consonant plus *y* form the plural by changing *y* to *ies*.

EXAMPLE   The store advertises prompt delivery, but places a limit on the number of *deliveries* in one day.

Some nouns ending in *o* add *es* to form the plural; others add only *s*.

EXAMPLES   One tomato plant produced twelve *tomatoes*.
We installed two *dynamos* in the plant.

Some nouns ending in *f* or *fe* add *s* to form the plural; others change the *f* or *fe* to *ves*.

EXAMPLES   cliff/cliffs, fife/fifes, knife/knives

Some nouns require an internal change to form the plural.

EXAMPLES   woman/women, man/men, mouse/mice, goose/geese

Some nouns do not change in the plural form.

EXAMPLE   *Fish* swam lazily in the clear brook while a few wild *deer* mingled with the *sheep* in a nearby meadow.

Compound nouns form the plural in the main word.

EXAMPLES   sons-in-law, high schools

Compound nouns written as one word add *s* to the end.

EXAMPLE   Use seven *tablespoonfuls* of freshly ground coffee to make seven cups of coffee.

If in doubt about the plural form of a word, look up the word in a good **dictionary.** Most dictionaries give the plural form if it is made in any way other than by adding *s* or *es*.

noun clauses

A noun clause is a subordinate **clause** that functions as a **noun.**

EXAMPLE   *Why the report took three months to prepare* is a mystery to me.

Noun clauses are frequently introduced by **interrogative** and **relative pronouns** and **adverbs** (*which, who, what, when, where,* and *why*). The **conjunction** *that* (not the relative pronoun) is also commonly used.

A noun clause can fit any **sentence** position that can be occupied by a noun.

EXAMPLES   *That we had succeeded* pleased us. (subject)
The treasurer admitted *that he had made a mistake.* (direct object of verb)
Upon retirement, she's going back to *where she came from.* (object of preposition)
Help out by doing *what you can.* (direct object of gerund)
Give *whoever comes* a ticket. (indirect object)
The main thing we should remember, *that we are in business to make a profit,* is the thing we seem to be forgetting. (appositive)
He made himself *what he wanted to be.* (objective complement)

Noun clauses appear often in definitions and explanations as **subjective complements.**

EXAMPLE   A common theory is *that competition keeps prices low.*

nowhere near

The **phrase** *nowhere near* is colloquial and should be avoided in writing.

CHANGE   His ability is *nowhere near* Jim's.
TO   His ability does *not approach* Jim's.
OR   His ability is *not comparable to* Jim's.
OR   His ability is *far inferior* to Jim's.

number

Number is the grammatical property of **nouns, pronouns,** and **verbs** that signifies whether one thing (singular) or more than one (plural) is being referred to. Nouns normally form the plural by simply adding *s* or *es* to their singular forms.

EXAMPLES   Many new business *ventures* have failed in the past ten *years.*
Partners in successful *businesses* are not always personal *friends.*

But some nouns require an internal change to form the plural.

EXAMPLES   woman/women, man/men, goose/geese, mouse/mice

All pronouns except *you* change internally to form the plural.

EXAMPLES   I/we
he, she, it/they

Most verbs show the singular of the third **person,** present **tense,** indicative **mood,** by adding an *s* or *es.*

EXAMPLES   he *stands,* she *works,* it *goes*

The verb *to be* normally changes form to indicate the plural.

EXAMPLES   I *am* ready to begin work. (singular)
We *are* ready to begin work. (plural)

See also **agreement.**

numbers

The general rule is to write numbers from zero to ten as words and numbers above ten as figures. There are, however, a number of exceptions.

One exception is that page numbers of books, as well as figure and **table** numbers, are expressed as figures.

> EXAMPLE   Figure 4 on page 9 and Table 3 on page 7 provide pertinent information.

Another exception is that units of measurement are expressed in figures.

> EXAMPLES   3 miles, 45 cubic feet, 9 meters, 27 cubic centimeters, 4 picas

Numbers that begin a **sentence** should always be spelled out, even if they would otherwise be written as figures.

> EXAMPLE   One hundred and fifty people attended the meeting.

If spelling out such a number seems awkward, rewrite the sentence so that the number does not appear at the beginning.

> CHANGE   One hundred and fifty people attended the meeting.
> TO   There were 150 people in attendance at the meeting.
> CHANGE   Two hundred seventy-three defective products were returned last month.
> TO   Last month, 273 defective products were returned.

When several numbers appear in the same sentence or **paragraph,** they should be expressed alike regardless of other rules and guidelines.

> EXAMPLE   The company owned 150 trucks, employed 271 people, and rented 7 warehouses.

When numbers appear together in the same **phrase,** write one as a figure and the other as a word.

> CHANGE   The order was for *12* 6-inch pipes.
> TO   The order was for *twelve* 6-inch pipes.

Approximate numbers are normally spelled out.

> EXAMPLE   More than two hundred people attended the conference.

Percentages are normally given as figures, and used with the percent **symbol.**

EXAMPLE Approximately 85 % of the area has been seeded with ponderosa pine.

In typed manuscript, page numbers are written as figures, but chapter or volume numbers may appear either way.

EXAMPLES page 37
Chapter 2 or Chapter Two
Volume 1 or Volume One

The plural of a written number is formed by adding *s* or *es* or by dropping *y* and adding *ies,* depending on the last letter, just as the plural of any other noun is formed. (See **nouns.**)

EXAMPLES sixes, elevens, twenties

The plural of a figure is written with *s* alone or with *'s.*

EXAMPLES 5s, 12s/5's, 12's

Do not follow a word representing a number with a figure in **parentheses** representing the same number.

CHANGE Send five (5) copies of the report.
TO Send five copies of the report.

CARDINAL AND ORDINAL NUMBERS

A cardinal number expresses an exact quantity.

EXAMPLES *one* pencil, *two* typewriters, *three* airplanes

An ordinal number expresses degree or sequence.

EXAMPLES *first* quarter, *second* edition, *third* degree

In most writing, an ordinal number should be spelled out only if it is a single word (*tenth, 312th*). Ordinal numbers can also function as **adverbs** (John arrived **first**). See also **first/firstly.**

DATES

The year and day of the month should be written as figures. Dates are usually written in month-day-year sequence, but businesses and industrial corporations are increasingly using the military day-month-year sequence.

EXAMPLES   August 24, 1975
24 August 1975

The **slash** form of expressing dates (8/24/75) is used in informal writing only.

TIME

Hours and minutes are expressed as figures when A.M. and P.M. follow.

EXAMPLES   11:30 A.M., 7:30 P.M.

When not followed by A.M. or P.M., however, time should be spelled out.

EXAMPLES   four o'clock, eleven o'clock

FRACTIONS

Fractions are expressed as figures when written with whole numbers.

EXAMPLES   27½ inches, 4¼ miles

Fractions are spelled out when they are expressed independently.

EXAMPLES   one-fourth, seven-eighths

Numbers with decimals are always written in figures.

EXAMPLE   5.21 meters

ADDRESSES

Numbered streets from one to ten should be spelled out except where space is at a premium.

EXAMPLE   East Tenth Street

Building numbers are written as figures. The only exception is the building number *one*.

EXAMPLES   4862 East Monument Street
One East Tenth Street

Highways are written as figures.

EXAMPLES   U.S. 70, Ohio 271, I 94

numeral adjectives

Numeral adjectives identify quantity, degree, or place in a sequence. They always modify **count nouns.** Numeral adjectives are divided into two subclasses: cardinal and ordinal.

A cardinal **adjective** expresses an exact quantity.

EXAMPLES   *one* pencil, *two* typewriters, *three* airplanes

An ordinal adjective expresses degree or sequence.

EXAMPLES   *first* quarter, *second* edition, *third* degree, *fourth* year.

In most writing, an ordinal adjective should be spelled out if it is a single word *(tenth)* and written in figures if it is more than one word *(312th).* Ordinal numbers can also function as **adverbs.**

EXAMPLE   John arrived *first.*

See also **first/firstly.**

O

objective

What do you want your **readers** to know or be able to do when they have read your finished writing project? When you have answered this question, you have determined the *objective* of your writing project. Too often, however, beginning writers state their objectives in broad terms that are of no practical value to them. Such an objective as "to write about the Model 6000 Accounting Machine" is too general to be of any real help. "To explain how to operate a Model 6000 Accounting Machine" is a specific objective that will help keep the writer on the right track.

The writer's objective is rarely simply to "explain" something, although on occasion it may be. You must ask yourself, "Why do I need to explain it?" In answering this question, you may find, for example, that your objective is also to persuade your reader to change his or her attitude toward the thing you are explaining.

A writer for a company magazine who has been assigned to write an article on the firm's new computer installation, in answer to the question *what,* could state the objective as "to explain the way the computer will increase efficiency and ultimately benefit employees." In answer to the question *why,* the writer might state "to ease the fear of automation that has become evident in employees."

If you answer these two questions *exactly,* and put your answers in writing as your stated objective, not only will your job be made easier but you will be considerably more confident of ultimately reaching your goal. As a test of whether you have adequately formulated your objective, try to state it in a single **sentence.** If you find that you cannot, continue to formulate your objective until you *can* state it in a single sentence.

Even a specific objective is of no value, however, unless you keep it in mind as you work. Guard against losing sight of your objective as you become involved with the other steps of the writing process.

objective complements

An objective complement is a **noun** or **adjective** that completes the meaning of a **sentence** by revealing something about the **direct object** of a **transitive verb;** it may either describe (adjective) or rename (noun) the direct object.

EXAMPLES   We painted the house *white*. (adjective)
I like my coffee *hot*. (adjective)
They call her a *genius*. (noun)

The objective complement is one of the four kinds of **complements** that complete the subject-verb relationship; the others are the direct object, the **indirect object,** and the **subjective complement.**

objects

There are three kinds of objects: **direct object, indirect object,** and object of a **preposition.** All objects are **nouns** or noun equivalents (**pronoun, gerund, infinitive,** noun phrase, **noun clause**) and can be replaced by a pronoun in the objective **case.**

DIRECT OBJECTS

The direct object answers the question "what?" or "whom?" about a **verb** and its **subject.**

EXAMPLES   Sheila designed a new *circuit*. (*Circuit,* the direct object, answers the question, "Sheila designed *what?*")
George telephoned the *Chief Engineer*. (*Chief Engineer,* the direct object, answers the question, "George telephoned *whom?*")

A verb whose meaning is completed by a direct object is called a **transitive verb.**

INDIRECT OBJECTS

An indirect object answers the questions "to whom?" or "to what?" or "for whom?" or "for what?" about a transitive verb, its subject, and its direct object. The indirect object always precedes the direct object.

EXAMPLES   We sent the *General Manager* a full report. (*Report* is the direct object. The indirect object, *General Manager,* answers the question, "We sent a full report *to whom?*")
The General Manager gave the *report* careful consideration. (*Consideration* is the direct object. The indirect object, *report,* answers the question, "The General Manager gave careful consideration *to what?*")
The Purchasing Department bought *Sheila* a new oscilloscope. (*Oscilloscope* is the direct object. The indirect object, *Sheila,* answers the question, "The Purchasing Department bought a new oscilloscope *for whom?*")

OBJECTS OF PREPOSITIONS

For a discussion of objects of prepositions, see **prepositional phrases.**

observance/observation

An *observance* is the "performance of a duty, custom, or law"; it is sometimes confused with *observation,* which is the "act of noticing or recording something."

EXAMPLES   The *observance* of Veteran's Day as a paid holiday varies from one organization to another.

The laboratory technician made careful *observations* during the experiment.

OK/okay

The expression *okay* (also spelled *OK*) is common in **informal writing** but should be avoided in more formal **correspondence** and **reports.**

CHANGE   Mr. Sturgess gave his *okay* to the project.
TO   Mr. Sturgess *approved* the project.

CHANGE   That is *okay* with me.
TO   That is *acceptable* to me.

on account of

As a substitute for *because,* the **phrase** *on account of* should be avoided in writing.

CHANGE   He felt that he had lost his job *on account of* the company's switch to automated equipment.

TO   He felt that he had lost his job *because* the company switched to automated equipment.

on/onto

On is normally a **preposition** meaning "supported by," "attached to," or "located at."

EXAMPLE   Install the telephone *on* the wall.

Onto implies movement to a position on or movement up and on.

EXAMPLE   The union members surged *onto* the platform after their leader's defiant speech.

on the grounds that/of

The **phrase** *on the grounds that*—or *on the grounds of*—is a wordy substitute for *because*.

CHANGE   She left *on the grounds that* the NASA position offered a higher salary.
TO   She left *because* the NASA position offered a higher salary.

one

When used as an **indefinite pronoun,** *one* may help you avoid repeating a **noun.**

EXAMPLE   We need a new plan, not an old *one*.

One is often redundant in **phrases** where it restates the noun, and it may take the proper **emphasis** away from the **adjective.**

CHANGE   The computer program was not a unique *one*.
TO   The computer program was not unique.

One can also be used in place of a noun or **personal pronoun** in a statement such as the following.

EXAMPLE   *One* cannot ignore *one's* physical condition.

In statements of this kind, to avoid the tedious **repetition** of *one* it is permissible to use a third-person singular personal pronoun with *one* as the antecedent.

CHANGE   *One* must do *one's* best to correct the injustices *one* sees.
TO   *One* must do *his* best to correct the injustices *he* sees.
OR   *One* must do *her* best to correct the injustices *she* sees.

If you use *one* in this manner, be careful not to shift back and forth between *one* and *you*. However, to use *one* in this way at all is distinctly formal and impersonal; in any but the most formal writing you are better advised to address your **reader** directly and personally as *you*.

CHANGE   *One* cannot be too careful about planning for leisure time. The cost of *one's* equipment can, for example, force *one* to work more and therefore reduce *one's* leisure time.

TO   *You* cannot be too careful about planning for leisure time. The cost of *your* equipment can, for example, force *you* to work more and thus reduce *your* leisure time.

See also **point of view.**

one of those . . . who

A **dependent clause** beginning with *who* or *that* and preceded by *one of those* takes a plural **verb.**

EXAMPLES   She is *one of those* executives *who are* concerned about their writing.
This is *one of those* policies *that make* no sense when you examine them closely.

In the preceding examples *who* and *that* are subjects of dependent clauses and refer to plural antecedents (*executives* and *policies*) and thus take plural verbs, *are* and *make.*

Because people so often use singular verbs in such constructions in everyday speech and very **informal writing,** this rule confuses many writers. The principle behind the rule becomes clearer if *among* is substituted for *one of.* Compare the following examples with the preceding ones.

EXAMPLES   She is *among* those executives *who are* concerned about their writing. (You would not write, ''She is among those executives who *is* concerned . . .'')
This is *among* those policies *that make* no sense when you examine them closely. (You would not write, ''This is among those policies that *makes* no sense . . .'')

Another way to clarify the principle behind the rule is to turn the **sentence** around. Compare the following examples with the examples given above.

EXAMPLES   Of those executives who *are* concerned about their writing, she is one.
Of those policies that *make* no sense when you examine them closely, this is one.

When the sentences are seen turned around in this rather unnatural but perfectly logical way, you would not be likely to use *is* for *are* (in the first sentence) or *makes* for *make* (in the second sentence). It may also be helpful to contrast these sentences with others in which the singular verb is the correct form.

EXAMPLES She is one executive *who is* concerned about her writing. (*Who* refers to a singular antecedent, *executive,* and is therefore singular and takes a singular verb, *is.*)

This is one policy *that makes* no sense when it is examined closely. (*That* refers to a singular antecedent, *policy,* and thus requires a singular verb, *makes.*)

There is one exception to the plural-verb rule stated at the very beginning of this entry: If the phrase *one of those* is preceded by *the only (the only one . . .),* the verb in the following dependent clause should be singular.

EXAMPLES She is *the only one* of those executives *who is* concerned about her writing. (The verb is singular because its subject, *who,* refers to a singular antecedent, *one.* If the sentence were reversed, it would read,"Of those executives, she is the only *one who is* concerned about her writing.")

This is *the only one* of those policies *that makes* no sense when you examine it closely. (If the sentence were reversed, it would read, "Of those policies, this is the only *one that makes* no sense when you examine it closely.")

Finally, it should be repeated that the plural-verb rule given at the beginning of this entry is often ignored in everyday speech and very informal writing. Indeed, in spoken English the singular form of the verb, though technically illogical, is probably far more common than the plural. In work-related writing, where preciseness is crucial, you should use a plural verb after *one of those who* or *one of those that.*

only

In writing, the word *only* should be placed immediately before the word or **phrase** it modifies.

CHANGE We *only* lack financial backing; we have determination.
TO We lack *only* financial backing; we have determination.

A speaker can place *only* before the **verb** and avoid **ambiguity** by stressing the word being modified; in writing, only correct placement of the word can ensure **clarity.** Incorrect placement of *only* can change the meaning of a sentence.

EXAMPLES *Only* he said that he was tired. (He alone said.)
He *only* said that he was tired. (He actually was not tired although he said he was.)

He said *only* that he was tired. (He said nothing except that he was tired.)

He said that he was *only* tired. (He was nothing except tired.)

openings

If your **reader** is already familiar with your subject, or if what you are writing is short, you may not need to begin your writing project with a full **introduction**. You may simply want to focus the reader's attention with a brief opening.

For many types of on-the-job writing, openings that simply get to the point are quite adequate, as shown in the following examples.

Correspondence
Mr. George T. Whittier
1720 Old Line Road
Thomasbury, WV 26401

Dear Mr. Whittier:

You will be happy to know that we have corrected the error in your bank balance. The new balance shows . . .

Progress Report Letter
William Chang, M.D.
Phelps Building
9003 Shaw Avenue
Parksville, MD 29099

Dear Dr. Chang:

To date, eighteen of the twenty specimens you submitted for analysis have been examined. Our preliminary analysis indicates . . .

Longer Progress Report
PROGRESS REPORT ON REWIRING
THE SPORTS ARENA

The rewiring program at the Sports Arena is continuing on schedule. Although the costs of certain equipment are higher than our original bid had indicated, we expect to complete the project without exceeding our budget, because the speed with which the project is being completed will save labor costs.

Work Completed
As of August 15th, we have . . .

Memorandum

To: Jane T. Meyers, Chief Budget Manager
From: Charles Benson, Assistant to the Personnel Director
Date: June 12, 19—
Subject: Budget Estimates for Fiscal Year 19—

The personnel budget estimates for fiscal year 19— are as follows
. . .

You may also use an intriguing or interesting opening, either by itself or in conjunction with an introduction, to stimulate your reader's interest. Such types of writing as **annual reports, journal articles,** and **house organ articles** often *require* an interesting opening because you do not have a captive audience. Such openings have two essential purposes: to indicate the subject and to catch the interest of the reader. Several types of openings achieve both purposes. The opening should be natural, not forced; an obviously "tacked on" opening will only puzzle your reader, who will be unable to establish any meaningful connection between the opening and the body of your article or report.

STATEMENT OF THE PROBLEM

One way to provide the reader with the perspective of your report is to present a brief account of the problem that led to the study or project being reported.

EXAMPLE   Several weeks ago a brewmaster noticed a discoloration in the grain supplied by Acme Farms, Inc. He immediately reported his discovery to his supervisor. After an intensive investigation, we found that Acme . . .

However, if the reader is familiar with the problem, or if the particular problem has been solved for a long period, a problem opening is likely to be boring to your reader.

DEFINITION

Although a definition can be useful as an opening, do not define something with which the reader is familiar or provide a definition that is obviously a contrived opening (such as "Webster defines business as . . ."). A definition should be used as an opening only if it provides insight into what follows.

EXAMPLE   *Risk* is a loosely defined term. It is used here in the sense of physical risk as a qualitative combination of the probability of an event and the severity of the consequences of that event. *Risk assessment* is the process of estimating the probabilities and consequences of events and of establishing the accuracy of these estimates. Another necessary term is *risk appraisal.* This goes far beyond assessment and involves judgment about people's perception of risk and their reactions to this perception; it extends to the final process of making decisions. Risk appraisal is thus an essential part of the work of a licensing and regulatory body in formulating safety policy and applying this policy to individual plants. It is part of the everyday work of inspectors. However, only in recent years have attempts been made to quantify the appraisal aspects of risk.

—H.J. Dunster and W. Vanek, "The Assessment of Risk—Its Values and Limitations," *Nuclear Engineering International* 24 (August 1979), p. 23.

INTERESTING DETAIL

Often an interesting detail of your subject can be used to gain readers' attention and arouse their interest. This requires, of course, that you be aware of your readers' interests. The following opening, for example, might be especially interesting to members of a personnel department.

EXAMPLE   The small number of graduates applying for jobs at Acme Corporation is disappointing, particularly after many years of steadily increasing numbers of applicants. There are, I believe, several reasons for this development . . .

Sometimes it is possible to open with an interesting statistic.

EXAMPLE   From asbestos sheeting to zinc castings, from a chemical analysis of the water in Lake Maracaibo (to determine its suitability for use in steam injection units) to pistol blanks (for use in testing power charges), the Purchasing Department attends to the company's material needs. Approximately 15,000 requisitions, each containing from one to fourteen separate items, are processed each year by this department. Every item or service that is bought . . .

BACKGROUND

The background or history of a certain subject may be quite interesting and may even put the subject in perspective for your reader. Consider the following example from a house organ article describing the process of oil drilling.

EXAMPLE From the bamboo poles the Chinese used when the pyramids were young to today's giant rigs drilling in a hundred feet of water, there has been a lot of progress in the search for oil. But whether four thousand years ago or today, in ancient China or a modern city, in twenty fathoms of water or on top of a mountain, the object of drilling is and has always been the same—to manufacture a hole in the ground, inch by inch. The hole may be either for a development well . . .

This type of opening is easily overdone, however; use it only if the background information is of some value to your reader. Never use it just as a way to get started.

QUOTATION

Occasionally, you can use a **quotation** to stimulate interest in your subject. To be effective, however, the quotation must be pertinent—not some loosely related remark selected from a book of quotations simply because it was listed under the subject heading that fits your writing project. Often a good quotation to use is one that predicts a new trend or development.

EXAMPLE Richard Smith, president of P. R. Smith Corporation, recently said, "I believe that the Photon projector will revolutionize our industry." His statement represents a growing feeling among corporate . . .

OBJECTIVE

In reporting on a project or activity of some kind, you may wish to open with a statement of the project or activity's objective. Such an opening gives the reader a basis for judging the actual results as they are presented.

EXAMPLE The primary objective of this project was to develop new techniques to measure heat transfer in a three-phase system. Our first step was to investigate . . .

SUMMARY

You can provide a summary opening by greatly compressing the results, conclusions, or recommendations of your article or report. Do not start a summary, however, by writing "This report summarizes . . ."

CHANGE   This report summarizes the advantages offered by the photon as
a means of examining the structural features of the atom. The
photon is a specially designed laser used for examining . . .

TO   As a means of examining the structure of the atom, the photon
offers several advantages. Since the photon is especially designed for examining . . .

FORECAST

Sometimes you can use a forecast of a new development or trend to
arouse the reader's interest.

EXAMPLE   In the very near future, we may be able to call our local library
and have a video tape of *Hamlet* replayed on our wall television.
This project and others are now being developed at Acme
Industries . . .

SCOPE

At times you may want to present the **scope** of your article or report in
your opening. By providing the parameters of your material—the limitations of the subject or the amount of detail to be presented—you enable
your readers to determine whether they want or need to read your article
or report. Be on guard against letting this kind of opening become a
narrated **table of contents** or an **abstract;** if either of those is appropriate, include it—but not as an artificial opening. The scope type of
opening is often useful for small writing projects where an abstract or
table of contents would be inappropriate.

EXAMPLE   This pamphlet provides a review of the requirements for obtaining an FAA pilot's license. It is not designed as a textbook to
prepare you to take the examination itself; rather, it gives you
an idea of the steps you need to take and the costs involved. . . .

oral reports

Preparing an oral presentation is much like preparing to write. You must
analyze your audience, determine your **objective,** and then develop
your **topic** and prepare an **outline.**

KNOWING YOUR AUDIENCE

To communicate effectively with your listeners you must know who they
are. Are they coworkers, customers, clients, management, or fellow
members of a professional organization? The answer to this question will
help determine how much knowledge of your subject your listeners

already possess. If you are speaking to those in your own field, you can assume that they understand the terminology of that field—so you can use it freely. If you are speaking to those outside your field, on the other hand, you should carefully define any special terms.

Knowing your audience will also help you recognize other factors that might improve your communication with your listeners. An audience composed of management is likely to be interested in the practical aspects of the material you are presenting: costs, scheduling, staff needs, and so on. If you are speaking to fellow technicians, on the other hand, they are likely to be concerned with the technical details of your presentation.

In preparing an oral presentation, you must pay particular attention to the size of your audience. Facial expressions and gestures that would be effective if you were speaking to half a dozen people sitting around a conference table, for example, would be lost on an audience of a hundred workers watching you demonstrate cardiopulmonary resuscitation. With such a large audience, you probably could not rely at all on facial expressions to convey an idea; and instead of pointing with your finger at a chart, you'd have to use a pointer. The size of your audience can also influence the approach you take to your presentation. For a small group of listeners, for instance, you might plan a presentation that is mainly discussion, composed of a short talk and a long question-and-answer period. For a large audience, on the other hand, you would be more likely to prepare a longer, more formally organized presentation with only a brief question-and-answer period.

DEFINING YOUR OBJECTIVE

Determine the purpose of your presentation by asking yourself what you want your listeners to know, to be able to do, or to believe after you finish. You must establish a clear purpose to know what to include in your presentation to make it effective.

DEVELOPING YOUR REPORT

Speaking requires a different approach than writing, for all audiences have limited attention spans. A presentation crammed with difficult ideas or numerous statistics is sure either to confuse listeners or to put them to sleep. Cite facts and figures sparingly, and use them in such a way that they directly support the major points you wish to make. Put your most important ideas into your **opening** and closing, where your listeners are more likely to remember them. Unless your presentation is short (under ten minutes), your **conclusion** should summarize the information presented in the rest of the talk.

There are as many ways to develop a topic in speaking as there are in writing. Several **methods of development** are especially good for oral presentations:

Increasing order of importance develops a topic from the least important point to the most important point and thus leaves the audience with the most important points fresh in their minds at the end.

The problem-solution pattern first describes and analyzes the problem, then presents the criteria for evaluating possible solutions, and finally explains the advantages and disadvantages of each possible solution.

Cause-and-effect explains why something happened or why you predict that something will happen. This approach can be effective in a speech because it can hold the audience's interest much as a detective story does.

Specific-to-general begins with specific details such as statistics, expert opinions, and examples and then uses them to make a general statement.

General-to-specific presents a general statement and then follows it with supporting details—statistics, expert opinions, and examples.

When you have determined the best method of development for your subject, create an **outline** for your presentation. The following type of outline is common.

OUTLINE

Background Information About the Autoclave (Sterilizer)

I. Opening
   A. I am Steve Philisandro, Assistant Manager of the Supply Department.
   B. Ms. Cynthia Lipanski has given you an overview of the department—now I would like to give you some background about our autoclave sterilizer units.
   C. The points I will cover are explained in greater detail in our department manual, which each of our employees uses.

II. Materials
   A. Materials that are sterilized include cloth, metal, glass, liquid, rubber, and plastic.
   B. Items that are sensitive to steam must be sterilized with the ethylene oxide unit: rubber goods, electric cords, telescopic lenses, and delicate instruments, for example.

   C. Materials, other than liquids, are wrapped in muslin or placed in *peel* packages and heat sealed *(show instrument and "peel" packages)*.
   D. Liquids, such as saline, are placed in pyrex bottles with closures that allow steam to penetrate.
   E. Each type of material requires precise settings for time, pressure, and temperature *(show operators' manual)*.

III. Procedure
   A. Operator checks that items are properly wrapped and spaced on the sterilizer cart *(point to cart)*.
   B. Chemical indicators are placed inside the packages, and heat sensitive tape is used on the outsides.
   C. Operator loads the cart (cooled to prevent condensation) into the sterilizer units.
   D. Although settings are usually pre-set, operator checks time, temperature, and pressure *(show gauges)*.
   E. Operator first closes the door, and then pushes the "lock" button; operation proceeds automatically from that point.
   F. Lights on the front of the door will indicate completion of cycle.
   G. Operator removes the sterilizer cart, using asbestos gloves.
   H. Packages are allowed to cool before they are placed on storage shelves in the next room.
   I. Westdale's process is more automated than those at smaller hospitals.

IV. Closing
   A. I hope I've given you some idea of how our sterilizer units work.
   B. Are there any questions?
   C. Thank you!
   D. Mrs. Sanches will now demonstrate Unit #2.

The amount of detail you should include in your outline depends upon both your confidence in front of a group and your familiarity with the subject. The more confident and familiar with the subject you are, the less detail you will need in your outline. However, you may need to write out and read verbatim any portions that you feel need precise wording—for example, important policy statements.

Write your notes either in a three-ring binder or on 4 × 6 inch note cards. Be careful not to cram the cards with too many notes. You should be able to grasp the information on a card at a glance so that you can maintain eye contact with your listeners. Number the cards so you don't get lost during your presentation.

VISUAL AIDS

Some speakers find it helpful to use visual aids such as **graphs, drawings, tables, photographs,** or models. You can present them by means of chalkboards, posters, flip charts, or projectors. Well-planned visual aids can add interest and emphasis to a presentation and can also clarify and simplify your message by reinforcing its key points.

Visual aids can be overdone, of course. Do not attempt to use visual aids to flesh out a skimpy presentation or to form the focal point of a disorganized one. Use aids only if they clarify a point or make your presentation more vivid and concise. If they are necessary, be certain to check the physical arrangement and facilities of the room where you will speak. Does it have a chalkboard? a projector screen? a flip chart? Will a slide or overhead projector be available? Also consider room size. If you plan to use a chalkboard, will the writing be visible to those seated in the back? Everything you use must be tailored to scale. Check to see that nothing obstructs the view of any member of your audience. Keep in mind the following points about using visual aids:

1. Keep the amount of information on each visual aid to a minimum. It is usually a good idea to present only one point on each visual aid. Avoid too many figures or equations.
2. Do not write out **sentences;** use only words and **phrases.**
3. During your presentation emphasize a visual aid by pointing to it— with a physical gesture *and* with words.
4. As you discuss the material on the visual aid, be careful not to block it with your body.
5. Be sure to talk to your listeners, not to the visual aid.
6. Do not simply read from the visual aid; after all, your audience can read. It is sometimes helpful, though, to list major points in key words on the visual aid to help your listeners keep them in mind.

Following are commonly used visual aids. Each has advantages based on the size of the audience and the flexibility that your subject requires.

A *chalkboard* is easy to use and to control; it gives you complete flexibility to decide what goes on, what comes off, and when. It adds animation to your presentation because you must physically put the information on the board. If your information is too long to put on the board conveniently while you speak, put part or all of your message on ahead of time. A disadvantage is that you must interrupt your presentation to write on the board. Chalkboards are good with small and medium-size audiences (up to 40 people).

A *poster* can be used for medium-size and large audiences (40 to 100 people). It must be large enough for your audience to see: A common size is 2 × 3 feet. Prepare the poster ahead of time, and make your lettering large enough to be seen comfortably from the back of the room. Felt-tip markers are good writing instruments for posters. Put the poster on the easel only when you are ready to use it; otherwise, it may distract your audience.

A *flip chart* provides you with good control of what you want the audience to see—you flip it over only when you need it. You can draw your **illustration** ahead of time with light pencil lines and fill the lines in with the bold lines of a felt-tip pen during your presentation (sketching it during the presentation helps hold the audience's attention). Flip charts are good for small and medium-size audiences.

Slide and *overhead projectors* can be easy to use but you must learn how to handle them. As you prepare your presentation, decide which slides you will use and where each one will go. Have them made up far enough in advance to be able to rehearse with them. Slides and transparencies may be used for audiences of various sizes.

DELIVERING AN ORAL PRESENTATION

Your oral presentation is most likely to be on a job-related topic. Therefore, you should be able to speak with knowledge and conviction about your subject. This is important because your listeners will have greater confidence in what you say if you speak with conviction.

The key to delivering your presentation well is practice. Rehearse with your outline until you know exactly how you want to move from one idea to another. If you practice, you will have no problem with the words you want to use; they will come naturally, and you will have the confidence you need to face your listeners.

Generating Confidence and Enthusiasm. Speaking is like selling. If you believe in your topic, you will be enthusiastic about it, and your audience will find themselves thinking, "Hey, I didn't know that!"

You can expect to be nervous; but if you know what you are going to say and how you are going to say it you will probably relax as you begin speaking. Stand poised and erect, but not stiff. A neat appearance can also add to your self-confidence. Remember to project an image of friendliness; most people react favorably to a friendly approach. Friendliness cannot be faked, of course; it must be honest to be effective.

Getting and Holding Your Audience. The opening of a presentation can be crucial. It should arouse interest, stimulate curiosity, or impress your listeners with the importance of your subject. You can begin your presentation in a number of ways that will catch the attention of your listeners. Study the different ways presented in **openings.**

Delivery Techniques. When you stand up to speak, do not rush your opening. Begin by taking a good look at your listeners, and let them take a good look at you. Then begin your planned opening firmly and authoritatively. Never apologize for inconvenient conditions or tell your listeners how nervous you are.

Try to talk to your listeners about your subject just as you would talk to a friend. Maintain eye contact with most of your listeners, not with just one person. In addition to establishing a directness of communication with them, eye contact can provide you with feedback from your audience: You will be able to notice if they begin to look puzzled or confused and will know if you need to clarify a point.

Be careful not to talk too fast, but do not talk in a monotone. Enunciate carefully, saying "going" instead of "gonna," and "don't you think?" instead of "dontchathink?" Try to eliminate sounds like "ah," "like," "okay?" and "yaknow?" Also, learn to suppress nervous mannerisms, such as adjusting your clothing or smoothing your hair.

To emphasize a point, say it—then pause to let its significance register with your listeners—and then repeat the point. You can also achieve **emphasis** by varying the pitch or the volume of your voice, and by varying the pace of your presentation. Gestures can also provide emphasis. A closed fist, for example, can stress an idea dramatically. Be sure, though, to use gestures only when they come naturally; in other words, use the same gestures you would in ordinary conversation.

Refer to your notes openly; when you need to look at them, do so. Your listeners won't be impatient; they are human, too.

Concluding. When you are finished, conclude promptly but not abruptly. Don't end as though you were beating a hasty retreat. Plan the ending of your presentation the way you planned your opening. It's your last chance to get your message across, so don't waste it. You may find that it is most effective to conclude with a summary of your main points—so that your listeners will be more likely to remember them. Or, depending on your topic, you might end by making a recommendation, by asking a question, or by using any of the other techniques suggested in **conclusions.**

Question-and-Answer Period. Depending upon the occasion and your subject, you may want to allow for a question-and-answer session at the end of your presentation. If so, answer questions as completely as time allows or as the question deserves. Be sure that your answers are accurate. Don't be afraid to respond with "I don't know." Be polite and objective in responding to a hostile question, but be prepared to go on to the next question if it becomes evident that the questioner simply wants to argue. Don't play for laughs at a questioner's expense. You'll offend not only the questioner but the rest of your listeners as well.

For advice on chairing committee meetings and leading panel discussions, see **meetings.**

oral/verbal

Oral refers to that which is spoken.

EXAMPLE   He made an *oral* commitment to the policy.

Although it is sometimes used synonymously with *oral, verbal* literally means "in words" and can refer to that which is spoken or written. To avoid possible confusion, do not use *verbal* if you can use *written* or *oral.*

CHANGE   He made a *verbal* agreement to complete the work.
TO   He made a *written* agreement to complete the work.
OR   He made an *oral* agreement to complete the work.

When you must refer to something both written and spoken, you should use both *written* and *oral* to make your meaning clear.

EXAMPLE   He demanded either a *written* or an *oral* agreement before he would continue the project.

order letters

One of the most common reasons for writing a letter, especially if you work for a small organization or are self-employed, is to order supplies or equipment. The supplies may be anything from company letterhead to X-ray film. Obviously, an order letter must be specific and complete if you are to receive the exact item you want. But be careful not to clutter the letter with unnecessary details, such as why you need the items or

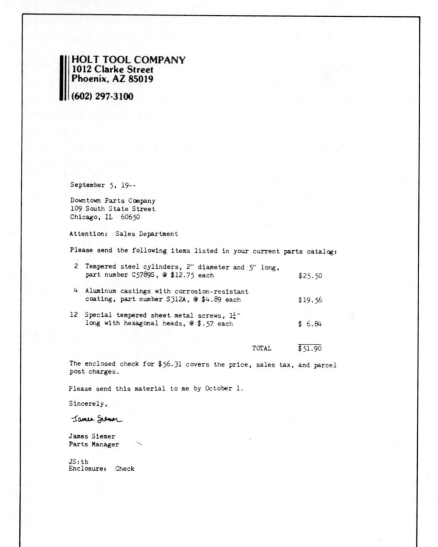

HOLT TOOL COMPANY
1012 Clarke Street
Phoenix, AZ 85019

(602) 297-3100

September 5, 19--

Downtown Parts Company
109 South State Street
Chicago, IL 60650

Attention: Sales Department

Please send the following items listed in your current parts catalog:

2	Tempered steel cylinders, 2" diameter and 5" long, part number C5789S, @ $12.75 each	$25.50
4	Aluminum castings with corrosion-resistant coating, part number S312A, @ $4.89 each	$19.56
12	Special tempered sheet metal screws, 1¼" long with hexagonal heads, @ $.57 each	$ 6.84

                                                    TOTAL $51.90

The enclosed check for $56.31 covers the price, sales tax, and parcel
post charges.

Please send this material to me by October 1.

Sincerely,

James Siemer

James Siemer
Parts Manager

JS:tb
Enclosure: Check

Letter 16

HOLT TOOL COMPANY
1012 Clarke Street
Phoenix, AZ 85019

(602) 297-3100

October 12, 19--

Downtown Parts Company
109 South State Street
Chicago, IL 60650

Attention: Sales Department

On September 5, I sent you a letter in which I placed an
order for a number of parts listed in your current catalog. I
also sent a check for $56.31.

I have enclosed a copy of that letter, and I request that you
please rush this material to me. If for some reason you
cannot send the material, please cancel my order and return
the check.

Thank you.

Sincerely,

James Siemer

James Siemer
Parts Manager

JS:tb
Enclosure: Copy of order letter

Letter 17

who will use them. Above all, be accurate. Since a misspelled word or misplaced decimal point could cause a staggering error, proofread carefully and double-check all price calculations. If you order several items, it is best to give them in list form.

Make sure that the order letter contains any of the following information that is appropriate.

1. The exact name and part number of the item
2. Any useful description of the item: size, style, color, and so on
3. The quantity of each item
4. The price of the item (both unit price and total price)
5. The shipping method: mail, air, express, etc.
6. The date of the order and the date by which you need the item (indicate "rush" or other instructions)
7. The place to which the material is to be shipped (including the exact shipping address)
8. The method of payment (either including a check or money order, indicating that you will pay c.o.d., or enclosing a purchase order number for the seller to bill your company)

Letter 16 is a typical order letter.

If you receive no response to an order letter within about ten days after the item was due, you may need to send a follow-up letter. If you do so, be courteous and avoid showing irritation. In your follow-up, identify the order by referring to the original letter by date. It is also wise to include a copy of the letter. Letter 17 is one follow-up letter. (See also **correspondence** and **lists.**)

organization

Organization is achieved by developing your **topic** in a way that will make it easiest for your **reader** to understand your message and then outlining your material on the basis of that **method of development.**

A logical method of development satisfies the reader's need for a controlling shape and structure for your subject. An appropriate method of development is the writer's tool for keeping things under control and the reader's means of following the writer's development of a theme. Many different methods of development are available to the writer; this book includes those that are likely to be used by technical people: **sequential, chronological, increasing-order-of-importance, decreasing-order-of-importance, division and classification, comparison,** spatial, specific-to-general, general-to-specific, and cause-and-effect. As the writer, you must choose the method of development that best suits your subject, your reader, and your **objective.**

Outlining provides structure to your writing by assuring that it has a beginning, a middle, and an end. It provides proportion to your writing by assuring that one step flows smoothly to the next without omitting anything important, and it enables you to emphasize your key points by placing them in the positions of greatest importance. The use of an outline makes larger and more difficult subjects easier to handle by breaking them into manageable parts. Finally, by forcing you to organize your subject and structure your thinking in the outline stage, creating a good outline releases you to concentrate exclusively on writing when you begin the rough draft.

organizational charts

An organizational chart shows how the various components of an organization are related to one another. It is useful where you want to give your **readers** an overview of an organization or where you want to show them the lines of authority within it.

The title of each organizational component (office, section, division) is placed in a separate box. These boxes are then linked to a central authority. (See Figure 33.) If your readers need the information, include the name of the person occupying the position identified in each box. (See Figure 34.)

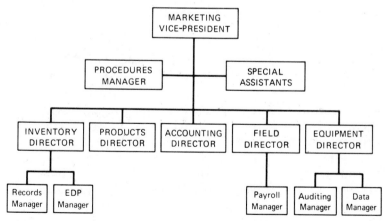

Figure 33

```
┌─────────────────────────┐
│    FIELD DIRECTOR        │
│     J. Ann Carlson       │
└─────────────────────────┘
```

Figure 34

As with all **illustrations,** place the organizational chart as close as possible to the text that refers to it.

orient/orientate

Orientate is merely a **long variant** of *orient,* meaning "locate in relation to something else." The shorter form should be used because it is simpler.

CHANGE   Let me *orientate* your group on our operation.
TO   Let me *orient* your group on our operation.

outlining

Outlining provides structure to your writing by assuring that it has a beginning **(introduction),** a middle (main body), and an end **(conclusion).** An outline gives your writing proportion so that one part flows smoothly to the next without omission of anything important. Outlining also enables you to emphasize your key points by placing them in the positions of greatest importance.

Like a road map, an outline indicates a starting point and keeps you moving logically so that you don't get lost before arriving at your conclusion. (Errors in logic are much easier to detect in an outline than in a draft.) The use of an outline makes larger and more difficult subjects easier to handle by breaking them into manageable parts; therefore, the less certain you are about your writing ability or about your subject, the fuller your outline should be. The parts of an outline are easily moved about, so you can experiment to see what arrangement of your ideas is most effective. Perhaps most important, creating a good outline frees you to concentrate on writing when you begin the rough draft (by forcing you to organize your subject and structure your thinking beforehand).

An outline is a tool that helps the writer to make his writing more logical and coherent. Two types of outlines are generally used: the topic outline and the sentence outline.

The topic outline consists of short **phrases** that are listed to show the sequential order and relative importance of ideas; in this manner, a topic outline provides order and establishes the relationships of topics to one another. The topic outline alone is generally not sufficient for a large or complex writing job; however, it may be used to good advantage for structuring the major and minor divisions of your **topic** in preparation for creating a sentence outline. (An outline for a small job is not as detailed as one for a larger job, but it is just as important; for example, a topic outline that lists your major and minor points can help greatly even in writing important letters.)

On a large writing project, it is wise to create a topic outline first and then use it as a basis for creating a sentence outline. In a sentence outline, the writer summarizes each idea in a single complete **sentence** that will become the **topic sentence** for a **paragraph** in the rough draft. The sentence outline begins with a main idea statement that establishes the subject, then follows with a complete sentence for each idea in the major and minor divisions of the outline. A sentence outline provides order and establishes the relationship of topics to one another to a considerably greater degree than a topic outline. A well-developed sentence outline offers a sure test of the validity of your arrangement of your material. If most of your notes can be shaped into controlling topic sentences for paragraphs in your rough draft, you can be relatively sure that your paper will be well organized before you begin its final composition. A good sentence outline, when stripped of its numbering symbols, needs only to be expanded with full details to become a rough draft.

Your outline should begin with a thesis statement, which is a statement of the main idea that you intend to develop in your report. This statement provides you with a goal that you can work toward as you create your outline. (Carried through to the final draft, the thesis statement summarizes your subject for your readers and lets them know what to expect.) A thesis statement often suggests a particular **method of development.**

EXAMPLES   The mission director must oversee a rigid chain of events that must occur prior to blast-off. (This thesis statement suggests a **sequential method of development.**)

Increased sales in the Chicago area appear to be a direct result of the new marketing techniques introduced there last year. (This thesis statement suggests a **cause-and-effect method of development.**)

The growth of the computer industry has closely paralleled the growth of conglomerates. (This thesis statement suggests a combination of the **chronological method of development** and **comparison method of development.**)

Although the following outlining technique is mechanical, it is easy to master and well suited to unraveling the complexities of large and difficult subjects.

The first step is to find the natural major divisions of your subject and write them down, then determine whether they are exactly what you need to meet the demands of your **objective,** your **reader,** and your **scope** of coverage. If they are—or when they are—arrange them in the proper order and label them with Roman numerals. For example, the major divisions for this discussion of outlining could be as follows.

        I. Advantages of outlining
       II. Types of outlining
      III. Effective outlining
       IV. Writing the draft from the outline

The second step is to establish your minor points by searching out the minor divisions within each major head. Arrange them in the proper sequence under their major **heads** and label them with capital letters. Keep major heads equal in importance and minor heads equal in importance; the purpose of this **parallel structure** is to keep your thinking in order. The major and minor heads for this discussion of outlining could be as follows.

        I. Advantages of outlining
       II. Types of outlining
           A. The topic outline
           B. The sentence outline
      III. Effective outlining
           A. Establish major and minor heads
           B. Sort note cards by major and minor heads
           C. Complete the sentence outline
       IV. Writing the draft from the outline

Of course, you will often need more than the two levels of heads illustrated here. If your subject is complicated, you may need four or five levels to keep all of your ideas straight and in proper relationships to one another. In that event, the following numbering scheme is recommended.

        I. First-level head
           A. Second-level head
              1. Third-level head
                 a. Fourth-level head
                    (i) Fifth-level head

The third step is to mark each of your note cards with the appropriate Roman numeral and capital letter. Sort the note cards by major and minor heads (Roman numerals and capital letters). Then arrange the cards within each minor head and mark each with the appropriate sequential Arabic number. Transfer your notes to paper, converting them to complete sentences. You now have a complete rough sentence outline, and the most difficult part of the writing job is over.

The final step is to polish your rough outline. Check to ensure that the **subordination** of minor heads to major heads is logical. Remember that all major heads should be parallel and all minor heads should be parallel.

CHANGE   A. Establish major and minor heads
         B. Note cards sorted by major and minor heads
         C. Complete the sentence outline
    TO   A. Establish major and minor heads
         B. Sort note cards by major and minor heads
         C. Complete the sentence outline

Make certain that your outline follows your method of development. Check for **unity.** Does your outline stick to the subject, or does it stray into unrelated or only loosely related topics? Stay within your established scope of coverage. Resist the temptation to tell all you know unless it is all pertinent to your objective. It is much easier to correct this problem in the outline stage than in the rough draft. If some information is of importance to only a minority of your readers, consider including that information in an **appendix;** those few readers can then pursue the subject further and the others will not be interrupted. Check your outline for completeness; scan it to see whether you need more information in any of your divisions, and plug in any information that is needed.

You now have a final, polished sentence outline. It isn't sacred, however; change it if you need to as you write the draft; the outline should be your point of departure and return. Return to it to find your place and your direction as your work.

outside of

In the **phrase** *outside of,* the word *of* is redundant.

CHANGE   Place the rack *outside of* the incubator.
    TO   Place the rack *outside* the incubator.

In addition, do not use *outside of* to mean "aside from" or "except for."

> CHANGE   *Outside of* his frequent absences, Jim has a good work record.
> TO   *Except for* his frequent absences, Jim has a good work record.

over with

In the expression *over with,* the word *with* is redundant; moreover, the word *completed* often expresses the thought better.

> CHANGE   You may enter the test chamber now that the experiment is *over with.*
> TO   You may enter the test chamber now that the experiment is *over.*
> OR   You may enter the test chamber now that the experiment has been *completed.*

P

pace

Pace is the speed at which the writer presents ideas to the **reader.** Your goal should be to achieve a pace that fits both your reader and your subject; at times you may need a fast pace, at other times a slow pace. The more knowledgeable the reader, the faster your pace can be—but be careful not to lose control of the pace. In the first passage of the following example, facts are piled on top of each other at a rapid pace. In the second passage, the same facts are spread out and presented in a more easily assimilated manner; in addition, a different and more desirable **emphasis** is achieved.

CHANGE   The generator is powered by a 90-horsepower engine, is designed to operate under normal conditions of temperature and humidity, produces 110 volts at 60 Hertz, is designed for use under emergency conditions, and may be phased with other units of the same type to produce additional power when needed.

TO   The generator, which is powered by a 90-horsepower engine, produces 110 volts at 60 Hertz under normal conditions of temperature and humidity. Designed especially for use under emergency conditions, this generator may be phased with other units of the same type to produce additional power when needed.

Check your draft to determine the pace at which you are presenting your facts and ideas to the reader. If you find that you are jamming your ideas too closely together, you have a problem with pacing. There are at least three solutions to such problems: **subordination, sentence construction,** and **transition.**

The best way to solve a pacing problem is to search out minor thoughts and subordinate them to the major points, just as, in the example, ". . . is designed for use under emergency conditions . . ." was subordinated to ". . . this generator may be phased with other units . . ."

Another solution is to consider a different type of **sentence** structure. A complex idea is normally better stated in a simple sentence structure; a series of simple thoughts is better stated in a more complex sentence structure. Notice in the example above that the first passage is one long,

432 **paragraphs**

complicated sentence and that the same ideas are expressed in shorter
sentences in the **revision.**

A final possible solution is to provide words and **phrases** of transition
to alter the pace.

CHANGE   The report was not finished. It was 6:15. We were all tired. We
          left the office and went home.

TO   The report was *still* not finished at 6:15. *However,* we were all tired,
      so we left the office and went home.

See also **subordination.**

paragraphs

A paragraph is a group of **sentences** that support and develop a single
idea; it may be thought of as an essay in miniature, for its function is to
expand upon the core idea stated in its **topic sentence.** A paragraph may
use a particular **method of development** to expand upon the idea stated
in the topic sentence. The following paragraph uses the **general-to-
specific method of development.** Its topic sentence is italicized.

> *The arithmetic of searching for oil is stark.* For all his scientific methods
> of detection, the only way the oilman can actually know for sure
> that there is oil in the ground is to drill a well. The average cost of
> drilling an oil well is over $300,000, and drilling a single well may
> cost over $8,000,000! And once the well is drilled, the odds
> against its containing any oil at all are 8 to 1!

The paragraph performs three essential functions: (1) it develops the unit
of thought stated in the topic sentence; (2) it provides a logical break in
the material; and (3) it creates a physical break on the page, which in turn
provides visual assistance to the **reader.**

TOPIC SENTENCE

A topic sentence states the subject of a paragraph; the rest of the
paragraph then supports and develops that statement with carefully
related details. The topic sentence may appear anywhere in the paragraph—a fact that permits the writer to achieve **emphasis** and variety—
and a topic statement may be more than one sentence if necessary.

The topic sentence is most often the first sentence of the paragraph
because it states the subject that the paragraph is to develop. The topic
sentence is effective in this position because the reader knows imme-

diately what the paragraph is about. In the following paragraph, the first sentence is the topic sentence.

> *Spot size, critical in specifying resolution, should be as uniform as possible across the screen* [or the cathode ray tube, or CRT]. If spots are significantly larger in the corners than at the screen's center, the imaging in the corners will lack sharp definition. Although resolution depends largely on spot size or line width, a well-focused electron beam lacks a definite edge and thus has no absolute size. Light distribution has a tendency to decrease from its center in a bell-shaped or Gaussian curve. The spot size specified will depend on its application and attendant variables.
> —James A. Geissinger, "Cathode-Ray-Tube Display," *Chemical Engineering* 87 (1980), p. 145.

On rare occasions, the topic sentence logically falls in the middle of a paragraph. In the following example, the topic sentence, italicized, is in the middle of the paragraph.

> While the principles and techniques of system coordination can be taught in a one- to two-week training course, actual expertise cannot be developed until one has coordinated several systems. The reasons for this is simple. *Coordination is at least 60% "art" and only 40% engineering.* For any given system, there are many different and acceptable solutions. The engineer performing coordination must have a depth of experience to select the optimum solution from the many choices.
> —Thad Brown, "Principles of Short-Circuit Analysis and Coordination," *Chemical Engineering* 87 (1980), p. 152.

Although the topic sentence is usually most effective early in the paragraph where the reader's attention is greatest, a paragraph can effectively be made to lead up to the topic sentence; this is sometimes done to achieve emphasis. When a topic sentence concludes a paragraph, it can also serve as a summary or **conclusion,** based on the details that were designed to lead up to it. Following is a paragraph whose topic sentence is at the end.

> Energy does far more than simply make our daily lives more comfortable and convenient. Suppose you wanted to stop—and reverse—the economic progress of this nation. What would be the surest and quickest way to do it? Find a way to cut off the nation's oil resources! Industrial plants would shut down; public utilities would stand idle; all forms of transportation would halt.

The country would be paralyzed, and our economy would plummet into the abyss of national economic ruin. *Our economy, in short, is energy-based.*

—*The Baker World* (Los Angeles: Baker Oil Tools, Inc., 1964), p. 5.

Because multiple paragraphs are sometimes used to develop different aspects of an idea, not all paragraphs have topic sentences. In this situation, **transition** between paragraphs is especially important so the reader knows that the same idea is being developed through several paragraphs.

Topic Sentence for all three paragraphs
To conserve valuable memory space, a large portion of the software package remains on disc; only the most frequently used portion resides in internal memory all of the time. The disc-resident software is organized into small modules that are called into memory as needed to perform specific functions.

Transition
The memory-resident portion of the operating system maintains strict control of processing. It consists of routines, subroutines, lists, and tables that are used to perform common program functions, such as processing input/output operations, calling other software routines from disc as needed, and processing errors.

Transition
The disc-resident portion of the operating system contains routines that are used less frequently in system operation, such as the peripheral-related software routines that are used for correcting errors encountered on the various units, and the log and display routines that record unusual operating conditions in the system log. The disc-resident portion of the operating system also contains Monitor, the software program that supervises the loading of utility routines and the user's programs.

—*NCR Century Operating Systems Manual* (Dayton: The NCR Corporation, 1974), p. 17.

In this example, the reason for breaking the development of the idea expressed in the topic sentence into three paragraphs is to help the reader assimilate the fact that the main idea has two separate parts.

PARAGRAPH LENGTH

Paragraph length should be tailored to aid the reader's understanding of ideas. A series of short, undeveloped paragraphs can indicate poor **organization** of material, in which case you should look for a larger idea to which the ideas in the short paragraphs relate and then make the larger idea the topic sentence for a single paragraph. A series of short paragraphs can also sacrifice **unity** by breaking a single idea into several pieces. A series of long paragraphs, on the other hand, can fail to provide the reader with manageable subdivisions of thought. A good rule of thumb is that a paragraph should be just long enough to deal adequately with the subject raised by its topic sentence. A new paragraph should begin whenever the subject changes significantly.

A short paragraph sometimes immediately follows a long paragraph to give emphasis to the thought contained in the short paragraph. Also, short paragraphs increase the pace of your writing, whereas long paragraphs slow it down.

> Because it has at times displaced some jobs, automation has become an ugly word in the American vocabulary. But the all-important fact that is so often overlooked is that it invariably creates many more jobs than it eliminates. (The vast number of people employed in the American automobile industry as compared with the number of people that had been employed in the harness- and carriage-making business is a classic example.) Almost always, the jobs that have been eliminated by automation have been menial, unskilled jobs, and those who have been displaced have been forced to increase their skills, which resulted in better and higher paying jobs for them.
>
> In view of these facts, is automation really bad? It has made our country the most wealthy and technologicaly advanced nation the world has ever known!

WRITING PARAGRAPHS

The paragraph is the basic building block of any writing effort. Careful paragraphing reflects the writer's accurate thinking and logical organization. Clear and orderly paragraphs help the reader follow the writer's thoughts more easily.

Outlining is the best guide to paragraphing. It is easy to group ideas into appropriate paragraphs when you follow a good working outline.

> *Outline*
> I. Advantages of Chicago as location for new plant
>   A. Transport facilities
>   1. Rail
>   2. Air
>   3. Truck
>   4. Sea (except in winter)
>   B. Labor supply
>   1. Engineering and scientific personnel
>   a. Many similar companies in the area
>   b. Several major universities
>   2. Technical and manufacturing personnel
>   a. Existing programs in community colleges
>   b. Possible special programs designed for us.

Resulting Paragraphs

Probably the greatest advantage of Chicago as a location for our new plant is its excellent transport facilities. The city is served by three major railroads. Both domestic and international air cargo service is available at O'Hare International Airport. Chicago is a major hub of the trucking industry, and most of the nation's large freight carriers have terminals there. Finally, except in the winter months when the Great Lakes are frozen, Chicago is a seaport, accessible through the St. Lawrence Seaway.

A second advantage of Chicago is that it offers an abundant labor force. An ample supply of engineering and scientific personnel is assured not only by the presence of many companies engaged in activities similar to ours but also by the presence of several major universities in the metropolitan area. Similarly, technicians and manufacturing personnel are in abundant supply. The seven colleges in the Chicago City College system, as well as half a dozen other two-year colleges in the outlying areas, produce graduates with associate degrees in a wide variety of technical specialties appropriate to our needs. Moreover, three of the outlying colleges have expressed an interest in establishing special courses attuned specifically to our requirements.

Consider not only the nature of the material you are developing but also the appearance of your page. An unbroken page looks forbidding.

PARAGRAPH COHERENCE AND UNITY

A good paragraph has unity, **coherence,** and adequate development. Unity is singleness of purpose, based on a topic sentence that states the core idea of the paragraph. When every sentence in the paragraph contributes to developing the core idea, the paragraph has unity. Coherence is holding to one **point of view,** one attitude, one **tense;** it is the joining of sentences into a logical pattern. Coherence is advanced by the careful choice of transitional words so that ideas are tied together as they are developed.

Topic
Sentence

Transitions

Transition

Any company which operates internationally today faces a host of difficulties. Inflation is worldwide. Most countries are struggling with other economic problems *as well. In addition,* there are many monetary uncertainties and growing economic nationalism directed against multinational companies. *Yet* there is ample business available in most developed countries if you have the right products, services, and marketing organization. To maintain the

growth NCR has achieved overseas, we recently restructured
Transition   our international operations into four major trading areas. *This*
will improve the services and support which the Corporation can
Transition   provide to its subsidiaries around the world. *At the same time* it
established firm management control, insuring consistent poli-
Transition   cies around the world. *So* you might say the problems of doing
business abroad will be more difficult this year but we are better
organized to meet those problems.

—*1974 Annual Report* (Dayton: The NCR Corporation, 1974), p. 3.

Simple enumeration (*first, second, then, next,* etc.) can also provide effective transition within paragraphs. Notice how the italicized words and **phrases** give coherence to the following paragraph.

Most adjustable office chairs have nylon hub tubes that hold metal spindle rods. To ensure trouble-free operation, lubricate these spindle rods occasionally. *First,* loosen the set screw in the adjustable bell. *Then* lift the chair from the base so that the entire spindle rod is accessible. *Next,* apply the lubricant to the spindle rod and the nylon washer, using the lubricant sparingly to prevent dripping. *When you have finished,* replace the chair and tighten the set screw.

Good paragraphs often use details from the preceding paragraph, thereby preserving and advancing the thought being developed. Appropriate **conjunctions** and the **repetition** of key words and **phrases** can help to provide unity and coherence among, as well as within, paragraphs, as the italicized words do in the following examples.

Six high power thyristors connected in a three-phase bridge configuration form the basic armature module. When necessary, the modules can be arranged in parallel order to meet higher power requirements. The basic armature module is constructed as a convenient *pull-out tray.* All *armature trays* are interchangeable.

For optimum shovel performance it is also desirable to use thyristors in the control of power fields. These smaller thyristors are also arranged in a *pull-out tray* arrangement called a *field tray.*

Should a fault ever occur in a tray, an *indicator tray* is provided, which by means of pilot lights indicates which tray is the source of the trouble. Through the *pull-out tray* concept, the mine electrician can quickly replace the faulty tray and need not troubleshoot.

Sometimes a paragraph is used solely for transition, as in the following example.

> . . . that marred the progress of the company.
>
> There were two other setbacks to the company's fortunes that year which contributed to its present shaky condition: the loss of many skilled workers through the early retirement program and the intensification of the devastating rate of inflation.
>
> The early retirement program . . .

parallel structure

Parallel sentence structure requires that **sentence** elements that are alike in function be alike in construction as well, as in the following example (in which three similar actions are stated in similar phrases).

EXAMPLES   The stream runs *under the culvert, behind the embankment,* and *into the pond.*

We need a supplementary work force *to handle* peak-hour activity, *to free* full-time employees from routine duties, *to relieve* operators during lunch breaks, and *to replace* vacationing employees.

Parallel structure achieves an economy of words, clarifies meaning, and pleases the **reader** aesthetically. In addition to adding a pleasing symmetry to a sentence, parallel structure expresses the equality of its ideas. This technique assists readers because they are able to anticipate the meaning of a sentence element on the basis of its parallel construction. When they recognize the similarity of word order or construction, readers know that the relationship between the new sentence element and the **subject** is the same as the relationship between the last sentence element and the subject. Because of this they can go from one idea to another more quickly and confidently.

EXAMPLES   The computer instruction contains *fetch, initiate,* and *execute* stages. (parallel words)

The computer instruction contains a *fetch stage, an initiate stage,* and *an execute stage.* (parallel phrases)

The computer instruction contains a fetch stage, it contains an initiate stage, and it contains an execute stage. (parallel clauses)

Parallel structure is especially important in creating your **outline,** your **table of contents,** and your **heads** because it is important for your reader to know the relative value of each item in your table of contents and each head in the body of your **report** (or other writing project).

In any type of writing, parallel structure channels the reader's attention and helps to draw together related ideas or to line up dissimilar ideas for comparison and contrast.

> EXAMPLE   Her book provoked much comment: *negative from her enemies, positive from her friends.*

The power of many of Lincoln's speeches comes from his use of parallel structure; in them, parallel structure harmonizes the elements within sentences and the relationships among sentences.

> EXAMPLE   But in a larger sense, *we cannot dedicate—we cannot consecrate—we cannot hallow*—this ground. The brave men, *living* and *dead,* who struggled here, have consecrated it far above our poor power to *add* or *detract. The world will little note nor long remember what we say here,* but *it can never forget what they did here.*
> —Abraham Lincoln, Gettysburg Address

A century later, John F. Kennedy used the same technique with striking effect.

> EXAMPLE   We shall *pay any price, bear any burden, meet any hardship, support any friend, oppose any foe* to assure the survival and the success of liberty.
> —John F. Kennedy, Inaugural Address

Although parallelism can be accomplished with words, phrases, or **clauses,** it is most frequently accomplished by the use of phrases.

> EXAMPLES   I was convinced of their competence *by their conduct, by their reputation,* and *by their survival in a competitive business.* (prepositional phrases)
> *Filling the gas tank, testing the windshield wipers,* and *checking tire pressure* are essential to preparing for a long trip. (gerund phrases)
> From childhood the artist had made it a habit *to observe people, to store up the impressions they made upon him,* and *to draw conclusions about mankind from them.* (infinitive phrases)

Correlative conjunctions *(either . . . or, neither . . . nor, not only . . . but also)* should always be followed by parallel structure. Both members of these pairs should be followed immediately by the same grammatical form: two words, two similar phrases, or two similar clauses.

EXAMPLES   Viruses carry either *DNA* or *RNA,* never both. (words)
Clearly, neither *serologic tests* nor *virus isolation studies* alone would
have been adequate. (phrases)
Either *we must increase our operational efficiency* or *we must decrease our
production goals.* (clauses)

To make a parallel construction clear and effective, it is often best to
repeat a **preposition,** an **article,** a **pronoun,** a **subordinating conjunction,** a **helping verb,** or the mark of an **infinitive.**

EXAMPLES   The Babylonians had *a* rudimentary geometry and *a* rudimentary astronomy. (article)
My father and *my* teacher agreed that I was not really trying.
(pronoun)
To run and be elected is better than *to* run and be defeated. (mark
of the infinitive)
The driver *must* be careful to check the gauge and *must* move
quickly when the light comes on. (helping verb)
New teams were being established *in* New York and *in* Hawaii.
(preposition)
The history of factories shows both *the* benefits and *the* limits of
standardization. (article)

When parallel constructions are closely related in thought and equal in
length and construction, they balance one another. Such balance is
useful for comparing and contrasting ideas.

EXAMPLE   A qualified staff without a sufficient budget is helpless; a sufficient budget without a qualified staff is useless.

FAULTY PARALLELISM

Faulty parallelism results when joined elements are intended to serve
equal grammatical functions but do not have equal grammatical form.
Avoid this kind of partial parallelism. Make certain that each element in
a series is similar in form and structure to all others in the same series.

CHANGE   We need a supplementary work force *to handle* peak-hour activity,
to free full-time employees from routine duties, *to relieve* operators at lunch break, and *for the replacement of* vacationing employees.
TO   We need a supplementary work force to *handle* peak-hour activity,
to free full-time employees from routine duties, *to relieve* operators at lunch break, and *to replace* vacationing employees.

In work-related writing, **lists** often cause problems with parallel structure. When you use a list of phrases or clauses, each phrase or clause in the list should begin with the same **part of speech.**

CHANGE   The following recommendations were made regarding the Cost Containment Committee's position statement:
1. *Stress* that this statement is for all departments.
2. *Start* the statement with "If the company continues to grow, the following steps should be taken."
3. *The statement* should emphasize that it applies both to department managers and staff.
4. *Such strong words* as *obligation, owe,* and *must* should be replaced with words that are less harsh.

TO   The following recommendations were made regarding the Cost Containment Committee's position statement:
1. *Stress* that this statement is for all departments.
2. *Start* the statement with "If the company continues to grow, the following steps must be taken."
3. *Emphasize* that it applies both to department managers and staff.
4. *Replace* such strong words as *obligation, owe,* and *must* with words that are less harsh.

In the first list the items are not grammatically parallel in structure: items 1 and 2 begin with verbs, but items 3 and 4 do not. Notice how much more smoothly the corrected version reads, with all items beginning with verbs.

Because faulty parallelism with correlative conjunctions *(either . . . or, neither . . . nor, not only . . . but also)* is one of the most common writing errors, the following examples are worth careful study:

CHANGE   You may travel to the new plant either *by train* or *there is a plane.* (different grammatical forms: a prepositional phrase, *by train;* and a clause, *there is a plane*)

TO   You may travel to the new plant either *by train* or *by plane.* (parallel grammatical forms: prepositional phrases)

OR   You may travel to the new plant by either *train* or *plane.* (parallel grammatical forms: nouns)

CHANGE   We are not only *responsible to our stockholders* but also *to our customers.* (different grammatical forms: an adjective with a modifying prepositional phrase, *responsible to our stockholders;* and a prepositional phrase, *to our customers*)

TO   We are responsible not only *to our stockholders* but also *to our customers.* (parallel grammatical forms: prepositional phrases)

OR   We are not only *responsible to our stockholders* but also *responsible to our customers.* (parallel grammatical forms: verb phrases)

CHANGE   We either *will pay the bill* or *we will return the shipment.* (different
          grammatical forms: a verb with its object, *will pay the bill,* and a
          clause, *we will return the shipment*)

TO   Either *we will pay the bill* or *we will return the shipment.* (parallel
          grammatical forms: clauses)

OR   We will either *pay the bill* or *return the shipment.* (parallel grammati-
          cal forms: verbs with their objects)

Be careful not to throw your reader off balance. If you make a statement
about two subjects and then follow with a further statement about one of
them, complete the thought and make a balancing statement about the
other subject.

CHANGE   The Commanche Commander and the Cessna Hawk are both
          small aircraft; *the Commander seats four.*

TO   The Commanche Commander and the Cessna Hawk are both
          small aircraft; *the Commander seats four* and *the Hawk seats six.*

paraphrasing

When you paraphrase a written passage, you rewrite it to state the
essential ideas in your own words. Because you do not quote your source
word-for-word when paraphrasing, it is unnecessary to enclose the
paraphrased material in **quotation marks.** However, paraphrased material should be footnoted because the ideas are taken from someone else
whether the words are identical or not.

Ordinarily, the majority of the notes you take during the **research**
phase of writing your **report** will paraphrase the original material (see
note-taking).

Original Material

One of the major visual cues used by pilots in maintaining
precision ground reference during low-level flight is that of object
blur. We are acquainted with the object-blur phenomenon experienced when driving an automobile. Objects in the foreground
appear to be rushing toward us while objects in the background
appear to recede slightly. There is a point in the observer's line of
sight, however, at which objects appear to stand still for a moment, before once again rushing toward him with increasing
angular velocity. The distance from the observer to this point
where objects appear stationary is sometimes referred to as the
"blur threshold" range.

—Wesley E. Woodson and Donald W. Conover, *Human Engineering Guide for Equipment
Designers,* 2nd ed. (Los Angeles: University of California Press, 1964), p. 139.

Paraphrased

Object blur refers to the phenomenon by which an observer in a moving vehicle looks out and sees foreground objects appear to rush at him, while background objects appear to recede. But objects at some point appear temporarily stationary. The distance separating the observer from this point is sometimes called the "blur threshold" range.

Note that the paraphrased version includes only the essential information from the original passage. Paraphrased material, consequently, will almost always be shorter than the original material. As with all summarized material, strive to put the original ideas into your words without distorting them.

parentheses

Parentheses ( ) are used to enclose words, **phrases,** or **sentences.** The material within parentheses can add **clarity** to a statement without altering its meaning. Parentheses de-emphasize (or play down) an inserted element. Parenthetical information may not be essential to a sentence, but it may be interesting or helpful to some **readers.**

EXAMPLE   Aluminum is extracted from its ore (called bauxite) in three stages.

Parenthetical material applies to the word or phrase immediately preceding it.

EXAMPLE   The development of International Business Machines (IBM) is a uniquely American success story.

Parentheses may be used to enclose figures or letters that indicate sequence. Enclose the figure or letter with two parentheses when it appears within a sentence, rather than using only one parenthesis.

EXAMPLE   The following sections deal with (1) preparation, (2) research, (3) organization, (4) writing, and (5) revision.

Parenthetical material does not affect the **punctuation** of a sentence. If a parenthesis closes a sentence, the ending punctuation appears after the parenthesis. Also, a **comma** following a parenthetical word, phrase, or **clause** appears outside the closing parenthesis.

EXAMPLE   These oxygen-rich chemicals, as for instance potassium permanganate ($KMnO_4$) and potassium chromate ($KCrO_4$), were oxidizing agents (they added oxygen to a substance).

However, when a complete sentence within parentheses stands independently, the ending punctuation goes inside the final parenthesis.

EXAMPLE   The new marketing approach appears to be a success; most of our regional managers report sales increases of 15% to 30%. (The only importance exceptions are the Denver and Houston offices.) Therefore, we plan to continue. . . .

In some **footnote** forms, parentheses enclose the publisher, place of publication, and date of publication.

EXAMPLE   [1] J. Lamarsh, *Introduction to Nuclear Reactor Theory* (Reading, Mass.: Addison Wesley, 1966), p. 9.

Use **brackets** to set off a parenthetical item that is already within parentheses.

EXAMPLE   We should be sure to give Emanuel Foose (and his brother Emilio [1812-1882] as well) credit for his part in founding the institute.

Do not overuse parentheses. Also be on guard against using parentheses where other marks of punctuation are more appropriate.

participial phrases

A participial phrase consists of a **participle** plus its **object** or **complement,** if any, and modifiers. Like the participle, a participial phrase functions as an **adjective** modifying a **noun** or **pronoun.**

EXAMPLES   The division *having the largest sales increase* wins the trophy.
Finding the problem resolved, he went to the next item.
Having begun, we felt that we had to see the project through.

The relationship between a participial phrase and the rest of the **sentence** must be clear to the **reader.** For this reason, every sentence containing a participial phrase must have a noun or pronoun that the participial phrase modifies; if it does not, the result is a dangling participial phrase that is misplaced in the sentence and so appears to modify the wrong noun or pronoun. Both of these problems are discussed below.

DANGLING PARTICIPIAL PHRASES

A dangling participial phrase occurs when the noun or pronoun the participial phrase is meant to modify is not stated but only implied in the sentence. (See also **dangling modifiers.**)

CHANGE   *Being unhappy with the job,* his efficiency suffered. (His *efficiency* was not unhappy with the job; what the participial phrase really modifies—*he*—is not stated but merely implied.)

TO   *Being unhappy with the job,* he grew less efficient. (Now what the participial phrase modifies—*he*—is explicitly stated.)

MISPLACED PARTICIPIAL PHRASES

A participial phrase is misplaced when it is too far from the noun or pronoun it is meant to modify and so appears to modify something else. This is an error that can sometimes make the writer look ridiculous indeed. (See also **modifiers.**)

CHANGE   *Rolling around in the bottom of the vibration test chamber,* I found the missing bearings.

TO   I found the missing bearings *rolling around in the bottom of the vibration test chamber.*

CHANGE   We saw a large warehouse *driving down the highway.*

TO   *Driving down the highway,* we saw a large warehouse.

participles

Participles are **verb** forms that function as **adjectives.**

EXAMPLES   The *waiting* driver raced his engine.
Here are the *revised* estimates.
The *completed* report lay on his desk.
Rising costs reduced our profit margin.

Participles belong to a larger class of verb forms called **verbals** or nonfinite verbs. (The other two types of verbals are **gerunds,** which always function as nouns, and **infinitives,** which have several functions.) Because participles are formed from verbs, they share certain characteristics of verbs even though they are adjectives: (1) they may take **objects** or **complements;** (2) they may be in the present, past, or perfect **tense;** and (3) they may be in the active or passive **voice.** But remember that a participle cannot be used as the verb of a sentence. Inexperienced writers sometimes make this mistake; the result is a **sentence fragment.**

CHANGE   The committee chairman was responsible. His vote *being* the decisive one.

TO   The committee chairman was responsible, his vote *being* the decisive one.

OR   The committee chairman was responsible. His vote *was* the decisive one.

TENSE

The present participle ends in *ing.*

EXAMPLE   *Declining* sales forced us to close one branch office.

Do not confuse present participles with gerunds, which are also verbals ending in *ing* but which always function as nouns, as in "*Running* is his favorite form of exercise."
The past participle may end in *ed, t, en, n,* or *d.*

EXAMPLES   Repair the *bent* lever.
What are the *estimated* costs?
Here is the *broken* calculator.
What are the metal's *known* properties?
The story, *told* many times before, was still interesting.

The perfect participle is formed with the present participle of the **helping verb** *have* plus the past participle of the main verb.

EXAMPLE   *Having gotten* (perfect participle) a large raise, the *smiling* (present participle), *contented* (past participle) employee worked harder than ever.

VOICE

Participles formed from **transitive verbs** may be in either the active or the passive voice. Some form of the helping verb *be* is used to form the passive voice of the perfect participle.

EXAMPLES   *Having finished* the job, we submitted our bill. (active)
The job *having been finished,* we submitted our bill. (passive)

MODIFIERS OF PARTICIPLES

Participles may be modified by **adverbs, prepositional phrases,** or **adverb clauses.**

EXAMPLES   *Rapidly* declining sales forced the shutdown. (adverb; the participle is *declining*)
Looking to the future, I felt that we should increase our research staff. (prepositional phrase; the participle is *looking*)
Having failed *because his laboratory was inadequate,* he requisitioned new equipment. (adverb clause; the participle is *having failed*)

OBJECTS AND COMPLEMENTS OF PARTICIPLES

Because participles are formed from verbs, they may take objects (direct or indirect) or complements.

EXAMPLES   *Having sold* (participle) *him* (indirect object) the new *model* (direct
object), we must now provide adequate repair service.
Having been *president,* I found the adjustment difficult. (complement; the participle *having been* is formed from a **linking verb**)
Striking *the table* for emphasis, he told us exactly what he thought.
(direct object; the participle *striking* is formed from a transitive
verb)

PARTICIPLE FORMS IN VERB PHRASES

The present-tense and past-tense participle forms are combined with
helping verbs to create **verb phrases.** The verb phrases function as main
verbs rather than as adjectives.

EXAMPLES   I *have worked* all day.
I *am working* every day.
We *had fought* for our plan, but we *had lost.*
We *will be fighting* for our plan.
He *has planned* this trip for weeks.
He *would have been working* in any case.

parts of speech

Part of speech is a term used to describe the class of words to which a
particular word belongs, according to its function in the **sentence;** that
is, each function in a sentence (naming, asserting, describing, joining,
modifying, exclaiming) is performed by a word belonging to a certain
part of speech.

If a word's function is to *name* something, it is a **noun** or **pronoun.** If a
word's function is to make an *assertion* about something, it is a **verb.** If its
function is to *describe* or *modify* something, the word is an **adjective** or an
adverb. If its function is to *join* or *link* one element of the sentence to
another, it is a **conjunction** or a **preposition.** If its function is to express
an exclamation it is an **interjection.** (See also **functional shift.**)

party

In legal language, *party* refers to an individual, group, or organization.

EXAMPLE   The injured *party* brought suit against my client.

The term is inappropriate in general writing; when an individual is being referred to, use *person*.

> CHANGE The *party* whose file you requested is here now.
> TO The *person* whose file you requested is here now.

Party is, of course, appropriate when it refers to a group.

> EXAMPLE Arrangements were made for the members of our *party* to have lunch after the tour.

per

This common business term means "by means of," "through," or "on account of,"and in these senses it is appropriate.

> EXAMPLES *per* annum, *per* capita, *per* diem, *per* head

When used to mean "according to" (*per* your request, *per* your order) the expression is business **jargon** at its worst and should be avoided. Equally annoying is the phrase *as per.*

> CHANGE *Per your request,* I enclose the production report.
> TO *As you requested,* I enclose the production report.

> CHANGE *As per our discussion,* I will send revised instructions.
> TO *As we agreed,* I will send revised instructions.

per cent/percent/percentage

Percent, which is replacing the two-word *per cent,* is used when **numbers** are written out.

> EXAMPLE Only five *percent* of our employees are covered by group health insurance.

Percentages are normally given as figures and used with the percent **symbol.**

> EXAMPLE Only 25% of the members attended the meeting.

Percentage, which is never used with numbers, indicates a general size.

> EXAMPLE Only a small *percentage* of the managers attended the meeting.

periods

A period (also called a full stop or end stop) usually indicates the end of a declarative **sentence.** Periods also link (when used as leaders) and indicate omissions (when used as **ellipses.**)

Although the primary function of periods is to end declarative sentences, periods also end imperative sentences that are not emphatic enough for an **exclamation mark.**

EXAMPLE   Send me any information you may have on the subject.

Periods may also end questions that are really polite requests and questions to which an affirmative response is assumed.

EXAMPLE   Will you please send me the specifications.

Periods end incomplete sentences when the meaning is clear from the context. These sentences are common in advertising. (See also **sentence faults.**)

EXAMPLE   Bell and Howell's new Double-Feature Cassette Projector will change your mind about home movies. *Because if you can press a button, now you can show movies. Instantly. Easily.*

IN QUOTATIONS

Do not use a period after a declarative sentence that is quoted in the context of another sentence.

CHANGE   "There is every chance of success." she stated.
TO   "There is every chance of success," she stated.

A period is conventionally placed inside **quotation marks.**

EXAMPLES   He liked to think of himself as a "tycoon."
He stated clearly, "My vote is yes."

WITH PARENTHESES

A sentence that ends in a **parenthesis** requires a period *after* the parenthesis.

EXAMPLE   The institute was founded by Harry Denman (1902-1972).

If a whole sentence (beginning with an initial **capital letter**) is in parentheses, the period (or any other end mark) is placed inside the final parenthesis.

EXAMPLE   The project director listed the problems facing her staff. (This was the third time she had complained to the board.)

CONVENTIONAL USES OF PERIODS

Use periods after initials in names.

EXAMPLES   W. T. Grant, J. P. Morgan

Use periods as decimal points with numbers.

EXAMPLES   109.2, $540.26, 6.9%

Use periods to indicate **abbreviations.**

EXAMPLES   Ms., Dr., Inc.

Use periods following the **numbers** in numbered **lists.**

EXAMPLE   1.
2.
3.

AS ELLIPSES

When you omit words in quoted material, use a series of three spaced periods—call **ellipsis** dots—to indicate the omission. Such an omission must not detract from or alter the essential meaning of the passage.

ORIGINAL   "Technical material distributed for promotional use is sometimes charged for, particularly in high-volume distribution to educational institutions, although prices for these publications are not uniformly based on the cost of developing them."

WITH OMISSION   "Technical material distributed for promotional use is sometimes charged for . . . although prices for these publications are not uniformly based on the cost of developing them."

When introducing a **quotation** that does not begin with the first word of a sentence (that is, when your quotation starts somewhere in the middle of a sentence of the material from which you are quoting), you do not need ellipsis dots; the lower-case letter with which you begin the quotation already indicates an omission.

ORIGINAL   "When the programmer has determined a system of runs, he must create a systems flowchart to provide a picture of the data flow through the system."

WITH OMISSION   The booklet states that the programmer "must create a systems flowchart to provide a picture of the data flow through the system."

If an ellipsis follows the end of a sentence, retain the period at the end of the sentence and add the three ellipsis dots.

ORIGINAL "During the year, every department participated in the development of a centralized computer system. The basic plan centered on the use of the computer as a cost reduction tool. At the beginning of the year, each department received a booklet explaining the purpose of the system."

WITH OMISSION "During the year, every department participated in the development of a centralized computer system. . . . At the beginning of the year, each department received a booklet explaining the purpose of the system."

AS LEADERS

When spaced periods are used in a **table** to connect one item to another, they are called leaders. The purpose of leaders is to help the reader align the data.

```
150 lbs ...................................... 1.7 psi
175 lbs ...................................... 2.8 psi
200 lbs ...................................... 3.9 psi
```

The most common use of leaders is in **tables of contents.**

PERIOD FAULTS

The incorrect use of a period is sometimes referred to as a period fault. When a period is inserted prematurely, the result is a sentence fragment.

CHANGE After a long day at the office in which we finished the report. We left hurriedly for home.

TO After a long day at the office in which we finished the report, we left hurriedly for home.

When a period is left out, the result is a "fused," or run-on, sentence. Be careful never to leave out necessary periods.

CHANGE Bill was late for ten days in a row Ms. Sturgess had to fire him.

TO Bill has late for ten days in a row. Ms. Sturgess had to fire him.

See also **sentence faults.**

person

Person refers to the form of a **personal pronoun** that indicates whether the **pronoun** represents the speaker, the person spoken to, or the person (or thing) spoken about. A pronoun representing the speaker is in the first person.

EXAMPLE   *I* could not find the answer in the manual.

If the pronoun represents the person or persons spoken to, the pronoun is in the second person.

EXAMPLE   *You* are going to be a good supervisor.

If the pronoun represents the person or persons spoken about, the pronoun is in the third person.

EXAMPLE   *They* received the news quietly.

The following table shows first, second, and third person pronouns.

Person	Singular	Plural
First	I, me, my	we, ours, us
Second	you, your	you, your
Third	he, him, his she, her, hers it, its	they, them, their

Identifying pronouns by person helps the writer avoid illogical shifts from one person to another. A very common error is to shift from the third person to the second person.

CHANGE   A *manager* should spend the morning hours on work requiring mental effort, for *your* mind is freshest in the morning.

TO   A *manager* should spend the morning hours on work requiring mental effort, for *his* mind is freshest in the morning.

OR   *You* should spend the morning hours on work requiring mental effort, for *your* mind is freshest in the morning.

See also **number** and **case**.

personal/personnel

Personal is an **adjective** meaning "of or pertaining to an individual person."

EXAMPLE   He left work early because of a *personal* problem.

Personnel is a **noun** meaning "a group of people engaged in a common job."

EXAMPLE   All *personnel* should pick up their paychecks on Thursday.

Be careful not to use *personnel* when the word you really need is *persons* or *people.*

CHANGE   The remaining two *personnel* will be moved next Thursday.
TO   The remaining two *persons* will be moved next Thursday.

personal pronouns

The personal pronouns are *I, me, my, mine; you, your, yours; he, him, his; she, her, hers; it, its; we, us, our, ours;* and *they, them, their, theirs.*
Don't attempt to avoid using the personal pronoun *I* where it is natural. The use of unnatural devices to avoid using *I* is more likely to call attention to the writer than would use of the pronoun *I.*

CHANGE   *The writer* wishes to point out that the proposed solution has several weaknesses.
TO   *I* wish to point out that the proposed solution has several weaknesses.
OR   The proposed solution has several weaknesses.

Also, avoid substituting the **reflexive pronouns** *myself* for *I* or *me,* or *ourselves* for *us.*

CHANGE   Joe and *myself* worked all day on it.
TO   Joe and *I* worked all day on it.
CHANGE   He gave it to my assistant and *myself.*
TO   He gave it to my assistant and *me.*
CHANGE   This decision cannot be made by managers like *myself.*
TO   This decision cannot be made by managers like *me.*
CHANGE   Nobody can solve this problem but *ourselves.*
TO   Nobody can solve this problem but *us.*

The use of *we* for general reference (''*We* are living in a time of high inflation.'') is acceptable. But using the editorial *we* to avoid *I* is pompous and stiff. It may also assume, presumptuously, that ''we'' all agree.

CHANGE   *We* must state unequivocally that *we* disapprove of such practices.
TO   *I* must state unequivocally that *I* disapprove of such practices.

Bear in mind that the use of *we* can be legally construed to commit the writer's company as well as the writer personally. If you are speaking for yourself—even in your role as an employee—it is better to use *I.*

One is also acceptable for general reference. Be aware, however, that it is impersonal and somewhat stiff and that overdoing it renders your writing awkward.

CHANGE   *One* must be careful about *one's* work habits lest *one* become sloppy.

TO   *People* must be careful about *their* work habits lest *they* become sloppy.

See also **one** and **point of view.**

persons/people

When we use *persons,* we are usually referring to individual people thought of separately.

EXAMPLE   We need to find three qualified *persons* to fill the vacant positions.

When we say *people,* we are identifying a large or anonymous group.

EXAMPLE   Many *people* have never even heard of our product.

persuasion

Persuasion in writing attempts to convince the **reader** to adopt the writer's point of view. The range of persuasive writing required on the job varies widely. Technical writers may use persuasive techniques in pleading for safer working conditions in a shop or factory, justifying the expense of a new program, or writing a **proposal** for a multimillion-dollar contract. Elements of persuasive writing are found in many types of **correspondence,** too.

In persuasive writing the way ideas are presented is as important as the ideas themselves. Of course, you must support your appeal with a sound presentation of facts, statistics, and examples. You must also consider any real or possible conflicting opinions. Only then can you demonstrate—and argue for—the merit of your point of view. But remember your audience, and take your readers' feelings into account, too. Always maintain a positive **tone.** And be very careful not to wander from your main point—avoid ambiguity and never make trivial, irrelevant, or false claims.

The following passage is taken from a guide on the selection and proper use of chainsaws. The intended readers, loggers, use chainsaws daily in their work. The author explains the necessity of maintaining the proper tension on the saw's chain. In explaining some of the technical problems that had to be overcome so that the proper tension could be maintained under all conditions, the author convincingly makes the case that chainsaws with a particular design are most effective.

> For maximum chain life, the chain should be properly tensioned. Maintaining the proper tension has always been difficult with solid-nosed bars because the chain develops a great deal of heat from friction as it slides over the nose of the bar. The heat causes the chain to expand and hence become too slack. If the chain is properly tensioned when cold, it will be too slack when running. If the chain is tensioned after it warms up, then stopping the saw long enough to refuel it might result in the chain shrinking to the bar so tight that it will not turn.
>
> The first attempt at solving these problems was the roller-nosed bar. This bar cuts down on the friction to a considerable extent, but is somewhat fragile, and since the chain is unsupported during part of its travel, it is not good practice to cut with the tip of the bar. The sprocket-nosed bar seems to have solved these problems. Here a thin sheet metal sprocket, roughly the thickness of the drive links, is all but hidden inside the nose of the bar. As the chain approaches the end of the bar, the sprocket lifts the drive link and therefore keeps the saw from running on the rails of the bar as it travels around the nose. This cuts friction while supporting the chain well enough so that the nose of the bar can be used for cutting. It is true that a saw with a sprocket-nosed bar will kick back harder than other saws because there is so little friction between the chain and the bar, and the sprocket nose may be somewhat more fragile, particularly if it is pinched in a cut. On the other hand, its advantages often outweigh its disadvantages. It gives faster cutting, better chain life, and better chain tension, which in itself is a safety factor. Most sprockets and bearings for bars can be replaced by a dealer, although if the nose portion of the bar is bent, the bar is ruined. Some sprocket-nosed bars are available with the entire tip of the bar made as a section that may be easily replaced by the user.
>
> —R. P. Sarna, *Chain Saw Manual* (Danville, Ill.: The Interstate Printers & Publishers, Inc., 1979), pp. 8-9.

Notice that the author acknowledges that this design is not entirely free of potentially harmful side effects: a saw with this bar will kick back harder than others and the bar itself is more fragile. In so doing, the author gains credibility. (See also **logic, methods of development,** and **forms of discourse.**)

phenomenon/phenomena

A *phenomenon* is an observable thing, fact, or occurrence. Its plural form is *phenomena*.

EXAMPLES   The natural *phenomenon* of earth tremors *is* a problem we must
anticipate in designing the California installation.
The *phenomena* associated with atomic fission *were* only recently
understood.

photographs

Photographs are vital to many publications. Pictures are the best way to show the superficial appearance of an object or to record an event or the development of a phenomenon over a period of time. Not all representations, however, call for photographs. They cannot depict the internal workings of a mechanism or below-the-surface details of objects or structures. Such details are better represented in drawings or diagrams.

HIGHLIGHTING PHOTOGRAPHIC SUBJECTS

Stand close enough to the object so that it fills your picture frame. To get precise and clear photographs, choose camera angles carefully. A camera will photograph only what it is aimed at; accordingly, select important details and the camera angles that will record these details. To show relative size, place a familiar object—such as a ruler, a book, or a person—near the object being photographed.

TIPS FOR USING PHOTOGRAPHS

Like all illustrative materials, photographs must be handled carefully. When preparing photographs for a **report,** observe these guidelines.

1. Mount photographs on white bond paper with rubber cement or another adhesive and allow ample margins.
2. If the photograph is the same size as the paper, type the caption, figure and page numbers, and any other important information on labels and fasten them with rubber cement or another adhesive to the photograph. (Photographs are given figure numbers in sequence with other **illustrations** in a publication; see Figure 35.)
3. Position the figure number and caption so that the reader can view them and the photograph from the same orientation.

As-received elbow section

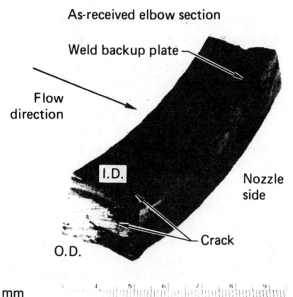

Weld backup plate

Flow
direction

I.D.

Nozzle
side

O.D.

Crack

mm

Figure 35.   Deep Crack of Counterbore Slope in Elbow Section

4. Do not draw crop marks (lines showing where the photo should be trimmed for reproduction) directly across a photograph. Draw them at the very edges of the photograph.
5. Do not write on a photograph, front or back. Tape a tissue-paper overlay over the face of the photograph, and then write very lightly on the overlay with a soft-lead pencil. Never write on the overlay with a ball-point pen.
6. Do not use paper clips or staples on photographs.
7. Do not fold or crease photographs.

phrases

Below the level of the **sentence,** there are two ways to combine words into groups: by forming **clauses,** which combine **subjects** and **verbs,** and by forming phrases, which are based on **nouns,** nonfinite verb forms, or verb combinations without subjects. A phrase is the most basic meaningful group of words; it cannot make a full statement as a clause can, however, because it does not contain a subject and a verb.

EXAMPLE   He encouraged his staff (clause) *by his calm confidence.* (phrase)

A phrase may function as an **adjective,** an **adverb,** a noun, or a verb.

EXAMPLES   The subjects *on the agenda* were all discussed. (adjective)
We discussed the project *with great enthusiasm.* (adverb)
Working hard is her way of life. (noun)
The Chief Engineer *should have been notified.* (verb)

Even though phrases function as adjectives, adverbs, nouns, or verbs, they are normally named for the kind of word around which they are constructed—**preposition, participle, infinitive, gerund,** verb, or noun.

PREPOSITIONAL PHRASES

A preposition is a word that shows the relationship of its **object** with another word; that is, it combines with its object (a noun or a **pronoun**) to form a modifying phrase; a **prepositional phrase,** then, consists of a preposition plus its object (the noun or pronoun) and its **modifiers.**

EXAMPLE   *After the meeting,* the regional managers adjourned *to the executive dining room.*

PARTICIPIAL PHRASES

A participle is any form of a verb that is used as an adjective. A **participial phrase** consists of a participle plus its object and its modifiers.

EXAMPLE   *Looking very pleased with himself,* the sales manager reported on the success of the policies he had introduced.

INFINITIVE PHRASES

An infinitive is the bare form of a verb *(go, run, talk)* without the restrictions imposed by **person** and **number;** an infinitive is generally preceded by the word *to* (which is usually a preposition but in this use is called the sign, or mark, of the infinitive). An **infinitive phrase** consists of the word *to* plus an infinitive and any objects or modifiers.

EXAMPLE   *To succeed in this field,* you must be willing *to assume responsibility.*

GERUND PHRASES

A gerund phrase, which consists of a gerund plus any objects or modifiers, always functions as a noun.

EXAMPLES   *Preparing an annual report* is a difficult task. (subject)
            She liked *running the department*. (direct object)

VERB PHRASES

A **verb phrase** consists of a main verb and its **helping verb.**

EXAMPLE   Company officials discovered that a computer *was emitting* more
           data than it *had been asked* for. After investigation, police sug-
           gested that an unauthorized program *had been run* through the
           computer at the coded command of another computer belong-
           ing to a different company. The police obtained a warrant to
           search for electronic impulses in the memory of the suspect
           machine. The case (involving trade secrets) and its outcome
           should make legal history.

NOUN PHRASES

A noun phrase consists of a noun and its modifiers.

EXAMPLES   *Many large companies* use computers.
            Have *the two new employees* fill out *these forms*.

plagiarism

To use someone else's exact words without **quotation marks** and appropriate credit or to use the unique ideas of someone else without acknowl-
edgement is known as plagiarism. In publishing, plagiarism is illegal; in
other circumstances, plagiarism is at the least unethical. You may quote
the words and ideas of another if you credit your source by using
footnotes. (However, if you intend to publish or reproduce and distribute material in which you have included quotations, you will often need
to obtain written permission from whoever holds the **copyright.**) You
may **paraphrase** a passage without citing the source as long as the idea is
not unique to one author. (See also **references** and **quotations.**)

point of view

Point of view is the attitude you exhibit toward your **reader** as reflected
in your use of **person** (the use of **personal pronouns** such as *I, we, you,*
and *they*). The writer's point of view may be impersonal.

EXAMPLE   It is regrettable that the material shipped on March 12 is unac-
           ceptable.

460 point of view
Or the writer's point of view may be personal.

EXAMPLE   I regret that we cannot accept your shipment of March 12.

Although the meaning of both **sentences** is the same, the sentence with the personal point of view (using *I, we,* and *your*) reflects the fact that two *people* are involved in the communication. Years ago technical people preferred the impersonal point of view because they thought it made writing sound more objective and professional. Unfortunately, however, the impersonal point of view often prevented clear and direct communication and actually caused misunderstanding at times. Today, good technical writing adopts a more personal point of view wherever it is appropriate.

Using the first person *(I, we)* makes the point of view more personal since the writer speaks of himself.

EXAMPLE   *I* can finish the project by August 1.

When a writer tries to avoid *I* by using *one*—when he is really talking about himself—he does not increase objectivity but merely makes the statement impersonal.

CHANGE   *One* can only conclude that the absorption rate is too fast.
TO   *I* can only conclude that the absorption rate is too fast.

Never use *the writer* to replace *I* in a mistaken attempt to sound very formal or dignified.

CHANGE   *The writer* feels that this project will be completed by the end of June.
TO   *I* feel that this project will be completed by the end of June.

Do not use the personal point of view when an impersonal point of view would be more appropriate or more effective.

CHANGE   I am inclined to think that each manager should attend the final committee meeting to hear the committee's recommendations.
TO   Each manager should attend the final committee meeting to hear the committee's recommendations.

Notice in the above examples that when you write in the third person, the point of view shifts from the writer to the person or thing being discussed. If you want the subject matter to receive more **emphasis** than either you (the writer) or the reader, you should establish an impersonal point of view. This is a fairly common practice in **formal writing** because the subject matter is given greatest emphasis.

In a letter on company stationery, be aware that use of the pronoun *we* may be interpreted as reflecting company policy, while *I* clearly reflects personal opinion. You should decide which pronoun to use on the basis of whether the matter discussed in the letter is in fact a corporate or individual concern.

EXAMPLES   *I* appreciate your suggestion regarding our need for more community activities.

         We appreciate your suggestion regarding our need for more community activities.

Remember, however, that the pronoun *we* may commit your organization to what you say.

CHANGE   *We* believe that your proposal was the best submitted.

TO   *I* believe that your proposal was the best submitted.

One way of reflecting a personal point of view is to write directly to the reader by use of the second person pronoun *(you)*. This approach aims your message directly at the reader.

EXAMPLE   If *you* finish *your* section of the report by the 1st, *we* can submit the whole report on the 5th.

When a technical writer tries to avoid using the pronoun *you*, the result is often awkward and sometimes even unclear.

CHANGE   *The operator* should check the cycle gauge before engaging the gear.

TO   *You* should check the cycle gauge before engaging the gear.

OR   Check the cycle gauge before engaging the gear.

Whether you adopt a personal or an impersonal point of view should depend upon your **objective** and what you know about your reader. In a business letter to an associate, you will most likely adopt a personal point of view. In a report to a large group, you will probably choose to emphasize the subject by adopting an impersonal point of view.

positive writing

Presenting positive information as though it were negative is a trap that technical writers fall into quite easily because of the complexity of the information they must write about. It is a practice that confuses the **reader,** however, and one that should be avoided.

CHANGE   If the error does *not* involve data transmission, the special function is *not* used.

TO   The special function is used only if the error involves data transmission.

In the first sentence, the reader must reverse two negatives to understand the exception that is being stated; the second sentence presents an exception to a rule in a straightforward manner.

On the other hand, negative facts or conclusions should be stated negatively; stating a negative fact or conclusion positively can mislead the reader.

CHANGE   For the first quarter of this year, employee exposure to airborne lead has been maintained to within 10% of acceptable state health standards.

TO   For the first quarter of this year, employee exposure to airborn lead continues to be 10% above acceptable state health standards.

possessive adjectives

Because possessive adjectives *(my, your, our, his, her, its, their)* directly modify **nouns,** they function as **adjectives,** even though they are **pronoun** forms. Anything that modifies a noun functions as an adjective.

EXAMPLE   The *proposal's* virtues outweighed *its* defects.

The possessives *mine, yours, hers, ours,* and *theirs* normally function as pronouns rather than as adjectives. (Expressions like *mine host* are archaic.) They often function as **subjective complements.**

EXAMPLE   That desk is *mine.*

They can, however, serve as **subjects.**

EXAMPLE   My desk is in the back; *yours* is by the window.

In either case, they function as pronouns.

practicable/practical

Practicable means that something is possible or feasible. *Practical* means that something is both possible and useful.

EXAMPLE   The program is *practical;* but, considering the company's recent financial problems, is it *practicable?*

Practical, not *practicable,* is used to describe a person, implying common sense, a commitment to what works rather than to theory or abstractions.

predicates

The predicate is that part of a **sentence** that contains the main **verb** and any other words used to complete the thought of the sentence (the verb's **modifiers** and **complements**). The principal part of the predicate is the verb, just as a **noun** (or noun substitute) is the principal part of the **subject.**

EXAMPLE   Bill *piloted the airplane.*

The simple predicate is the verb (or **verb phrase**) alone; the complete predicate is the verb and its modifiers and complements.

A compound predicate consists of two or more verbs with the same subject. It is an important device for economical writing.

EXAMPLE   The company *tried* but *did not succeed* in that field.

predicate adjectives

A predicate adjective is an **adjective** that follows a **linking verb** and describes the **subject.**

EXAMPLES   He is *sick.*
That flower is *artificial.*
The milk smells *sour.*
The proposal looks *good.*

The predicate adjective is one kind of **subjective complement;** the other is the **predicate nominative.** (See also **predicates.**)

predicate nominatives

A predicate nominative is a **noun** construction that follows a **linking verb** and renames the **subject.**

EXAMPLES   His is my *lawyer.* (noun)
His excuse was *that he had been sick.* (noun clause)

The predicate nominative is one kind of **subjective complement;** the other is the **predicate adjective.**

prefixes

A prefix is a letter, or letters, placed in front of a root word that changes the meaning, often causing the new word to mean the opposite of the root word.

Root Word	With Prefix
enter	*re*enter (enter again)
acceptable	*un*acceptable (not acceptable)
operate	*co*operate (operate together)
symmetrical	*a*symmetrical (not symmetrical)
honest	*dis*honest (not honest)
science	*meta*science (beyond science)

When a prefix ends with a vowel and the root word begins with the same vowel, the prefix may be separated from the root word with a **hyphen** (re-enter, co-operate, re-elect), the second vowel may be marked with a dieresis (reënter, coöperate, reëlect), or the word may be written with neither hyphen nor dieresis (reenter, cooperate, reelect). The hyphen and the dieresis are visual aids that help the **reader** recognize that the two vowels are pronounced differently. Since typewriters do not have the dieresis mark, the hyphen is the more practical way to help the reader.

Except between identical vowels, the hyphen rarely appears between a prefix and its root word. At times, however, a hyphen is necessary for clarity of meaning; for example, *reform* means ''correct'' or ''improve,'' while *re-form* means ''change the shape of.''

preparation

To write clearly and effectively does not require innate creative talent. It does require a systematic approach, however, which should logically begin with proper preparation.

Preparation consists of determining the **objective** of your writing project, its **readers,** and the **scope** within which you will cover your subject. Although these fundamentals of the writing process are often overlooked, you must come to grips with them if you hope to write well. They are the foundations upon which all the other steps of the writing process are built.

OBJECTIVE

What *exactly* do you want your readers to know or be able to do when they have finished reading what you have written? When you can answer this question—again, *exactly*—you will have determined the objective of your writing. A good test of whether you have formulated your objective adequately is to state it in a single **sentence.** This statement may also function as the thesis statement in your **outline,** although such a function should not be a requirement of your stated objective.

AUDIENCE

Know who your readers are and learn certain key facts about them, such as their educational level, technical knowledge, and needs relative to your subject. Their level of knowledge, for example, should determine whether you need to cover the fundamentals of your subject and which terms you must define.

SCOPE

If you know the objective of your writing project and your readers, you will know the type and amount of detail you will need to include in your writing. This is called scope, which may be defined as the depth and breadth to which you cover your subject. If you do not determine your scope of coverage you will not know how much or what kind of information to include, and therefore you are likely to invest unnecessary work during the **research** phase of your writing project.

One of the most common—and damaging—mistakes technical writers make is to begin writing too soon. The preparation stage of the writing process is analogous to focusing a camera before taking a picture. By determining who your reader is, what your objective is, and what your scope of coverage should be, you are bringing your whole writing effort into focus before beginning to write the draft. As a result, your **topic** is much more clearly focused for your reader. (See also the Checklist of the Writing Process.)

prepositions

A preposition is a word that links a **noun** or **pronoun** (its **object**) to another **sentence** element by expressing such relationships as direction *(to, into, across, toward),* location *(at, in, on, under, over, beside, among, by, between, through),* time *(before, after, during, until, since),* or figurative locations *(for, against, with).* Although only about seventy prepositions exist in the English language, they occur frequently. Together the preposition, its **object,** and the object's modifiers form a **prepositional phrase,** which acts as a **modifier.**

The object of a preposition, the word or phrase following the preposition, is always in the objective **case.** This situation gives rise to a problem in such constructions as "between you and *me,*" a phrase that is frequently and incorrectly written as "between you and *I.*" *Me* is the objective form of the pronoun, and *I* is the subjective form.

Many words that function as prepositions also function as **adverbs.** If the word takes an object and functions as a **connective,** it is a preposition; if it has no object and functions as a modifier, it is an adverb.

EXAMPLES   The manager sat *behind* the desk *in* his office. (prepositions)
The customer lagged *behind;* then she came *in* and sat *down.* (adverbs)

AVOIDING ERRORS WITH PREPOSITIONS

Do not use redundant prepositions, such as "off *of,*" "in back *of,*" and "inside *of.*"

CHANGE   The client arrived *at about* four o'clock.
TO   The client arrived *at* four o'clock. (to be exact)
OR   The client arrived *about* four o'clock. (to be approximate)

Do not omit necessary prepositions

CHANGE   He was oblivious and not distracted by the view from his office window.
TO   He was oblivious *to* and not distracted *by* the view from his office window.

Avoid adding the preposition *up* to **verbs** unnecessarily.

CHANGE   Call *up* and see if he is in his office.
TO   Call and see if he is in his office.

USING A PREPOSITION AT THE END OF A SENTENCE

If a preposition falls naturally at the end of a sentence, leave it there.

EXAMPLE   I don't remember which file I put it *in.*

Be aware, however, that a preposition at the end of a sentence can be an indication that the sentence is awkwardly constructed.

CHANGE   The branch office is where he was *at.*
TO   He was at the branch office.

COMMON USES OF PREPOSITIONS

Certain verbs, adverbs, and **adjectives** are used with certain prepositions. For example, we say "interested *in*," "aware *of*," "devoted *to*," "equated *with*," "abstain *from*," "adhere *to*," "conform *to*," "capable *of*," "comply *with*," "object *to*," "find fault *with*," "inconsistent *with*," "independent *of*," "infer *from*," and "interfere *with*." (See also **idioms.**)

USING PREPOSITIONS IN TITLES

When a preposition appears in a title, it is capitalized if it is the first word in the title or if it is four letters or more. (See also capitalized letters.)

EXAMPLES   *Composition for the Business and Technical World* was reviewed recently by the *Journal of Technical Communications.*
The book *Managing Like Mad* should be taken seriously in spite of its frothy title.

prepositional phrases

A **preposition** is a word that shows relationship and combines with a **noun** or **pronoun** (its **object**) to form a modifying **phrase.** A prepositional phrase, then, consists of a preposition plus its object and the object's **modifiers.**

EXAMPLE   *After the meeting,* the district managers adjourned *to the executive dining room.*

Prepositional phrases, because they normally modify nouns or **verbs,** usually function as **adjectives** or **adverbs,** although occasionally they function as nouns.

A prepositional phrase may function as an adverb of motion.

EXAMPLE   Turn the dial four degrees *to the left.*

A prepositional phrase may function as an adverb of manner.

EXAMPLE   Answer customers' questions *in a courteous fashion.*

A prepositional phrase may function as an adverb of place.

EXAMPLE   We ate lunch *in the company cafeteria.*

A prepositional phrase may function as an adjective.

EXAMPLE   Garbage *with a high paper content* has been turned into a protein-rich animal food.

When functioning as adjectives, prepositional phrases follow the nouns they modify.

EXAMPLE   Here are the results *of our study.*

When functioning as adverbs, prepositional phrases may appear in different places.

EXAMPLES   *In residential and farm wiring,* one of the wires must always be grounded.
One of the wires must always be grounded *in residential and farm wiring.*

Be careful when using prepositional phrases; separating a prepositional phrase from the noun it is modifying can cause **ambiguity.**

CHANGE   *The man* standing by the drinking fountain *in the gray suit* is our president.
TO   *The man in the gray suit* who is standing by the drinking fountain is our president.

See also **misplaced modifiers.**

principal/principle

Principal, meaning an "amount of money on which interest is earned or paid" or a "chief official in a school or court proceeding," is sometimes confused with *principle,* meaning a "basic truth or belief."

EXAMPLES   The bank will pay 6½% per month on the *principal.*
She is a person of unwavering *principles.*
He sent a letter to the *principal* of the high school.

Principal is also an adjective, meaning "main" or "primary."

EXAMPLE   My *principal* objection is that it will be too expensive.

progress reports

A progress report provides information about the status of a project—its current emphasis, whether it is on schedule, whether it is within budget, and so on. Progress reports are issued at regular intervals throughout the life of a project; they state what has been done in a specified interval and what has yet to be done before the project can be completed. The progress report is used primarily with projects that involve many steps over a period of weeks, months, or even years. Progress reports help keep projects running smoothly because on the basis of such reports, management may assign workers, adjust schedules, allocate budget, or schedule supplies and equipment.

All reports in a series of progress reports should be uniform in **format.** Although a progress report could be either a **formal report** or an informal report, it is commonly informal. If the progress report is an external report, it might use the **format** of a letter; if it is an internal report, it might use the format of a **memorandum.**

The **introduction** to the first progress report should identify the project, any specified methods and necessary materials, and the date by which the project is to be completed. Subsequent reports should then include only brief introductions that review the previous report and then sum up the progress since that report.

The body of the progress report should describe in detail the present status of the project.

The report should end with any conclusions and recommendations about changes in the schedule, materials, techniques, and so on. The **conclusion** may also include an estimate of future progress.

The following example is a first progress report submitted by a construction firm.

Introduction   This is a progress report, as agreed to in our contract, on the rewiring program at the Sports Arena. Although costs of certain equipment are higher than our original bid indicated, we expect to complete the project without going over cost; the speed with which the project is being completed will save labor costs.

WORK COMPLETED

As of August 15 we have finished the installation of the circuit breaker panels and meters of level I service outlets and of all sub-

Body   floor rewiring. Lighting fixture replacement, level II service outlets, and the upgrading of stage lighting equipment are in the preliminary stages (meaning that the wiring has been completed but installation of the fixtures has not yet begun).

COSTS

Equipment used up to this point has cost $10,800, and labor has cost $31,500 (including some subcontracted plumbing). My estimate for the rest of the equipment, based on discussions with your lighting consultant, is $11,500; additional labor costs should not exceed $25,000.

WORK SCHEDULE

I have scheduled the upgrading of stage lighting equipment from August 16 to October 5, the completion of level II service outlets from October 6 to November 12, and the lighting fixture replacement from November 15 to December 17.

CONCLUSION

Conclusion Although my original estimate on equipment ($20,000) has been exceeded by $2,300, my original labor estimate ($60,000) has been reduced by $3,500; so I will easily stay within the limits of my original bid. In addition, I see no difficulty in having the arena finished for the December 23 Christmas program.

pronouns

A pronoun is a word that is used as a substitute for a **noun** (the noun for which a pronoun substitutes is called its antecedent). Using pronouns in place of nouns relieves the monotony of repeating the same noun over and over. Pronouns fall into several different categories: personal, demonstrative, relative, interrogative, indefinite, reflexive, intensive, and reciprocal.

PERSONAL PRONOUNS

The **personal pronouns** refer to the person or persons speaking *(I, me, my, mine; we, us, our, ours);* the person or persons spoken to *(you, your, yours);* or the person or thing spoken of *(he, him, his; she, her, hers; it, its; they, them, their, theirs).* (See also **person.**)

EXAMPLE *I* wish *you* had told *me* that *she* was coming with *us.*
If *their* figures are correct, *ours* must be in error.

DEMONSTRATIVE PRONOUNS

The **demonstrative pronouns** *(this, these; that, those)* indicate or point out the thing being referred to.

EXAMPLES *This* is my desk.
These are my co-workers.
That will be a difficult job.
Those are incorrect figures.

RELATIVE PRONOUNS

The **relative pronouns** are *who* (or *whom*), *which,* and *that.* A relative pronoun performs a dual function: (1) it takes the place of a noun; and

(2) it connects, and establishes the relationship between, a **dependent clause** and its main **clause.**

EXAMPLE   The personnel manager decided *who* would be hired.

The pronoun *that* is normally used with restrictive clauses and the pronoun *which* with nonrestrictive clauses. (See also **restrictive and nonrestrictive elements.**)

EXAMPLES   Companies *that adopt the plan* nearly always show profit increases. (restrictive clause)
The annual report, *which was distributed yesterday,* shows that sales increased 20% last year. (nonrestrictive clause)

INTERROGATIVE PRONOUNS

Interrogative pronouns (*who* or *whom, what,* and *which*) ask questions. They differ from relative pronouns in two ways: They are used only to ask questions and they may introduce independent interrogative **sentences,** while relative pronouns connect or show relationship and introduce only dependent clauses.

INDEFINITE PRONOUNS

Indefinite pronouns specify a class or group of persons or things rather than a particular person or thing. *All, any, another, anyone, anything, both, each, either, everybody, few, many, most, much, neither, nobody, none, several, some,* and *such* are indefinite pronouns.

EXAMPLE   Not *everyone* liked the new procedures; *some* even refused to follow them.

REFLEXIVE PRONOUNS

A **reflexive pronoun,** which always ends with the **suffix** *-self* or *-selves,* indicates that the **subject** of the sentence acts upon itself.

EXAMPLE   The electrician accidentally shocked *herself.*

The reflexive and intensive pronouns are *myself, yourself, himself, herself, itself, oneself, ourselves, yourselves,* and *themselves.*
Myself is not a substitute for *I* or *me* as a personal pronoun.

CHANGE   John and *myself* completed the report on time.
TO   John and *I* completed the report on time.
CHANGE   The assignment was given to Wally and *myself.*
TO   The assignment was given to Wally and *me.*

INTENSIVE PRONOUNS

The intensive pronouns are identical in form with the reflexive pronouns (see page 471), but they perform a different function: to give **emphasis** to their antecedents.

EXAMPLE   I *myself* asked the same question.

RECIPROCAL PRONOUNS

The **reciprocal pronouns** (*one another* and *each other*) indicate the relationship of one item to another. *Each other* is commonly used when referring to two persons or things and *one another* when referring to more than two.

EXAMPLES   They work well with *each other.*
The crew members work well with *one another.*

GRAMMATICAL PROPERTIES OF PRONOUNS

Case. Pronouns have forms to show the subjective, objective, or possessive **case,** as the following chart shows.

	Subjective	*Objective*	*Possessive*
1st person singular	I	me	my, mine
2nd person singular	you	you	your, yours
3rd person singular	he	him	his
	she	her	her, hers
	it	it	its
1st person plural	we	us	our, ours
2nd person plural	you	you	your, yours
3rd person plural	they	them	their, theirs

A pronoun that is used as the subject of a clause or sentence is in the subjective case *(I, we, he, she, it, you, they, who).* The subjective case is also used when the pronoun follows a **linking verb.**

EXAMPLES   *He* is my boss.
My boss is *he.*

A pronoun that is used as the **object** of a **verb** or of a **preposition** is in the objective case *(me, us, him, her, it, you, them, whom).*

EXAMPLES   Mr. Davis hired Tom and *me.* (object of verb)
Between *you* and *me,* he's wrong. (object of preposition)

A pronoun that is used to express ownership is in the possessive case *(my, mine, our, ours, his, her, hers, its, your, yours, their, theirs, whose).*

EXAMPLE   He took *his* notes with him on the business trip.

Only the possessive form of a pronoun should ever be used with a **gerund.**

EXAMPLE   The boss objected to *my* arriving late every morning.

Several indefinite pronouns (*any, each, few, most, none,* and *some*) form the possessive only in "of" phrases (*of any, of some*). Others, however, use the **apostrophe** (*anyone's*).

CHANGE   *Each's* opinion was solicited.
TO   The opinion *of each* was solicited.

A pronoun **appositive** takes the case of its antecedents.

EXAMPLES   Two systems analysts, Joe and *I,* were selected to represent the company. (*Joe and I* is in apposition to the subject, *systems analysts,* and must therefore be in the subjective case.)
The systems analysts selected two members—Joe and *me.* (*Joe and me* is in apposition to *two members,* which is the object of the verb *selected,* and therefore must be in the objective case.)

How to Determine the Case of a Pronoun. To determine the case of a pronoun, try it with a **transitive verb,** such as *resembled.* If the pronoun can precede the verb, it is in the subjective case; if it must follow the verb, it is in the objective case.

EXAMPLES   *She* resembled her father. (subjective)
Her father resembled *her.* (objective)

If compound pronouns cause problems in determining case, try using them singly.

EXAMPLES   In his letter, John mentioned *you* and *me.*
In his letter, John mentioned *you.*
In his letter, John mentioned *me.*

They and *we* must discuss the terms of the merger.
They must discuss the terms of the merger.
We must discuss the terms of the merger.

When a pronoun modifies a noun, try it without the noun to determine its case.

EXAMPLES   *(We/Us)* pilots fly our own planes.
We fly our own planes. (You would not write "*Us* fly our own planes.")

He addresses his remarks directly to *(we/us)* technicians.
He addressed his remarks directly to *us.* (You would not write "He addressed his remarks directly to *we.*")

To determine the case of a pronoun that follows *as* or *than,* try mentally adding the words that are normally omitted.

EXAMPLES   The director does not have as much formal education *as he* [does]. (You would not write "*him* does.") His friend was taller than *he* [was]. (You would not write "*him* was.")

Number. **Number** is a frequent problem only with a few indefinite pronouns (*each, either, neither;* and those ending with *-body* or *-one,* such as *anybody, anyone, everybody, everyone, nobody, no one, somebody, someone*), which are normally singular and so require singular verbs and are referred to by singular pronouns.

EXAMPLES   As *each arrives* for the meeting and takes *his* seat, please hand *him* a copy of the confidential report. *Everyone is* to return *his* copy before *he* leaves. *No one* should be offended by these precautions when the importance of secrecy has been explained to *him.* I think *everybody* on the committee *understands* that *neither* of our major competitors *is* aware of the new process that we have developed.

Person.   Third person personal pronouns usually have antecedents.

EXAMPLE   John presented the report to the board of directors. *He* (John) first read *it* (the report) to *them* (the board of directors) and then asked for questions.

First and second person personal pronouns do not normally require antecedents.

EXAMPLES   *I* like my job. *You* were there at the time. *We* all worked hard on the project.

pronoun reference

Avoid vague and uncertain references between a **pronoun** and its antecedents.

CHANGE   Studs and thick treads make snow tires effective. *They* are implanted with an air gun.

TO   Studs and thick treads make snow tires effective. *The studs* are implanted with an air gun.

CHANGE   We made the sale and delivered the product. *It* was a big one.

TO   We made the sale, which was a big one, and delivered the product.

The **noun** to which a pronoun refers must be unmistakably clear. Three basic problems are encountered in regard to pronoun references. One is an ambiguous reference, or one that can be interpreted in more than one way.

CHANGE   Jim worked with Tom on the report, but *he* wrote most of it. (Who wrote most of it, Jim or Tom?)

TO   Jim worked with Tom on the report, but *Tom* wrote most of it.

A general (or broad) reference, or one that has no real antecedent, is another basic problem.

CHANGE   He deals with personnel problems in his work. *This* helps him in his personal life.

TO   He deals with personnel problems in his work. This experience helps him in his personal life.

A hidden reference, or one that has only an implied antecedent, is the third basic problem.

CHANGE   A high lipid, low carbohydrate diet is "ketogenic" because it favors *their* formation.

TO   A high lipid, low carbohydrate diet is "ketogenic" because it favors the formation of *ketone bodies*.

For the sake of **coherence,** pronouns should be placed as close as possible to their antecedents. The danger of creating an ambiguous reference increases with distance. Don't force your **reader** to go back too far to find out what a pronoun stands for.

CHANGE   The *house* in the meadow at the base of the mountain range was resplendent in *its* coat of new paint.

TO   The house, resplendent in *its* coat of new paint, was nestled in a meadow at the base of the mountain range.

Do not repeat an antecedent in **parentheses** following the pronoun. If you feel that you must identify the pronoun's antecedent in this way, you need to rewrite the **sentence.**

CHANGE   The senior partner first met Bob Evans when he (Evans) was still a trainee.

TO   Bob Evans was still a trainee when the senior partner first met him.

When making reference to a noun that includes both sexes (student, teacher, clerk, everyone), the pronouns *he* and *his* are traditionally used.

EXAMPLE　Each employee is to have *his* annual X-ray taken by Friday.

However, many people feel that this conventional use of the masculine **personal pronoun** implies sexual bias. If there is danger of offending, it is often best to avoid the problem by substituting an **article** or changing the statement from singular to plural. (See also **he/she.**)

EXAMPLES　Each employee is to have *his* annual X-ray taken by Friday.
　　　　　　All employees are to have *their* annual X-rays taken by Friday.

proofreaders' marks

Publishers have established **symbols,** called proofreaders' marks, which writers and editors use to communicate with printers in the production of publications. A familiarity with these symbols makes it easy for you to communicate your changes to others. See page 477.

proofreading

The final step in the **revision** of any document is proofreading. The effort spent on proofreading will depend on the complexity and the importance of the document. A short, simple **memorandum** to a close business associate may not require the same amount of proofreading as would a complex set of guidelines distributed to 1,000 people throughout a large organization. However, even a short, relatively simple letter to a potential client could cause you embarrassment if it contained errors.

To proofread effectively, you must be critical and alert. Much like revising, proofreading requires that you look at your writing objectively. In her article "36 Aids to Successful Proofreading," Ruth H. Turner offers useful tips for proofreading three types of material: short narratives (letters or memorandums, etc.); long narratives (manuals or **reports** of ten or more pages), and technical material and statistical tables.*

---

*Adapted, with permission, from *The Secretary,* Vol. 38, no. 2, p. 12, Feb. 1978 by the National Secretaries Association (International), Kansas City, MO 64108.

Mark in Margin	Instruction	Mark on Manuscript	Corrected Type
e	Delete	the ~~lawyer's~~ bible	the bible
lawyers	Insert	The ⌃bible	The lawyer's bible
stet	Let stand	the ~~lawyer's~~ bible	the lawyer's bible
Cap	Capitalize	the b̲ible	the Bible
lc	Make lower case	the /Law	the law
ital	Italicize	the lawyer's bible	the *lawyer's* bible
tr	Transpose	the ⌣bible⌣lawyer's⌣	the lawyer's bible
⌒	Close space	the Bi◡ble	the Bible
sp	Spell out	②bibles	two bibles
#	Insert space	The⌃Bible	The Bible
¶	Start paragraph	¶ The lawyer's...	The lawyer's...
run in	No paragraph	...marks⤾ ⌐Below is a....	....marks. Below is a....
sc	Set in small capitals	The b̲i̲b̲l̲e̲	The BIBLE
rom	Set in roman type	The⟨bible⟩	The bible
bf	Set in boldface	The b̰ḭb̰l̰ḛ	The **bible**
lf	Set in lightface	The⟨**bible**⟩	The bible
⊙	Insert period	The lawyers have their own bible⌃	The lawyers have their own bible.
⋏	Insert comma	However⌄ we cannot....	However, we cannot....
⌢=⫽⌢⌢	Insert hyphens	half⌃and⌃half	half-and-half
○	Insert colon	We need the following⌃	We need the following:
⌃	Insert semicolon	Use the law⌄ don't....	Use the law; don't....
⌄	Insert apostrophe	John⌄s law book	John's law book
⌣/⌣	Insert quotation marks	The ⌄law⌄ is law.	The "law" is law.
(⫽)	Insert parentheses	John's⌃law⌃book	John's (law) book
[⫽]	Insert brackets	John⌃1920-1962⌃went....	John [1920-1962] went....
$\frac{\vert}{N}$	Insert en dash	1920⌃1962	1920-1962
$\frac{\vert}{M}$	Insert em dash	Our goal⌃victory	Our goal—victory
⌄	Insert superior type	$3^{\lor}= 9$	$3^2 = 9$
⌃	Insert inferior type	$H\underset{\land}{S}O_4$	H_2SO_4
⌣	Insert asterisk	The law$^{\lor}$	The law*
†	Insert dagger	The law⌃	The law†
‡	Insert double dagger	The bible⌃	The bible‡
§	Insert section symbol	⌃Research	§Research

SHORT NARRATIVES

1. To concentrate specifically on typographical errors, read backwards (from right to left). Of course, since this method will not reveal omissions or duplications you must read again for content.
2. Arrange with another person to proofread each other's material. However, do not allow this procedure to delay your progress.
3. Concentrate as you slowly read for errors. Speed-reading does not help.
4. Pay attention to dates; do not assume they are correct. Check the **spelling** of months and even the correctness of the year.
5. Do not overlook names, addresses, subject or reference lines, signature lines, even the copies list. Errors can sneak into these items as well as into the body of the letter or memorandum.
6. Examine the end and beginning of lines in the body of the letter or memorandum to be sure little words like *that* or *and* have not been unnecessarily repeated.
7. If you have duplicated typed material and have ended each line at the same place, check your copy by laying it over the original and holding them both up to a strong light. This procedure will reveal errors without the need for reading.
8. Proofread personalized letters or memorandums that are identical except for names, addresses, or short insertions by carefully examining the personalized portions. Then hold them both up to a light, as described in the previous suggestion.

LONG NARRATIVES

1. Make a style sheet to assure consistency throughout a manuscript regarding spelling of specialized terminology or **jargon,** the capitalization and **punctuation,** the **heads, formats** for **outlines** or **tables,** and the names and titles. Any items that are unique to your project can be jotted down on this style sheet. Keep this sheet for proofreading later revisions or additional manuals or reports in a series.
2. Proofread in steps. For example, check all the heads and titles first. Be sure that you have consistently followed the style sheet.
3. Scan page numbers. Have you missed or duplicated any?
4. Read the body of the material against the original, **sentence** by sentence. This is a tedious, slow task and can be speeded up only if you know the material well enough to catch duplications, omissions, and typing errors with a continuous, concentrated reading.
5. Make sure any references to other parts of the work are correct. For

example, "See page 6 for a similar list." Possibly, in the final typing, the list moved to page 7.

6. Verify title and page numbers of other reference materials if you have that responsibility. Look up each item if at all possible.
7. When proofreading from a draft that has handwritten inserts, be sure none has been overlooked in the final draft.

TECHNICAL MATERIAL AND STATISTICAL TABLES

1. If possible, use two people—one reading aloud, the other checking copy—to verify itemized data, codes, figures, and so on.
2. When the columns of a table have been typed by tabulating across the page, proofread your copy down the columns, folding the original from top to bottom along the column and laying it next to the corresponding column of the copy.
3. If the draft is typewritten, lay a ruler under each line. This technique will help you keep your place on both the original and the copy. It also helps when the inevitable interruptions occur.
4. Check for both vertical and horizontal alignment of table columns. Proofread across the page, folding the original beneath each line and laying it under the corresponding lines of typed entries so that you can easily read the original against your copy.
5. Count the number of entries in each table column, and compare the total with the number in your copy. If there is a difference, hunt down the culprit.
6. Proofread outlines by breaking the task into components: check headings, then check the mechanics (**numbers,** letters, **indentations**), and finally perform a continuous reading of each item against the original.

A CHECKLIST FOR PROOFREADING

1. Make the **dictionary** your best friend. Is it *privilege, privelege* or *priviledge; gauge* or *guage; discression* or *discretion; chief* or *cheif?* Never be embarrassed to look up a word, for you might be more embarrassed at an error in completed work. Watch closely for omission of *-ed* or final *-s.*
2. Study related words until you understand differences in their meaning and use: **affect/effect, imply/infer, diagnosis/prognosis, advice/advise, insure/ensure/assure.** (For a list of such words, check the List of Usage Entries in the front of this book.)
3. Mentally repeat each syllable of long words with many vowels: *evacuation, responsibilities, continuously, individual.* It is easy to omit one.

4. Be careful with double letters. They are hard to remember, especially in words like *accommodation* and *occurrence.*

5. Remember that *supersede* is the only word in English ending in *-sede,* and *succeed, proceed,* and *exceed* are the only ones ending in *-ceed.* All others (*cede, recede, precede,* etc.) end in *-cede.*

6. Be careful with silent letters: *rhythm, rendezvous, malign.*

7. Check all punctuation marks after completing all other proofreading. Be careful not to omit a closing **parenthesis** or **quotation mark.**

8. Refer to the dictionary to determine whether to spell a compound word as separate words, hyphenated words, or one word. If the dictionary does not list the word as hyphenated or separate words, write it as one word.

9. Be sure that **capital letters** are used consistently. Follow any established style for the type of document or subject matter. If an established style does not exist, adopt a style of your own.

10. One last suggestion—proofread tomorrow what you worked on today.

For further information on proofreading and editing, you may wish to consult the following books.

Butcher, Judith. *Copy-Editing: The Cambridge Handbook*

Lasky, Joseph. *Proofreading and Copy-Preparation*

A Manual of Style. University of Chicago Press

Skillin, M. E., and Gay, R. M., eds. *Words Into Type*

proper nouns

A proper noun names a specific person, place, thing, concept, action, or quality. Therefore, proper nouns are always capitalized.

Proper Nouns	*Common Nouns*
Jane Jones	person
Chicago	city
General Electric	company
Tuesday	day
Declaration of Independence	document

For the **case** and **number** of proper nouns, see **nouns.**

proposals

Proposals, which suggest a plan or course of action, may be internal or external. An **internal proposal** usually recommends change or improvement within an organization. The recommendation might be to expand cafeteria service, to combine multiple manufacturing operations, or to introduce new working procedures. An internal proposal is normally prepared by an employee or department and then sent to a higher-ranking person in the organization for approval.

External proposals include both **sales proposals** and **government proposals.** A sales proposal is a company's offer to provide goods or services to a potential buyer within a certain amount of time and at a specified cost. The purpose of the proposal is to present a product or service in the best possible light and to offer reasons why a buyer should choose it over competitive products or services. The very survival of some companies depends upon their ability to produce effective proposals. Each type of proposal is explained in detail in its own entry.

proved/proven

Both *proved* and *proven* are acceptable past **participles** of *prove,* although *proved* is currently in wider use.

EXAMPLES   They had *proved* more obstinate than expected.
They had *proven* more obstinate than expected.

Proven is more commonly used as an **adjective.**

EXAMPLE   She was hired because of her *proven* competence as a manager.

pseudo/quasi

As a **prefix,** *pseudo,* meaning "false or counterfeit," is joined to a word without a **hyphen** unless the word begins with a **capital letter.**

EXAMPLES   *Pseudo*science (false science)
pseudo-Americanism (pretended Americanism)

Pseudo is sometimes confused with *quasi,* meaning "somewhat" or "partial." Unlike *semi, quasi* does not mean half. *Quasi* is usually hyphenated in combinations.

EXAMPLE   *quasi*-scientific literature.

See also **bi-/semi-.**

punctuation

Punctuation is a system of **symbols** that help the **reader** understand the structural relationship within (and the intention of) a **sentence.** Marks of punctuation may link, separate, enclose, indicate omissions, terminate, and classify sentences. Most of the thirteen punctuation marks can perform more than one function. The use of punctuation is determined by grammatical conventions and the writer's intention—in fact, punctuation often substitutes for the writer's facial expressions. Misuse of punctuation can cause your reader to misunderstand your meaning. Detailed information on each mark of punctuation is given in its own entry. The following are the thirteen marks of punctuation.

apostrophe	,
brackets	[ ]
colon	:
comma	,
dash	—
exclamation mark	!
hyphen	-
parentheses	( )
period	.
question mark	?
quotation marks	" "
semicolon	;
slash	/

Q

question marks

The question mark (?) has the following distinct uses.

Use a question mark to end a **sentence** that is a direct question.

EXAMPLE   Where did you put the specifications?

Use a question mark to end any statement with an interrogative meaning (a statement that is declarative in form but asks a question).

EXAMPLE   The report is finished?

Use a question mark to end an interrogative **clause** within a declarative sentence.

EXAMPLE   It was not until July (or was it August?) that we submitted the report.

Retain the question mark in a title that is being cited even though the sentence in which it appears has not ended.

EXAMPLE   *Should Engineers Be Writers?* is the title of her book.

When used with **quotations,** the question mark may indicate whether the writer who is doing the quoting or the person being quoted is asking the question. When the writer doing the quoting asks the question, the question mark is outside the **quotation marks.**

EXAMPLE   Did she say, "I don't think the project should continue"?

If, on the other hand, the quotation itself is a question, the question mark goes inside the quotation marks.

EXAMPLE   She asked, "When will we go?"

If the writer doing the quoting and the person being quoted both ask questions, use a single question mark inside the quotation marks.

EXAMPLE   Did she ask, "Will you go in my place?"

Question marks may follow a series of separate items within an interrogative sentence.

EXAMPLE   Do you remember the date of the contract? its terms? whether you signed it?

A question mark should never be used at the end of an indirect question.

CHANGE   He asked me whether sales had increased this year?
TO   He asked me whether sales had increased this year.

When a directive or command is phrased as a question, a question mark is usually not used. However, a request (to a customer or a superior, for instance) would almost always require a question mark.

EXAMPLES   Will you make sure that the machinery is operational by August 15.

Will you please telephone me collect if your entire shipment does not arrive by June 10?

questionnaires

A questionnaire—a series of questions on a particular topic, sent out to a number of people—is a sort of interview on paper. It offers several advantages over the personal interview but has some disadvantages. A questionnaire allows you to sample many more people than personal interviews can. It enables you to obtain responses from people in different parts of the country. Even people who live near you may be easier to reach by mail than in person. Those responding to a questionnaire do not face the constant pressure posed by someone jotting down their every word—a fact that could result in more thoughtful answers from questionnaire respondents. And the questionnaire reduces the possibility that the interviewer might influence an answer by tone of voice or facial expression. Finally, the cost of a questionnaire is lower than the cost of numerous personal interviews.

Questionnaires have drawbacks too. People who have strong opinions on a subject are more likely to respond to a questionnaire than those who do not. This factor could slant the results. An interviewer can follow up on an answer with a pertinent question; at best, a questionnaire can be designed to let one question lead logically to another. Furthermore, mailing a batch of questionnaires and waiting for replies take considerably longer than a personal interview does.

A questionnaire must be properly designed. Your goal should be to obtain as much information as possible from your recipients with as little effort on their part as possible. The first rule to follow is to keep the questionnaire brief. The longer the questionnaire is, the less likely the recipient will be to complete and return it. Next, the questions should be easy to understand. A confusing question will yield confusing results, whereas a carefully worded question will be easy to answer. Ideally, questions should be answerable with a yes or no.

> Would you be willing to work a four-day work week, ten hours a day, with every Friday off?
>
> Yes _____
>
> No _____
>
> No opinion _____

When it is not possible to phrase questions in such a straightforward style, provide an appropriate range of answers. Questions must be phrased neutrally; their wording must not lead respondents to a particular answer.

> How many hours of overtime would you be willing to work each week?
>
> 4 hours _____   10 hours _____
>
> 6 hours _____   More than 10 hours _____
>
> 8 hours _____   No overtime _____

When preparing your questions, remember that you must eventually tabulate the answers; therefore, try to formulate questions whose answers can be readily computed. The easiest questions to tabulate are those for which the recipient does not have to compose an answer. Any questions that require a comment for an answer take time to think about and write and thus lessen your chances of obtaining a reponse. They are also difficult to interpret. Questionnaires should include a section for additional comments, though, where recipients may clarify their overall attitude toward the subject. If the information will be of value in interpreting the answers, include questions about the recipient's age, education, occupation, and so on. Include your name, your address, the purpose of the questionnaire, and the date by which an answer is needed.

A questionnaire sent by mail must be accompanied by a letter explaining who you are, the purpose of the questionnaire, how it will be used, and the date by which you would like to receive a reply. If the information provided will be kept confidential or if the recipient's identity will not be disclosed, say so in the letter.

May 18, 19--

To: All Company Employees

From: Nelson Barrett, Director of Personnel

Subject: Review of Flexible Working Hours Program

     Please complete this questionnaire regarding Luxwear Corporation's trial program of flexible working hours. Your answers will help us to decide whether the program should continue. Return the questionnaire to the Personnel Department by May 28.
     If you want to discuss any item in more detail, call Tania Peters in Personnel at extension 8812.

1. What kind of position do you hold?

   _____   _____
   supervisory   nonsupervisory

2. Indicate your exact starting time under flexitime.

   _____

3. Where do you live?

   Talbot County _____   Greene County _____

   Montgomery County ____   Other, specify _____

4. How do you usually travel to work?

   Drive alone _____   Walk _____

   Car pool _____   Bus _____

   Train _____   Motorcycle _____

   Bicycle _____   Other, specify _____

5. Has flexitime affected your commuting time? If so, please indicate the approximate number of minutes.

   _____   _____   _____
   increase   decrease   no change

6. If you drive, has flexitime affected the amount of time it takes to find a parking space?

   _____   _____   _____
   increase   decrease   no change

Sample Questionnaire

7. Do you think that flexitime has affected your productivity?

_____   _____   _____
increase   decrease   no change

8. Have you had difficulty getting in touch with colleagues whose work schedules are different from yours?

Yes_____   No_____

9. Have you had trouble scheduling meetings?

Yes_____   No_____

10. Has flexitime affected the way you feel about your job?

_____   _____   _____
feel better   feel worse   no change

11. How important is it for you to have flexibility in your working hours?

Very_____   Not very_____

Somewhat_____   Not at all_____

12. If you have children, has flexitime made it easier or more difficult for you to obtain babysitting or day-care services?

_____   _____   _____
easier   more difficult   no change

13. Do you recommend that the flexitime program be made permanent?

Yes_____   No_____

14. Do you have suggestions for any changes in the program? If so, please specify.

_____

_____

Select the recipients for your questionnaire carefully. If you want to survey the opinions of all the employees in a small laboratory, simply send each worker a questionnaire. To survey the members of a professional society, mail questionnaires to everyone on the membership list. But to survey the opinions of large groups in the general population—for example, all medical technologists working in private laboratories—is not so easy. Since you cannot include everybody in your survey, you have to choose a representative cross-section.

The questionnaire on pages 486–487 was sent to employees in a large organization who had participated in a six-month program of flexible working hours. Under the program, employees worked a forty-hour, five-day week, with flexible starting and quitting times. Employees could start work between 7 and 9 A.M. and leave between 3:30 and 6:30 P.M., provided that they worked a total of eight hours each day and took a one-half-hour lunch period midway through the day.

quid pro quo

Quid pro quo, which is Latin for "one thing for another," can suggest mutual cooperation or "tit for tat" in a relationship between two groups or individuals. The term may be appropriate to business and legal contexts if you are sure your **reader** understands its meaning.

EXAMPLE   It would be unwise to grant their request without insisting on some *quid pro quo.*

quotation marks

Quotation marks (" ") are used to enclose direct **repetition** of spoken or written words. They should not be used as a means of **emphasis** under normal circumstances. There are a variety of guidelines for using quotation marks.

Enclose in quotation marks anything that is quoted word for word (direct quotation) from speech.

EXAMPLE   She said clearly, "I want the progress report by three o'clock."

Do not enclose indirect **quotations**—usually introduced by *that*—in quotation marks. Indirect quotations are **paraphrases** of a speaker's words or ideas.

EXAMPLE   She said that she wanted the progress report by three o'clock.

Handle quotations from written material the same way: place direct quotations within quotation marks, but not indirect quotations.

EXAMPLES   The report stated, "The potential in Florida for our franchise is as great as in California."
The report indicated that the potential for our franchise is as great in Florida as in California.

Quotations longer than four types lines (at 50 characters per line) are normally indented (*all* the lines) five spaces from the left margin, single-spaced, and *not* enclosed in quotation marks.

Unless it is indented as just described, a quotation of more than one **paragraph** is given quotation marks at the beginning of each new paragraph, but at the end of only the last paragraph.

Use single quotation marks (on a typewriter use the **apostrophe** key) to enclose a quotation that appears within a quotation.

EXAMPLE   John said, "Jane told me that she was going to 'hang tough' until the deadline is past."

Slang, colloquial expressions, and attempts at humor, although infrequent in technical writing in any case, seldom rate being set off by quotation marks.

CHANGE   Our first six months in the new office amounted to little more than a "shakedown cruise" for what lay ahead.
TO   Our first six months in the new office amounted to little more than a shakedown cruise for what lay ahead.

Use quotation marks to set off special words or technical terms only to point out that the term is used in context for a unique or special purpose (used, that is, in the sense of *the so-called*).

EXAMPLES   Typical of deductive analyses in real life are accident investigations: What chain of events caused the sinking of an "unsinkable" ship such as the Titanic on its maiden voyage?

Use quotation marks to enclose titles of short stories, articles, essays, radio and television programs, short musical works, paintings, and other art works.

EXAMPLE   Did you see the article "No-Fault Insurance and Your Motorcycle" in last Sunday's *Journal?*

Titles of books and periodicals are underlined (to be typeset in **italics**).

EXAMPLE   Articles in the *Business Education Forum* and *Scientific American* quoted the same passage.

Some titles, by convention, are neither set off by quotation marks nor underlined, although they are capitalized.

EXAMPLES   the Bible, the Constitution, Lincoln's Gettysburg Address, the Montgomery Ward Catalog

Commas and **periods** always go inside closing quotation marks.

EXAMPLE   "Reading *Space Technology* gives me the insider's view," he says, adding "it's like having all the top officials sitting in my office for a bull session."

Semicolons and **colons** always go outside closing quotation marks.

EXAMPLES   He said, "I will pay the full amount"; this sure surprised us.
The following are her favorite "sports": eating and sleeping.

All other **punctuation** follows the logic of the context: if the punctuation is a part of the material quoted, it goes inside the quotation marks; if the punctuation is not part of the material quoted, it goes outside the quotation marks.

Quotation marks may be used as ditto marks, instead of repeating a line of words or numbers directly beneath an identical set. In **formal writing,** this use is confined to **tables** and **lists.**

EXAMPLE   A is at a point equally distant from L and M.
B " " " "   "   "   " S and T.
C " " " "   "   "   " R and Q.

quotations

When you have borrowed words, facts, or ideas of any kind from someone else's work, acknowledge your debt by giving your source credit in a **footnote.** Otherwise, you are guilty of **plagiarism.** Be sure that you have represented the original material honestly and accurately.

DIRECT QUOTATIONS

Direct word-for-word quotations are enclosed in **quotation marks.** They are usually, although not always, separated from the rest of the **sentence** in which they occur either with a **comma** or a **colon.**

EXAMPLES   The noted economist says, "If monopolies could be made to behave as if they were perfectly competitive, we would be able to enjoy the benefits both of large-scale efficiency and of the perfectly working price mechanism."

The noted economist pointed out: "There are three options available when technical conditions make a monopoly the natural outcome of competitive market forces: private monopoly, public monopoly, or public regulation."

INDIRECT QUOTATIONS

Indirect quotations, which are essentially **paraphrases** and are usually introduced by *that,* are not set off from the rest of the sentence by **punctuation** marks.

DIRECT He said in a recent interview, "Regulation cannot supply the dynamic stimulus that in other industries is supplied by competition." (direct quotation)

INDIRECT In a recent interview he said that regulation does not stimulate the industry as well as competition does. (indirect quotation)

Where a quotation is divided, the material which interrupts the quotation is set off, before and after, by commas, and quotation marks are used around each part of the quotation.

EXAMPLE "Regulation," he said in a recent interview, "cannot supply the dynamic stimulus that in other industries is supplied by competition."

At the end of a quoted passage, commas and **periods** go inside the quotation marks, and colons and **semicolons** go outside the quotation marks.

DELETIONS OR OMISSIONS

Deletions or omissions from quoted material are indicated by three **ellipsis** dots within a sentence and four ellipsis dots at the end of a sentence, the first of the four dots being the period that ends the sentence.

EXAMPLE "If monopolies could be made to behave . . . we would be able to enjoy the benefits of . . . large-scale efficiency. . . ."

When a quoted passage begins in the middle of a sentence rather than at the beginning, however, **ellipsis** dots are not necessary; the fact that the first letter of the quoted material is not capitalized tells the **reader** that the quotation begins in midsentence.

CHANGE He goes on to conclude that ". . . coordination may lessen competition within a region."

TO He goes on to conclude that "coordination may lessen competition within a region."

Such quotations should be worked into one of your sentences rather than left standing alone. Where quoted material is worked into a sentence, be sure that it is related logically, grammatically, and syntactically to the rest of the sentence.

INSERTING MATERIAL INTO QUOTATIONS

When it is necessary to insert a clarifying comment within quoted material, use **brackets.**

> EXAMPLE   "The industry is organized as a relatively large, integrated system serving an extensive [geographic] area, with smaller systems existing as islands within the larger system's sphere of influence."

When quoted material contains an obvious error, or might in some other way be questioned, the expression *sic,* enclosed in brackets, follows the questionable material to indicate that the writer has quoted the material exactly as it appeared in the original. (See also **sic.**)

> EXAMPLE   In *Basic Astronomy,* Professor Jones notes that the "earth does not revolve around the son [sic] at a constant rate."

INCORPORATING QUOTATIONS INTO TEXT

Depending on the length, there are two mechanical methods of incorporating quotations into your text. Quotations of four lines or less are incorporated into your text and enclosed in quotation marks. (Many writers use this method regardless of length. When this method is used with multiple **paragraphs,** quotation marks appear at the beginning of each new paragraph, but at the end of only the last paragraph.) Material longer than five lines is now usually inset; that is, it is set off from the body of the text by being indented five spaces from the left margin and by triple spacing above and below the quotation. The quoted passage is single-spaced and not enclosed in quotation marks.

OVERQUOTING

Do not rely too heavily on the use of quotations in the final version of your **report** or paper. If you do, your work will appear merely derivative. The temptation to overquote during the **note-taking** phase of your **research** can be avoided if you concentrate on summarizing what you read. Quote word-for-word only where your source concisely sums up a great deal of information or points to a trend or development important to your subject. As a rule of thumb, avoid quoting anything over one paragraph.

R

raise/rise

Both words mean "move to a higher position." However, *raise* is a
transitive verb and always takes an object (*raise* crops), whereas *rise* is an
intransitive verb and never takes an object (heat rises).

EXAMPLES   *Raise* the lever to the ON position.
            Turn off the motor if the internal temperature *rises* above 110°F.

re

Re (and its variant form *in re*) is business and legal **jargon** meaning "in
reference to" or "in the case of." *Re* is sometimes used in **memorandums.**

EXAMPLE   To: Elaine Barr
           From: Edgar Roden
           Re: Revised Office Procedures

Re, however, is now giving way to *subject.*

EXAMPLE   To: Elaine Barr
           From: Edgar Roden
           Subject: Revised Office Procedures

readers

The first rule of effective writing is to help the reader. The responsibility
that technical writers have to their readers is nicely expressed in the
Society of Technical Communications's *Code for Communicators,* which
calls for technical writers to hold themselves responsible for how well
their audience understands their message and to satisfy the audience's
need for information rather than the writer's need for self-expression.
This code reminds writers to concentrate on writing *for* the needs of
specific readers rather than merely *about* certain subjects. Too often,
writers are blinded by their own familiarity with the technical subject and
therefore tend to overlook their readers' lack of knowledge.

As a technical writer, you must usually assume that your readers are
less familiar than you with the subject. You have to be careful, for
example, in writing **instructions** for operators of equipment that you

designed or in writing a **report** about a technical system for executives whose training is in such nontechnical areas as management, accounting, law, and so on. Such readers are unlikely to have had training in your technical specialty and thus need definitions of technical terms as well as extremely clear explanations of principles that you, as a specialist, take for granted. Even if you write a **journal article** for others in the field, you must remember to explain new or special uses of standard terms and principles.

Before starting to write, you must determine exactly who your readers are and then adjust the amount of detail and technical vocabulary to their training and experience. For example, to describe a new technique for repairing a vibration dampener to an experienced auto mechanic, you could use standard auto repair terms and procedures. To describe the same technique in a training manual, however, the terms and procedures would require a much fuller explanation.

As the previous example implies, when you write for many readers (hundreds of auto mechanic trainees, for example), try to visualize a single, typical member of that group and write for that reader. You might also make a list of characteristics (experience, training, and work habits for example) of that reader. This technique, used widely by professional writers, enables you to decide what should or should not be explained according to that reader's needs. When you are writing to a group, it is usually best to aim at those with the least training and experience. The extra information and explanation will not insult those who know more; they can simply read or follow the instructions more quickly, perhaps skipping sections with which they are thoroughly familiar.

When your readership includes people with widely varied backgrounds, however, such as in a report or **proposal,** consider aiming various sections of the document at different sets of readers. Recommendations, **executive summaries,** and **abstracts** can be aimed at executives, who will be reading to understand the general implications of projects or technical systems. **Appendixes** containing **tables, graphs,** and raw technical data can be aimed at specialists who wish to examine or use such supporting data. A **glossary,** of course, must be written for readers who are not familiar with standard technical terminology. The body of a report or proposal should be aimed at readers with the most serious interest or who need to make decisions based on the details in the contents.

Routine **memorandums** and short reports that are written for one individual reader do not require such elaborate design. Simply remember that person's exact needs as you write.

Letters written to readers outside your organization, however, require a special **tone.** In such letters, you must concern yourself not only with the readers' understanding of your **topic,** but with their reaction to it as well. You must be as careful in representing your organization as you are in explaining your subject matter. (See **correspondence** for a discussion of the extra courtesy required in such business letters.)

No matter what your topic or format, never forget that your readers are the most important consideration, for the most eloquent prose is meaningless unless it reaches its audience.

reason is because

The **phrase** *reason is because* is a colloquial expression to be avoided in writing. *Because,* which in this phrase only repeats the notion of cause, should be replaced by *that.*

CHANGE   Sales have increased more than 20%. The *reason is because* our sales force has been more aggressive this year.

TO   Sales have increased more than 20%. The *reason is that* our sales force has been more aggressive this year.

OR   Sales have increased more than 20% this year *because* our sales force has been more aggressive.

reciprocal pronouns

The reciprocal pronouns *one another* and *each other* indicate the relationship between two or more persons or things. *Each other* is normally used when referring to two persons or things.

EXAMPLE   The acid and metal reacted with *each other* in the beaker.

One another is normally used when referring to more than two persons or things.

EXAMPLE   The researcher observed the several chemicals reacting with *one another* in the beaker.

reference books

The following list provides many basic sources for library research and general reference.

ABSTRACTS

Agricultural Engineering Abstracts, 1976--. (monthly)

American Statistics Index (ASI), 1974--. (monthly)
    Index and abstracts of government publications containing statistics.

Astronomy and Astrophysics Abstracts, 1969--. (semiannually)

Biological Abstracts, 1926--. (every two weeks)

Biological Abstracts/RRM, 1962--. (monthly)
    Formerly *Bioresearch Index.*

Chemical Abstracts (CA), 1907--. (every two weeks)

Computer Abstracts, 1967--. (monthly)
    This is section C of *Science Abstracts.*

Dissertation Abstracts International, 1938--. (monthly)
    Part B contains science and engineering; Part C, European abstracts.

Ecological Abstracts, 1974--. (every other month)

Electrical and Electronic Abstracts, 1898--. (monthly)
    This is section B of *Science Abstracts.*

Energy Information Abstracts, 1976--. (monthly)

Energy Research Abstracts, 1976--. (every two weeks)

Engineering Index, 1906--. (monthly)

Environment Abstracts, 1971--. (monthly)

INIS Atomindex, 1970--. (every two weeks)
    Formerly *Nuclear Science Abstracts.*

International Aerospace Abstracts, 1961--. (every two weeks)

Key Abstracts: Electrical Measurement and Instrumentation, 1976--. (monthly)

Key Abstracts: Physical Measurement and Instrumentation, 1976--. (monthly)

Key Abstracts: Solid State Devices, 1978--. (monthly)

Mathematical Reviews, 1940--. (monthly)

Metals Abstracts, 1968--. (monthly)

Oceanic Abstracts, 1966--. (every other month)

Physics Abstracts (PA), 1898--. (every two weeks)
    This is section A of *Science Abstracts.*

Pollution Abstracts, 1970--. (every two months)

Review of Plant Pathology, 1922--. (monthly)
   Formerly *Review of Applied Mycology.*

ATLASES

Moore, Patrick. *The Atlas of the Universe*

National Geographic Atlas of the World

Rand McNally Commercial Atlas and Marketing Guide (updated annually)

The Rand McNally Atlas of the Oceans

Times Atlas of the World: Comprehensive Edition

United States Geological Survey. *National Atlas of the United States of America*

BIBLIOGRAPHIES

Alred, Gerald, Diana Reep and Mohan Limaye. *Business and Technical Writing: An Annotated Bibliography of Books 1880-1980*

Besterman, Theodore. *A World Bibliography of Bibliographies*

Bibliographic Index: A Cumulative Bibliography of Bibliographies, 1937--. (updated three times annually)

Philler, Theresa Ammennito, et al., eds. *An Annotated Bibliography on Technical Writing, Editing, Graphics, and Publishing 1950-1965*

Sheehy, Eugene P. *Guide to Reference Books*

Walford, A. J. *Walford's Guides to Reference Material*
   Volume 1 contains science and technology.

BIOGRAPHIES

American Men and Women of Science

Biography Index, 1946-1973

Current Biography, 1940--. (monthly)

Dictionary of Scientific Biography, 1970-1980

Ireland, Norma Olin. *Index to Scientists of the World from Ancient to Modern Times: Biographies and Portraits*

Who's Who in America, 1899-- (updated every two years)

BOOK GUIDES

Books in Print. (issued annually) (lists author, title, publication data, edition, and price.

Cumulative Book Index, 1928--. (lists all books published in English, with author, title, and subject listings; issued annually with monthly supplements)

New Technical Books, 1915--. (annotated selections, arranged by subject and updated every two months)

Proceedings in Print, 1964--. (announcements of conference proceedings with citation, subject, agency, and price; updated every two months)

BOOK REVIEWS

Book Review Digest, 1905--. (monthly, with annual accumulation)

Current Book Review Citations, 1976--.

Technical Book Review Index, 1935--. (guide to reviews in scientific, technical, and trade journals; issued monthly except for July and August)

COMMERCIAL GUIDES AND PROFESSIONAL DIRECTORIES

Encyclopedia of Associations (annual guide to professional groups and publications with a quarterly supplement, *New Associations and Projects*)

MacRae's Blue Book (annually updated buying directory for industrial material and equipment)

National Trade and Professional Associations of the United States and Canada and Labor Unions, 1966--. (annual)

Thomas Register of American Manufacturers, 1905--. (monthly)

CURRENT AWARENESS SERVICES

Chemical Titles, 1960--. (lists current journal articles, indexed by keyword and author, issued every two weeks)

Conference Papers Index, 1973--. (monthly)

Current Contents, 1961--. (weekly listing of current journals' tables of contents; sections include agriculture, biology and environment services, engineering and technology; physical, chemical and earth sciences; life sciences)

Current Papers in Electrical and Electronics Engineering, 1969--. (monthly)

Current Papers in Physics, 1966--. (every two weeks)

Current Papers on Computer and Control, 1969--. (monthly)

Environmental Periodicals Bibliography, 1972--. (monthly)

Index to Scientific and Technical Proceedings, 1978--. (monthly)

DICTIONARIES

For unabridged and desk dictionaries, see **dictionary.** For selected specialized subject dictionaries, see **library research.**

DISSERTATIONS

Dissertation Abstracts International, 1938--. (monthly)
Since 1966, Part B has been devoted to science and engineering.

Masters Abstracts, 1962--. (quarterly)

ENCYCLOPEDIAS

Encyclopedia Americana

Encyclopedia of Computer Science and Technology

Gray, Peter, ed. *The Encyclopedia of the Biological Sciences*

International Encyclopedia of Statistics

McGraw-Hill Encyclopedia of Science and Technology

The New Encyclopaedia Britannica

Newman, James R., ed. *Harper Encyclopedia of Science*

Sills, David L., ed. *International Encyclopedia of the Social Sciences*

Van Nostrand's Scientific Encyclopedia

FACTBOOKS AND ALMANACS

The Budget in Brief (a condensed version of the annual U.S. federal budget)

Demographic Yearbook, 1979-- (annual collection of information on world economics and trade published by the United Nations)

Facts on File, 1974--. (weekly digest of world news, collected annually)

Statistical Abstract of the United States (annual publication of the U.S. Bureau of the Census that summarizes data on industrial, political, social, and economic organizations in the United States)

GOVERNMENT PUBLICATIONS

Government Reports Announcements (GRA), 1946--. (twice-monthly listing of federally sponsored research reports arranged by subject with abstracts)

Government Reports Index, 1965--. (semimonthly index of reports abstracted in *GRA* and arranged by subject, author, and report number)

Index to U.S. Government Periodicals, 1970--. (quarterly)

Monthly Catalog of U.S. Government Publications, 1895--. (lists nonclassified publications of all federal agencies by subject, author, and catchword; also gives ordering instructions)

Monthly Checklist of State Publications, 1940--. (state government publications arranged by state with an annual subject and title index)

Schmeckebier, Laurance F., and Roy B. Eastin. *Government Publications and Their Use* (guide to government publications)

GRAPHICS

Arkin, Herbert, and Raymond R. Colton. *Statistical Methods*

Enrick, Norbert L. *Handbook of Effective Graphic and Tabular Communication*

Kepler, Harold B. *Basic Graphical Kinematics*

LANGUAGE USAGE

Bernstein, Theodore M. *The Careful Writer: A Guide to English Usage*

Copperud, Roy H. *American Usage and Style: The Consensus*

Morris, William and Mary Morris. *Harper Dictionary of Contemporary Usage*

Nicholson, Margaret. *A Dictionary of American English Usage*

LOGIC

Beardsley, Monroe C. *Thinking Straight: Principles of Reasoning for Readers and Writers*

Brennan, Joseph G. *A Handbook of Logic*

Copi, Irving M. *Introduction to Logic*

Weddle, Perry. *Argument: A Guide to Critical Thinking*

PARLIAMENTARY PROCEDURE

Deschler, Lewis. *Deschler's Rules of Order*

Robert, Henry M. et al. *Robert's Rules of Order*

PERIODICAL INDEXES AND DIRECTORIES

Applied Science and Technology Index, 1958--. (alphabetical subject list of more than 50,000 periodicals; issued monthly)

Bibliography of Agriculture, 1942--. (every two months)

General Science Index, 1978--. (monthly)

Index Medicus, 1880--. (monthly)

New York Times Index, 1851--. (every two weeks)

Reader's Guide to Periodical Literature, 1900--. (every two weeks)

Science Citation Index, 1962--. (quarterly)

Ulrich's International Periodicals Directory: A Classified Guide to Current Periodicals, Foreign and Domestic (annually)

Wildlife Review, 1935--. (quarterly)

Zoological Record, 1864--. (annually)

SPEECH AND GROUP DISCUSSION

Capp, Glen R. and G. Capp, Jr., *Basic Oral Communication*

Harnack, R. Victor, et al. *Group Discussion: Theory and Technique*

Maier, Norman. *Problem-Solving Discussions and Conference: Leadership Methods and Skills*

Wilcox, Roger P. *Oral Reporting in Business and Industry*

(See also **oral reports**).

STYLE MANUALS AND GUIDES

For a selected listing of style manuals and guides, see **references.**

SUBJECT GUIDES

Jenkins, Frances B. *Science Reference Sources*

Lasworth, Earl J. *Reference Sources in Science and Technology*

Malinowsky, Harold R. *Science and Engineering Literature: A Guide to Reference Sources*

Mount, Ellis. *Guide to Basic Information Sources in Engineering*

Owen, Dolores B. *Abstracts and Indexes in Science and Technology: A Descriptive Guide*

SYNONYMS

Rodale, J. L. *The Synonym Finder*

Roget, Peter M. *Roget's International Thesaurus*

Webster's Collegiate Thesaurus

reference letters

Almost everyone is called upon at some time to provide a recommendation or reference for a colleague, an acquaintance, or an employee. Reference letters may range from completing an admission form for a prospective student to composing a detailed description of professional accomplishments and personal characteristics for someone seeking employment.

In order to write an effective letter of recommendation, you must be familiar enough with the applicant's abilities or actual performance to offer an evaluation. Then you must *truthfully*, without embellishment, communicate that evaluation to the inquirer. For the reference letter to serve as a valid selection device, you must address specifically the applicant's skills, abilities, knowledge, and personal characteristics in relation to the requested objective. You may also be asked to say what you know about the person's former employment and education or even his or her military discharge and personnel records.

In a reference letter, always respond directly to the inquiry, being careful to address the specific questions asked. For the record, you must identify yourself: name, title or position, employer, and address. You will also have to state how long you have known the applicant and the circumstances of your acquaintance. You should mention with as much substantiation as possible, one or two outstanding characteristics of the applicant. Organize the details in your letter in order of importance; put the most important details first. Conclude with a statement of recommendation and a brief summary of the applicant's qualifications.

The Privacy Act of 1974 has had a major impact on reference letters. Under the Act applicants have a legal right to examine the materials in an organization's files that concern them—unless they sign a waiver of their right to do so. Organizations seeking to avoid the privacy issue may

2511 Raintree Drive
West Lafayette, IN 47906
May 20, 19--

Mr. Phillip Lester
Personnel Director
Thompson Enterprises
201 State Street
Springfield, IL 62705

Dear Mr. Lester:

As her employer and her former professor, I am happy to have
the opportunity to recommend Kerry Hawkins. I've known
Kerry for the last four years, first as a student in my class
and for the last year as a research assistant.

I have found Kerry to be an excellent student, with above
average grades in our program. On the basis of a GPA of 5.6
(A=6), Kerry was offered a research assistantship to work on
a grant under my supervision. In every instance, Kerry
completed her library search assignment within the time
agreed upon. The material provided in the reports Kerry
submitted met the requirements for my work and more. These
reports were always well written. While working 15 hours a
week on this project, Kerry has maintained a class load of
12 hours per semester.

I strongly recommend Kerry for her ability to work
independently, to organize her time efficiently, and for her
ability to write clearly and articulately. Please let me
know if I can be of further service.

Sincerely yours,

Michael Paul

Michael Paul
Professor of Business

MP:mt

Letter 18

offer applicants the option of signing a waiver against reading letters of reference written about them. When you are requested to serve as a reference or asked to supply a letter of reference, you should thus inquire about the confidentiality status of your letter—whether the applicant's file will be open or closed. Letter 18 is a typical reference letter.

references

In scientific and technical writing, references identify the sources of facts or ideas and direct readers to further information on the **topic.**

A list of references, entitled *Literature Cited, References Cited,* or simply *References,* is placed either at the end of a brief work or at the end of each chapter or section of a longer work. The reference section at the end of a brief work usually begins on a new page. Reference sections at the end of chapters or sections in longer works can begin on the last page of text, provided that there is room enough for the title of the reference section and at least one complete reference citation. Otherwise, begin the reference section on a new page. References that appear in more than one section or chapter should be repeated in full in each appropriate references section.

The references list cites only those works that are specifically referred to in the text. Sources not cited in the text but used for background information are listed in the **bibliography.** The following reference guidelines also apply to bibliography citations. Whereas references are arranged in a numerical sequence according to where they fall in the text, bibliography citations are arranged alphabetically by author's name.

The citations in a references list are arranged in numerical sequence (1, 2, 3), according to the order in which they are first mentioned in the text. The most common method for directing readers from your text to specific references is to place the reference number in **parentheses** after the work cited. Thus, in the text, the number one in parentheses (1) after a reference to a book, article, or other work refers the **reader** to the first citation in the references list. The number five in parentheses (5) refers the reader to the fifth citation in the reference list. A second number in the parentheses, separated from the first by a colon (3:27), refers to the page number of the source from which the information was taken. There are several other common methods for directing readers from your text to specific references. You can place the word *Reference,* or the abbreviation *Ref.,* within parentheses with the numbers: (Reference 1) or (Ref.

1). Or you can write the reference number as a superscript: text[1]. Sometimes you may cite your sources within a sentence in the text; then, you should spell out the word *Reference:* "The data in Reference 3 include . . . ." For subsequent references in the text, simply repeat the reference number of the first citation.

References systems vary so widely from one field of study to another that it is impossible to provide comprehensive guidelines for all possible alternatives. Most publications and organizations provide guidelines for their own materials. The most comprehensive guidelines for preparing references and bibliographies have been established by the American National Standards Institute. The reference style that follows is based on the guidelines in the Institute's *American National Standard for Bibliographic References.* References should usually contain the author's name, the full title, and the publication information.

PREPARING REFERENCES

When typing the actual references section, single-space within a citation and double-space between citations. Type the reference number flush with the left margin, followed by a period. Begin the first line of the reference two spaces from the period, and indent second and subsequent lines to align with the first line. For references identified by superscript, type the reference number slightly above the reference line, flushed left with the margin and not followed by a period. Begin the first line of the reference two spaces from the number and indent second and following lines to align with the first line.

EXAMPLE   [1]Thomas, Willard. Interrelated video factors and a basic model for system selection. Technical Communication. 27(1): 16–18; 1980.

List the author's name last name first *(Doe, John Q.)*. Give the name exactly as it is given in the original work. For a work with more than one author, list all names separated from one other by a **semicolon** *(Doe, John, Q.; Roe, Jane R.)*. Separate the author's name from the work's title with a period. List any editor's name, last name first, followed by the abbreviation *ed. (Coe, Harry M., ed.)*. For works by a corporate author— that is, by an organization rather than by one author—list the organization's name in the author position, capitalizing the first and all major words in the name *(U.S. Comptroller General, American Chemical Society)*. If you cannot determine even a corporate author, begin the citation with the title. Standard reference works that are mainly known by title rather than by editor (encyclopedias, **dictionaries,** and some handbooks) should usually be cited by title.

Capitalize only the first letter of the first word and of any **proper nouns.** Conclude the title with a period. For titles of journals always spell out the complete title.

Citing Books. Include the pertinent publication facts in this order: (1) author (or authors), editor, or corporate author; (2) complete title; (3) title of the series, if it is part of a series; (4) edition, if it is not the first edition; (5) city of publication; (6) publisher; (7) year of publication; and (8) volume number for multivolume works.

EXAMPLE   1. Marsh, W. D. Economics of electric utility power generation. New York: Oxford University Press; 1980.

Citing Periodical Articles. Include the pertinent publication facts in this order: (1) author (or authors); (2) complete title of the article; (3) name of the periodical; (4) volume number; (5) issue number (in **parentheses**); (6) page numbers; and (7) date of publication.

EXAMPLE   2. Thomas, Willard. Interrelated video factors and a basic model for system selection. Technical Communications, 27(1): 16–18; 1980.

Citing Technical Reports. Include publication facts in this order: (1) author (or authors); (2) complete title; (3) year of publication; and (4) information about how to obtain a copy.

EXAMPLE   3. U.S. Comptroller General. Waste disposal practices— a threat to health and the nation's water supply, report to Congress. 1978; available from U.S. General Accounting Office.

EXAMPLES OF REFERENCE CITATIONS

Following are examples of different kinds of reference citations. Notice and follow the punctuation in each citation.

BOOK, ONE AUTHOR

1. Hanley, Wayne. Natural history in America. New York: Quadrangle Press / New York Times Books; 1977.

BOOK, TWO OR MORE AUTHORS

2. Jelley, Herbert M.; Herrmann, Robert O. The American consumer: issues and decisions. New York: McGraw-Hill; 1973.

3. Young, Hugo; Silcock, Bryon; Dunn, Peter. Journey to tranquillity. Garden City, NY: Doubleday; 1970.

BOOK EDITION, IF NOT THE FIRST

4. Silverman, S.R.; Davis, Hollowell. Hearing and deafness. wnd ed. New York: Holt, Rinehart and Winston; 1978.

MULTIVOLUME BOOK

5. Bartholomew, John, ed. Times atlas of the world. Mid-century edition. London: Times Publishing Co., Ltd.; 1955-1959. 5 V.

BOOK IN A SERIES

6. Cooper, E. L., ed. Invertebrate immunology. Contemporary topics in immunology. New York: Plenum Publishing Corp.; 1974. V. 4.

EDITOR

7. Fowles, Jib, ed. Handbook of futures research. Westport, Conn.: Greenwood Press; 1978.

JOURNAL ARTICLE

8. Thomas, Willard. Interrelated video factors and a basic model for system selection. Technical Communication. 27(1): 16-18; 1980.

MAGAZINE ARTICLE

9. It's natural! it's organic! or is it? Consumer Reports. 45(7): 410-415; 1980 July.

NEWSPAPER ARTICLE

10. Allan, John H. Fixed-income securities ease. The New York Times. 1979 June 15; Late City ed. Sec. D:7.

SYMPOSIUM OR CONFERENCE PAPER IN A PROCEEDING

11. Ballard, T.M. Role of soil reconnaissance surveys in predicting use-impact susceptibility of wildlands. Ittner, R.; Potter, D. R.; Agee, J. K.; Auschell, A., eds. Conference proceedings on the recreational impact on wildlands; 1978 October 27-29; Seattle, Washington. Seattle: U.S. Forest Service, Pacific Northwest Region, 1979: 45-49.

THESIS OR DISSERTATION

12. Johnson, M. G. Infiltration capacities and surface erodibility associated with forest harvesting activities in the Oregon Cascades. Corvallis, Oregon: Oregon State University; 1978. 172 p. Master's Thesis.

MOTION PICTURE

13. Computing the weather Motion Picture . U.S. Naval Weather Service. National Audiovisual Center, 1975. 28 min.

FILMSTRIP WITH SOUND RECORDING

14. Prison Filmstrip . U.S. Bureau of Prisons. Urbes Affiliates, Inc., 1975. Accompanied by two audio cassettes.

Many professional societies and other organizations publish style manuals that provide reference **formats** for publications in their fields. Following is a selected listing of guides.

A SELECTED LIST OF STYLE MANUALS AND GUIDES

American Chemical Society. Handbook for authors. Washington, DC: American Chemical Society Publications; 1978.

American Institute of Physics. Style manual. New York: American Institute of Physics; 1978.

American Mathematical Society. Manual for authors of mathematical papers. 6th ed. Providence, RI: American Mathematical Society; 1978.

American Medical Association, Scientific Publications Division. Style book: editorial manual. 6th ed. Acton, Mass.: Publishing Sciences Group, Inc.; 1976.

American national standard for bibliographic references, 239.29-1977. New York: American National Standards Institute; 1977.

American Psychological Association. Publication manual. 2nd ed. Washington, DC: American Psychological Association; 1979.

American Society of Mechanical Engineers. An ASME paper. New York: American Society of Mechanical Engineers; 1976.

Cochran, W.; Hill, M.; Feuner, P., eds. Geowriting: a guide to writing, editing, and printing in earth science. 2nd ed. Falls Church, VA: American Geological Institute; 1974.

Council of Biology Editors style manual. 4th ed. Arlington, VA: American Institute of Biological Sciences; 1978.

Turabian, K. L. A manual for writers of term papers, theses, and dissertations. 4th ed. Chicago: Univ. of Chicago Press; 1974.

University of Chicago Press. A manual of style. 12th ed. rev. Chicago: Univ. of Chicago Press; 1969.

U.S. Geological Survey. Suggestions to authors of the reports of the United States Geological Survey. 6th ed. Washington, DC: Government Printing Office; 1978.

reflexive pronouns

A reflexive pronoun, which always ends with the **suffix** -self or -selves, indicates that the **subject** of the sentence acts upon itself.

EXAMPLE   The lathe operator cut *himself.*

The reflexive form of the **pronoun** may also be used as an intensive to give **emphasis** to its antecedent. The reflexive and intensive pronouns are *myself, yourself, himself, herself, itself, oneself, ourselves, yourselves, themselves.*

EXAMPLE   I collected the data *myself.*

Never use a reflexive pronoun as a subject or as a substitute for a subject.

CHANGE   Joe and *myself* worked all day on it.
TO   Joe and *I* worked all day on it.

A common mistake is substituting a reflexive pronoun for *I, me* or *us.*

CHANGE   John and myself wrote the report.
TO   John and *I* wrote the report.

CHANGE   He gave it to my assistant and *myself.*
TO   He gave it to my assistant and *me.*

CHANGE   Nobody can solve the problem but *ourselves.*
TO   Nobody can solve the problem but *us.*

refusal letters

When you receive a **complaint letter** or an **inquiry letter** to which you must give a negative reply, you may need to write a refusal letter. For example, a professional organization may have asked you to speak at their regional meeting, but they require your firm to pay your expenses. Because your firm has declined the request, you must write a refusal letter turning down the invitation. The refusal letter is difficult to write because it contains bad news; however, you can tactfully and courteously convey the bad news. In this kind of **correspondence,** that should be your goal.

In your letter you should lead up to the refusal. To state the bad news in your **opening** would certainly affect your **reader** negatively. The ideal refusal letter says *no* in such a way that you not only avoid antagonizing your reader but keep his goodwill. To achieve such an **objective,** you must convince your reader of the justness of your refusal *before refusing.* The following pattern is an effective way to deal with this problem:

1. A buffer beginning
2. A review of the facts
3. The bad news, based on the facts
4. A positive and pleasant closing

The primary purpose of a buffer beginning is to establish a pleasant and positive **tone.** You want to convince your reader that you are a reasonable person. One way is to indicate some form of agreement with, or approval of, the reader or the project. If your refusal is in response to a complaint letter, do not begin by recalling the reader's disappointment ("We regret your dissatisfaction . . ."). Keep your buffer **paragraph** positive and pleasant. You can express appreciation for your reader's time and effort, if appropriate. Stating such appreciation will soften the disappointment and pave the way for future good relations.

> EXAMPLE   Thank you for your cooperation, your many hours of extra work answering our questions, and your patience with us as we struggled to make a very difficult decision. We believe our long involvement with your company certainly indicates our confidence in your products.

Next you should analyze the circumstances of the situation sympathetically. Place yourself in your reader's position and try to see things from his or her point of view. Establish very clearly the reasons you cannot do what the reader wants—even though you have not yet said you cannot do it. A good explanation of the reasons should so thoroughly justify your refusal that the reader will accept your refusal as a logical conclusion.

> EXAMPLE   However, the Winton Check Sorter has all the features that your Abbott Check Sorter offers and, in fact, offers two additional features that your sorter does not. The more important of the two is a back-up feature that retains totals in memory, even if the power fails, so that processing doesn't have to start again from scratch following a power failure. The second additional feature that the Winton Sorter offers is stacked pockets, which takes less space than the linear pockets on your sorter.

State your refusal quickly, clearly, and as positively as possible.

> EXAMPLE   After much deliberation, we have decided to purchase Winton Check Sorters because of the extra features they offer.

Close your letter with a friendly remark, whether to assure the reader of your high opinion of his product or merely to offer him good success.

> EXAMPLE   We have enjoyed our discussions with your company and feel we have learned a great deal about check processing. Although we did not select your sorter, we were very favorably impressed with your systems, your people, and your company. Perhaps we will work together in the future on another project.

The following letter refuses an invitation to speak at a professional organization's regional meeting.

> I am honored to have been invited to address your regional meeting in St. Louis on May 17. To be considered one who might make a meaningful contribution to such a gathering of experts is indeed flattering.
>
> On checking with those who control the purse strings here at Watashaw Engineering Consultants, however, I learned that it would be contrary to company policy for Watashaw to pay my expenses for such a trip. Therefore, as much as I would enjoy attending your meeting, I must decline.
>
> I have been very favorably impressed over the years with your organization's contribution to the engineering profession, and I am proud to have received your invitation.

relative adjectives

Like **relative pronouns,** the relative adjectives (*whose, which,* and *what*) are **pronoun** forms that link **dependent clauses** to main **clauses.** The difference is that relative pronouns perform this linking function while taking the place of **nouns,** and relative adjectives perform this linking function while *modifying* (describing) nouns. Examples of the uses of relative adjectives follow.

EXAMPLES   The president commended the managers *whose* divisions were the most profitable. (*Whose* modifies "divisions.")
He also told us *which* divisions needed to show the greatest improvement. (*Which* modifies "divisions.")
Finally, he told us *what* steps we should take to improve our profits. (*What* modifies "steps.")

relative pronouns

Relative **pronouns** (the most common are *who, whom, which, what,* and *that*) perform a dual function: They substitute for a **noun** while linking the **dependent clause** in which they appear to a main **clause.** (Relative pronouns differ in use from **relative adjectives,** which are pronouns in form and have a similar linking function, but which *modify* [describe] nouns.) In each of the following examples, two words are italicized: the relative pronoun and the word it is substituting for.

EXAMPLES The specifications were sent to the *manager, who* immediately approved them.

The project was assigned to *Ed Cone, whom* everyone respects for his skill as a designer.

The customer's main *objection, which* can probably be overcome, is to the high cost of installation.

The *memo* from Martha Goldstein *that* you received yesterday sets forth the most important points.

Sometimes a relative pronoun does not stand for another specific noun, but does act as a noun in serving as the subject of a clause. In the following example, *what* acts as a subject in the dependent clause while also linking that clause to the main clause.

EXAMPLE We invited the committee to come to see *what* was being done about the poor lighting.

The relative pronoun *who (whom)* normally refers to a person, *that* and *which* normally refer to a thing and *that* can refer to either. The pronoun *which* is normally used with restrictive clauses and the pronoun *that* with nonrestrictive clauses. See also **restrictive and nonrestrictive elements.**

EXAMPLES The annual report, *which was distributed yesterday,* shows that sales increased 20% last year.

Companies *that adopt the plan* nearly always show profit increases.

Relative pronouns can often be omitted.

EXAMPLE I received the handcrafted ornament (that) I ordered.

Do not omit the word *that,* however, when it is used as a conjunction and its omission could cause the reader to misread a sentence.

CHANGE This time delay ensures the operator has time to load the card hopper.

TO This time delay ensures *that* the operator has time to load the card hopper.

See also **who/whom.**

repetition

EFFECTIVE USE

The deliberate use of repetition to build a sustained effect or emphasize a feeling or idea can be quite powerful.

> EXAMPLE   Similarly, atoms *come and go* in a molecule, but the molecule *remains;* molecules *come and go* in a cell, but the cell *remains;* cells *come and go* in a body, but the body *remains;* persons *come and go* in an organization, but the organization *remains*.
> —Kenneth Boulding, *Beyond Economics* (Ann Arbor: University of Michigan Press, 1968), p. 131.

Repeating key words from a previous **sentence** or **paragraph** can also be used effectively to achieve **transition.**

> EXAMPLE   For many years, *oil* has been a major industrial energy source. However, *oil* supplies are limited, and other sources of energy must be developed.

INEFFECTIVE USE

Purposeless repetition, however, makes a sentence awkward and hides its key ideas.

> CHANGE   He *said that* the customer *said that* the order was canceled.
> TO   He *said that* the customer canceled the order.

The harm caused to your writing by careless repetition is not limited to words and **phrases;** the needless repetition of ideas can be equally damaging. (See also **conciseness/wordiness.**)

> CHANGE   In this modern world of ours today, the well-informed, knowledgeable executive will be well ahead of the competition.
> TO   To succeed, the contemporary executive must be well informed.

reports

A report is an organized presentation of factual information that answers a request or supplies needed information. All reports fall into two broad categories: formal and informal.

Formal reports are for projects that require many months of work or involve large sums of money, or for projects that are done for government agencies. Formal reports can be quite elaborate, and usually include a cover, a title page, a table of contents, and often even appendixes and a **glossary.**

Informal reports, normally only a few pages long, are designed for circulation within an organization and include only the essential parts— an **introduction,** a body, a **conclusion,** and any recommendations.

There are different types of informal reports, many of which are described in separate entries in this handbook. Common reports include **feasibility reports, investigative reports, progress reports, trip reports, laboratory reports, test reports, trouble reports,** and **annual reports.** Many of these reports can be formal or informal, depending on their length and complexity.

research

Research is the process of investigation, the discovery of facts. As part of the writing process, research must be preceded by **preparation** (that is, by determining your **reader, objective,** and **scope**). During the preparation stage you establish the extent of the coverage needed in your work, then during the research stage you can logically determine what information you will need and how much detail. In short, you must do certain preparation so that your research is adequately focused and defined.

Most research is based on several different sources of information. The most common ones include written materials, interviews, and personal experience or observations.

Scan written material quickly before reading it in order to get an idea of the kind of information it includes. Then, if it seems useful, read it carefully, taking notes of any information that falls within the scope of your research. Be sure to note down your sources so that you can find them again if necessary. As you read, you may think of questions about the topic; record these questions and label them as such (for example, with a question mark). Some of your questions may eventually be answered in other written materials; those that remain unanswered can guide you in further research.

If, after studying all written sources, you still need more information, you might consider talking to an expert. Here is someone who may be able to answer your remaining questions and who can probably suggest additional sources. (See **interviewing.**)

You may also draw upon any personal experience that is applicable;

that is, you may make certain assumptions that are based on your own general knowledge of the field. You must, of course, verify that these assumptions are valid; but verifying your assumptions is much easier than doing original research. If you interview an expert, for example, you can simply ask whether your assumption is correct rather than asking for an open-ended explanation. Last, you can use your own observations. If you are writing about equipment or tools, for example, the most valuable source of information would be actually operating the equipment or using the tools.

See also **library research, note-taking,** and **questionnaires.**

respective/respectively

Respective is an **adjective** that means "pertaining to two or more things regarded individually."

> EXAMPLE The committee members prepared their *respective* reports.

Respectively is the **adverb** form of *respective,* meaning "singly, in the order designated."

> EXAMPLE The first, second, and third prizes in the sales contest were awarded to Maria Juarez, Gloria Hinds, and Margot Luce, *respectively.*

Respective and *respectively* are often unnecessary because the meaning of individuality is already clear.

> CHANGE The committee members prepared their *respective* reports.
> TO Each committee member prepared a report.

restrictive and nonrestrictive elements

Modifying **phrases** and **clauses** may be either restrictive or nonrestrictive. A nonrestrictive phrase or clause provides additional information about what it modifies, but it does not limit, or restrict, the meaning of what it modifies. Therefore, the nonrestrictive phrase or clause could actually be removed from its **sentence** without changing the essential meaning of the sentence. It is, in effect, a parenthetical element, and so it is set off by **commas** to show its loose relationship with the rest of the sentence.

EXAMPLES   This instrument, *called a backscatter gauge,* fires beta particles at an object and counts the particles that bounce back. (nonrestrictive phrase)

The annual report, *which was distributed yesterday,* shows that sales increased 20% last year. (nonrestrictive clause)

A restrictive phrase or clause limits, or restricts, the meaning of what it modifies. If it were removed from its sentence, the essential meaning of the sentence would be changed. Because a restrictive phrase or clause is so essential to the meaning of the sentence, it is never set off by commas.

EXAMPLES   All employees *wishing to donate blood* may take Thursday afternoon off. (restrictive phrase)

Companies *that adopt the plan* nearly always show profit increases. (restrictive clause)

It is extremely important for writers to distinguish carefully between nonrestrictive and restrictive elements. The same sentence can take on two entirely different meanings depending on whether a modifying element is set off by commas (because nonrestrictive) or not set off (because restrictive). The results of a slip by the writer can be not only misleading but downright embarrassing.

CHANGE   I think you will be impressed by our systems engineers who are thoroughly experienced in projects like yours.

TO   I think you will be impressed by our systems engineers, who are thoroughly experienced in projects like yours.

The problem with the first sentence is that it suggests that "you may not be so impressed by our other, less experienced, systems engineers."

résumés

The résumé is an essential element in any **job search.** Sent out with an **application letter,** a résumé is, in effect, a catalog of what you have to offer prospective employers. As such, it is the basis for their decision to invite you for a **job interview.** It tells them who you are, what you know, what you can do, what you have done, and what your job objectives are. It will serve as the focus for the interview because prospective employers can base specific questions on the information contained in it. It will also help them evaluate your interview after your departure by serving as an organized reminder of what you covered orally during the interview.

The résumé, then, is an important document that deserves careful planning and execution.

Three steps are involved in preparing and using a résumé effectively. The first is to analyze yourself and your background; the second is to organize and prepare the résumé; the third is to identify those to whom you will submit the résumé.

The starting point in preparing a résumé is a thorough analysis of your experience; your education and training; and your personal traits. List the major points for each category before actually beginning to organize and write your résumé. The following list indicates the kinds of things you should consider in each category.

Experience. List all your employment—full-time, part-time, vacation jobs, and freelance work—and analyze each job on the basis of the following list of questions.

1. What was the job title?
2. What did you do (in reasonable detail)?
3. What experience did you gain that you can apply to another job?
4. Why were you hired for the job?
5. What special skills did you develop on that job?
6. Why did you leave it?
7. When did you start and when did you leave the job?
8. Would your former employer give you a reference?
9. What special traits were required of you on the job (initiative, leadership, ability to work with details, ability to work with people, imagination, ability to organize, etc.)?

Education. For the applicant with little work experience, such as a recent graduate, the education category is of primary importance. For the applicant with extensive work experience, education is still quite important even though it is now secondary to work experience. List the following information about your education.

1. Colleges attended and the appropriate dates
2. Degrees and the dates they were awarded
3. Major and minor subjects
4. Courses taken or skills acquired that might be especially pertinent to the job now being applied for
5. Cumulative grade averages and any academic honors
6. Extracurricular activities
7. Scholarships and awards
8. Special training courses (academic or industrial)

Personal Characteristics. Listing your personal characteristics will help you find your personal strong points. Be as objective as possible in listing your personal characteristics. Weigh your strengths and weaknesses honestly, and determine which factors are pertinent to the job you are seeking. Some of the points to be considered are in the following list.

1. Age and marital status
2. Height, weight, health, and physical limitations
3. Appearance and grooming
4. Communication skills and speech characteristics
5. Social attitudes (aggressive, cooperative, cheerful, tactful, moody, courteous, etc.)
6. Career objectives and plans for the future

ORGANIZING AND WRITING THE RÉSUMÉ

After listing all items in the preceding three categories, you must analyze them in terms of the job you are seeking, evaluating all the items, rejecting some of them, and finally selecting those that you will include in your résumé. Base your decisions on the following questions.

1. What exact job are you seeking?
2. Who are the prospective employers?
3. What information is most pertinent to that job and those employers?
4. What details should be included, and in what order?
5. How should you present your qualifications most effectively to get an interview?

Résumés should be brief—preferably no more than one page—and should contain the following information.

1. Your name, address, and phone number
2. Your immediate and long-range job objectives
3. Your professional training
4. Your professional experience, including the firms where you've worked and your responsibilities there
5. Your specific skills and any professional registration
6. Pertinent personal information

FORMAT OF THE RÉSUMÉ

A number of different formats can be effective. Most important is that your résumé be attractive, well organized, easy to read, and free of error. One common format uses the following heads: Personal Data, Employ-

R É S U M É

CAROL ANN WALKER
273 East Sixth Street
Bloomington, IN 47401

Job Objective: Associate copywriter, with the ultimate goal
of working into advertising or public
relations management.

EDUCATION
19--   Will graduate in June from Indiana University with
a Bachelor of Science degree.
Major: Journalism Minor: Mass Communications
Grade avg.: 3.63
Honors: Dean's List, Senior Honor Society, 19--
Activities: Business Manager (19--) and Associate
Editor (19--) of college newspaper, President of
Alpha Phi sorority.

WORK EXPERIENCE
Summer 19--   Elwood, Dowd, and Harvey in Indianapolis,
Indiana. Intern editorial assistant.
Proofread copy and prepared preliminary
layouts for ads. Helped copywriters with
research. Wrote copy for two industrial
ads that were used in trade publications.

Summer 19--   Dayton Newspapers, Inc., in Dayton, Ohio.
Intern editorial assistant. Provided
research assistance to writers for the
editorial page. Prepared background
material for analysis of major news events.

Summer 19--   Oak Park Swim Club in Dayton, Ohio. Life
guard and instructor for pre-school
swimming classes.

Previously earned spending money by babysitting and house-
cleaning; had a paper route for two years (19-----).

REFERENCES

Dr. Walter Duncan   Mr. Peter Crowley
Professor, School of Journalism   4927 Far Oaks Lane
Indiana University   Dayton, OH 45420
Bloomington, IN 45401   (513) 299-3952

Mr. E.P. Dowd
President, Elwood, Dowd, and Harvey
Suite 412
Midstates Building
Indianapolis, IN 46206
(317) 533-4422

Student Résumé

```
                    R É S U M É
                CAROL ANN WALKER
                1436 W. Schantz Avenue
                Dayton OH  45429
                (513) 224-0773

Job Objective: Manager of Consumer or Public Relations for a
               consumer products company.
WORK EXPERIENCE

   January 19-- - Present

      Kerfheimer Associates in Dayton, Ohio

         Public Relations Group Leader (October 19-- to present).
         Creative director and project leader for staff of five
         public relations specialists.  Originated public rela-
         tions projects for fourteen client companies.  Most
         projects designed to respond to particular problems
         involving such things as consumer resistance to
         automated checkout systems in a small supermarket chain,
         press relations for an industrial firm involved in a
         labor dispute, and speech writing for the executives of
         a utility company to educate the public about the need
         for a price increase.  Named Kerfheimer Creator of the
         Year in 19-- for the supermarket project.

         Public Relations Specialist (September 19-- to October
         19--).  Responsible for creating a special public
         relations program to renew customer interest in a former
         shopping center that had been converted (at great
         expense) to an enclosed mall of boutiques and specialty
         shops.  Created press releases and promotional materials
         (printed and novelty pieces), developed slide programs
         for consumer groups, and wrote television and radio
         spots.  Published feature articles in the local papers,
         including a two-page article in the Sunday magazine.

         Copywriter (January 19-- to September 19--).  Created
         copy and preliminary layouts for advertising campaigns
         for clients that included home builders, a milk
         producers' association, fashion retailers, and new car
         dealers.  Named Kerfheimer Creator of the Month for
         May 19-- and September 19--.

   June 19-- - December 19--

      Elwood, Dowd, and Harvey in Indianapolis, Indiana

         Associate Copywriter (September 19-- to December 19--).
         Apprentice position working with copywriters and account
         executives in preparing advertising campaigns for
```

Résumé Organized by Job

clients. Assisted account executives and others in
preparing proposals to clients. Responsibilities
involved work in both editorial and art departments.

Editorial Assistant (June 19-- to September 19--).
Assisted copywriters, account executives, and administrative executives. Proofread copy, did marketing
research, worked in clipping service, wrote copy, key-
lined art, and created art work for ad campaigns.
Worked in some role in every creative department in the
firm. Intern editorial assistant the summer of 19--.

Summer 19--

   Dayton Newspapers, Inc., in Dayton, Ohio

   Intern Editorial Assistant Provided research assistance
   to writers for the editorial page. Prepared background
   papers for analysis of major news events.

Part-Time Work Experience

   Lifeguard and swimming instructor the summer of 19--.
   Earned spending money in high school by babysitting and
   house-cleaning. Had a paper route for two years (19-- -
   --). Business manager (19--) and associate editor
   (19--) of college newspaper.

EDUCATION

   19--   Wright State University
          Degree: M.B.A. in Management

   19--   Indiana University
          Degree: B.S. in Journalism magna cum laude

References and a portfolio of projects will be furnished on
request.

R É S U M É

CAROL ANN WALKER
1436 W. Schantz Avenue
Dayton, Ohio 45429
(513) 224-0773

Job Objective: Manager of Consumer or Public Relations for a
consumer products company.

SUMMARY OF EXPERIENCE

Public Relations
Prepared public and consumer relations programs for fourteen companies. Served as both creative director and
administrative head of group of five specialists. Wrote
speeches for executives, designed campaigns in all media
to overcome poor public image and create interest in a new
development. Work for eight of the fourteen companies was
done originally as a one-time, problem-oriented project.
Four of these companies were converted to full-time clients
as a result of the success of the original projects.
Clients included a supermarket chain, industrial companies,
home and apartment builders, and a milk producers'
association. Named Kerfheimer Creator of the Month for one
project, Creator of the Year for another campaign.

Publicity
Created campaigns to "spread the word" about new projects
in the community; largest campaign was to increase
interest in a refurbished small shopping center located in
a deteriorating neighborhood. The shopping center was
converted to a mall, featuring specialty shops and
boutiques--fashion, food, arts and crafts, and galleries--
with no major department stores. The mall, based on a
"world bazaar" theme, has been an outstanding commercial
success.

Press Relations
Placed news releases in local and regional newspapers,
feature articles in local magazine. Also sent news and
feature releases weekly to local newspapers and radio and
television stations. Secured effective participants and
had them placed on local business news and television
programs, including "Dayton Today," a program featuring
local people and places of interest. My time working for
a newspaper, although limited, exposed me to the media
point of view.

Client Relations
Work as project leader and as creative director has
exposed me to the selling aspects of business--selling

Résumé Organized by Function

campaigns, selling ideas, and selling services. Sold
long-term arrangements to four clients within a year
(billings of $200,000 per year) as a direct result of
effective campaigns.

Management
Developed budgets for press and advertising campaigns,
administered projects with a staff of five people. Coordinated activities of publications staff with art,
photographic, research, and media departments. Projects
administered showed a gross profit of 12 percent (average)
over an eighteen month period in which billings were
$920,000.

Meetings
Organized and developed programs for consumer meetings.
Created audio-visual materials to present controlled,
effective presentations. Worked with hotel convention
center and student union management in planning the
logistics for these meetings: facilities, meals, room
arrangements for attendees, hospitality programs, and
audio-visual equipment. Press rooms were established
when appropriate. Coordinated registration activities and
created publicity for these meetings.

Community Activities
Served as publicity chairman for Girl Scout Camping
campaign and for Country Memorial Hospital fund raising
campaign. Created campaigns that exceeded objectives,
directing a staff of hard-working volunteers for each
campaign.

EDUCATION

    19--   Wright State University
               Degree: M.B.A. in Management

    19--   Indiana University
               Degree: B.S. in Journalism magna cum laude

References and a portfolio of projects will be furnished on
request.

ment Objective, Education, Employment Experience, and References. Underline or capitalize the heads to make them stand out on the page. Whether you list education or experience first depends on which is stronger in your background. If you are a recent graduate, list education first; if you have one or more years of related job experience, list experience first. In both cases, list the most recent education or job experience first, the next most recent experience second, and so on.

The Heading. Center the title, "résumé," at the top of the page. Follow it with your name, address, and telephone number, also usually centered.

Personal Data. You may want to include your date of birth, your marital status and your hobbies. You need not feel compelled to include personal data, of course, if you prefer not to do so.

Employment Objective. State not only your immediate employment objective but the direction you hope your career will take.

Education. List the college or colleges you attended, the dates you attended each one, any degree or degrees received, your major field of study, and any academic honors you earned. Some people give the name of their high school, its location, and the dates they attended.

Employment Experience. List all your full-time jobs, starting with the most recent and working backward. If you have had little full-time work experience, list part-time and temporary jobs too. Give the details of your employment, including the job title, dates of employment, and the name and address of the employer. Provide a concise description of your duties only for those jobs whose duties are similar to those of the job you are seeking; otherwise, give only a job title and a brief description. Specify any promotions or pay increases you received. Do not, however, list your present salary. If you have been with one company a number of years, highlight your accomplishments during those years. List military service as a job: give the dates you served, your duty specialty, and your rank at discharge. Discuss the duties only if they relate to the job you are applying for.

Some résumés arrange employment experience by function rather than by time. Instead of starting with the most recent job, you would begin with the job closest in function to the one you seek.

References. You can include references as part of the résumé or provide a statement on the résumé that references will be furnished upon request. Either way, do not give anyone as a reference without first obtaining his or her permission to do so.

Style. As you write your résumé, use action **verbs** and state ideas concisely. There is no reason to avoid the *I* altogether, but it is best not to overuse it.

CHANGE   I was promoted to shop foreman in June 19--.

TO   Promoted to shop foreman in June 19--.

Be truthful in your résumé. If you give false data and are found out, the consequences could be serious. At the very least, you will have seriously damaged your credibility with your employer.

Your résumé must be flawless. If you are not a skilled typist, you may want to have it professionally typed. Printing usually produces more professional-looking copies than does photocopying.

The three sample résumés on pages 519–523 are all for the same person. The first one is a student's first résumé, and the remaining two show work experience later in her career (one is organized by job, the other by function).

revision

The more natural a work of writing seems to the **reader,** the more effort the writer has probably put into revision. Anyone who has ever had to say of his own writing, "Now I wonder what I meant by that," can testify to the importance of revision. The time you invest in revision will make the difference between clear writing and unclear writing.

If you have followed the appropriate steps of the writing process, you have a very rough draft that could hardly be considered finished writing. Revision is the obvious next step.

Allow a couple of days, if possible, to go by without looking at the draft before beginning to revise it. Without a cooling period, you are too close to the draft to be objective in evaluating it. Only when enough time has elapsed so that you can say, "How did I ever write that?" or "I wonder why I said that so awkwardly?" are you ready to start revising your rough draft.

A different frame of mind is required for revising than for **writing the draft.** Read and evaluate the draft, which should have been written quickly, with cool deliberation and objectivity—from the point of view of a reader or critic. Be anxious to find and correct faults, and be honest. In revising, consider your reader first.

Do not try to do all your revision at once. Read through your rough draft several times, each time searching for and correcting a different set of problems.

Check your draft for completeness. Your writing should give readers exactly what they need, but it should not burden them with unnecessary information or get sidetracked into insignificant or only loosely related subjects. Check your draft against your **outline** to make certain that you did, in fact, follow your plan. If not, now is the time to insert any missing points into your draft. This may mean inserting a line here and there; it may mean writing a **paragraph** or passage on a separate sheet and attaching it to the appropriate page; or it may mean major additions to your rough draft, depending upon how carefully you followed your outline.

Take time to examine the facts in your draft for accuracy. Keep in mind that no matter how careful and painstaking you may have been in conducting your **research,** compiling your notes, and creating your outline, you could easily have made errors when transferring your thoughts from the outline to the rough draft. Checking the facts in your draft for accuracy is quickly and easily accomplished (especially if you took complete notes) yet can be critically important if it catches but a single error. Be certain that contradictory facts have not crept into your draft.

Be consistent with names. Make sure, for example, that you have not called the same item a ''routine'' on one page and a ''program'' on another.

Your **introduction,** if you use one, should give your **objective** and provide a frame in which your reader can place the detailed information that follows in the body of your draft. Check your introduction to see that it does, in fact, provide the reader with such assistance. If you do not use an introduction, check your **opening** to see that it is useful and interesting to the reader.

To determine whether you have good **transition,** look for **unity** and **coherence** as key factors. If a paragraph has unity, all its sentences and ideas are closely tied together and contribute directly to the main idea expressed in the **topic sentence** of the paragraph. Writing that is coherent flows smoothly from one point to another, from one sentence to another, and from one paragraph to another. Where transition is missing, provide it; where transition is weak, strengthen it. Also look for unity in the entire report, and provide any transition that is needed between paragraphs or sections.

Check your draft for **pace.** If you find that you are jamming your ideas too closely together, space them out and slow down the pace.

Check your use of **jargon.** If you have any doubt that your entire reading audience will understand any jargon you may have used, eliminate it from your draft.

Check your draft for **conciseness.** Tighten your writing so that it says exactly what you mean by pruning unnecessary words, **phrases,** sentences, and even paragraphs.

Replace **clichés** with fresh **figures of speech** or with direct statements. Watch for words that could carry an undesired **connotation.** Also delete or replace vague or pretentious words, coined words, and unnecessary **intensifiers.** (See also **word choice.**)

Check your draft for **awkwardness,** determine the reason for any you may find, and correct the problem.

Check your draft for possible grammatical errors. Any such errors should be caught in your check for awkwardness, but a final review may save you embarrassment.

Finally, check your final draft for typographical errors by **proofreading.**

rhetorical questions

A rhetorical question is a question that a writer asks the **reader**—a question to which a specific answer is not necessarily needed or expected. The question is often intended to make the reader think about the subject from a different perspective; the writer then answers the question in the article or essay.

EXAMPLES   Is methanol the answer to the energy shortage?
Does advertising lower consumer prices?

The answer to a rhetorical question may not be a yes or no; in the above examples, it might involve a detailed explanation of the pros and cons of the value of methanol as a source of energy or of the effect of advertising on consumer prices.

The rhetorical question can be an effective **opening,** and it is often used for a **title.** By its nature, however, it is somewhat informal, and it should therefore be used judiciously in technical writing. For example, a rhetorical question would not be an appropriate opening for a **report** or **memorandum** addressed to a busy superior in your company. When you do use a rhetorical question, be sure that it is not trivial, obvious, or forced. More than any other writing device, the rhetorial question requires that you know your reader.

run-on sentences

A run-on **sentence,** sometimes called a fused sentence, is two or more sentences without **punctuation** to separate them. The term is also sometimes applied to a pair of **independent clauses** separated by only a **comma,** although this variation is usually called **a comma fault** or comma splice. Run-on sentences can be corrected by (1) making two sentences, (2) joining the two clauses with a **semicolon** (if they are closely related), (3) joining the two clauses with a comma and a **coordinating conjunction,** or (4) subordinating one clause to the other.

CHANGE   The new manager instituted several new procedures some were impractical. (run-on sentence)

TO   The new manager instituted several new procedures. Some were impractical. (period)

OR   The new manager instituted several new procedures; some were impractical. (semicolon)

OR   The new manager instituted several new procedures, but some were impractical. (comma plus coordinating conjunction)

OR   The new manager instituted several new procedures, some of which were impractical. (one clause subordinated to the other)

See also **sentence faults.**

S

sales proposals

Whether they sell products or services, business firms often prepare sales proposals. Sales proposals describe what the seller proposes to furnish to the customer and demonstrate that the customer's prospective purchase would be a wise investment.

Some sales proposals are small and fairly simple, covering a page or two; such a sales proposal might be used to bid on the construction of a single home. Other proposals are much larger and more elaborate; such a proposal might be used to sell a complete network of computer systems.

SIMPLE SALES PROPOSALS

This type of proposal should (1) state the facts of the proposal, such as total cost, estimated starting date, and completion date; (2) provide a breakdown of specific costs; and (3) express the reasons for your confidence that your company can do the job. The following example shows a simple sales proposal made in response to an invitation for bids sent to a nursery and several of its competitors.

PROPOSAL
TO LANDSCAPE THE NEW CORPORATE HEADQUARTERS
OF THE
WATFORD VALVE CORPORATION

Submitted to: Ms. Tricia Olivera, Vice-President
Submitted by: Jerwalted Nursery, Inc.
Date Submitted: February 1, 19--

Introduction states purpose and scope of proposal, indicates when project can be started and completed.   Jerwalted Nursery, Inc., proposes to landscape the new corporate headquarters of the Watford Valve Corporation, on 1600 Swason Avenue, at a total cost of $8,000. The lot to be landscaped is approximately 600 feet wide and 700 feet deep. Landscaping will begin no later than April 30, 19-- and will be completed by May 31.

The following trees and plants will be planted, in the quantities and sizes given and at the prices specified.

Body lists products to be provided, cost per item.

4 maple trees (not less than 7 ft.) @ $40 each—$160
41 birch trees (not less than 7 ft.) @ $65 each—$2,665
2 spruce trees (not less than 7 ft.) @ $105 each—$210
20 juniper plants (not less than 18 in.) @ $15 each—$300
60 hedges (not less than 18 in.) @ $7 each—$420
200 potted plants (various kinds) @ $2 each—$400

$$\begin{aligned} \text{Total Cost of Plants} &= \$4,155 \\ \text{Labor} &= \$3,845 \\ \hline \text{Total Cost} &= \$8,000 \end{aligned}$$

Conclusion specifies time limit of proposal, expresses confidence, and looks forward to working with prospective customer.

All trees and plants will be guaranteed against defect or disease for a period of 90 days, the warranty period to begin June 1, 19--.

The prices quoted in this proposal will be valid until June 30, 19--.

Thank you for the opportunity to submit this proposal. Jerwalted Nursery, Inc., has been in the landscaping and nursery business in the St. Louis area for thirty years, and our landscaping has won several awards and commendations, including a citation from the National Association of Architects. We are eager to put our skills and knowledge to work for you, and we are confident that you will be pleased with our work. If we can provide any additional information or assistance, please call.

ELABORATE SALES PROPOSALS

Before a large sales proposal can be prepared, you must conduct a study of the prospective customer's requirements (the relative importance of this study varies with the complexity of your product). Until you have made such a study, you cannot know either *whether* you can help the prospective customer or *how* you can help him. For a large project, you will need to review the customer's **specifications.** Once you have determined that you can do the job and the ways in which you can, however, you are ready to prepare the proposal.

The sales proposal may contain (1) a cover letter, (2) a summary of the proposal's conclusions (the ways the particular product being proposed solves the customer's particular problem), (3) a sales brochure or other description of your product that provides detailed information, (4) **illustrations** (charts, **graphs,** etc.) that demonstrate the benefits the customer will realize by using your product (cost savings, increased production, greater efficiency, etc.), (5) an explanation of the reasons why your company is well qualified to meet the customer's needs, (6) possibly **résumés** of key personnel, and (7) a cost analysis that specifies in detail all the costs of the transaction you are proposing. If the transaction includes training and post-sale support or maintenance, you must provide an explanation of the training and support you are proposing.

Such a proposal usually includes a **table of contents** at the front of it.

Cover Letter. The cover letter accompanying the proposal should (1) express appreciation for the opportunity to make your proposal and for any assistance that has been given you in the study of the customer's requirements, (2) summarize in a single **paragraph** the proposal's recommendations, and (3) offer a soft sales pitch for your company and its products. Letter 19 shows one example.

Summary. The purpose of the summary is to provide the customer with an overview of your proposal and recommendations that he can use after your departure in deliberating and reaching his final decision. The summary should be brief enough that the customer will read it, yet complete enough that he will not overlook any of the significant points.

SUMMARY

The OBR Acro system is a combination of specially designed application software and an online order entry configuration of hardware. This new system has been designed to handle your present applications more effectively while allowing for expansion and upward compatibility as your needs change in the future. The Acro software package consists of three application programs—Order Entry, Inventory Control, and Accounts Receivable. All related data-entry and data-inquiry functions can be performed through the CRT terminal with the option of entering data into the system through the card reader. This online processing system enables you to produce picking tickets, packing slips, and invoices while providing a wealth of management and control information within its many reports. In addition, Accounts Payable, Payroll, and General Ledger Applied Systems can be processed in subsequent processing runs.

Our recommendation consists of the Acro system with two CRT terminals (one located in the accounting office and one at the order desk). Operator lead-through techniques incorporated in the CRT units simplify their operation and tend to reduce the task of training order entry clerks. Use of CRT units will significantly reduce the amount of information that must be keyed into the system to process an order. Item descriptions, prices, and other types of information are retrieved from computer memory

March 15, 19--

Mr. Robert X. Bostick
Bostick Hardware Supply Co.
4042 South Plain Street
Plainsboro KS 45023

Dear Mr. Bostick:

Express your appreciation for the opportunity to make the proposal. —

We appreciate your cooperation and assistance in obtaining the details concerning the various operations within Bostick Hardware Supply Company. Information gathered during our survey of your operations on September 26 has been analyzed by our team of specialists.

Summarize the proposal's recommendations. —

As a result of this study we are recommending the installation of the OBR Acro at Bostick Hardware Supply Company. The reasons for recommending OBR Acro are two. First, the installation of the Acro system will enable you to automate your order entry, inventory control, and accounts receivable systems. This automation will provide a more efficient system that permits more effective use of warehouse space, personnel, and assets. Second, specific facts that were obtained during our survey were input to our Return-on-Investment analysis program, and it was discovered that your current return on investment is 3.84%. We feel that this can be appreciably increased by using the OBR Acro system as a management tool. Our recommendation offers Bostick Hardware Supply Company the highest degree of efficiency and economy by employing the most modern data processing system available today.

Make a soft Sales pitch. —

We would like to thank you for the opportunity to make these recommendations. We are confident that you, your staff, and OBR will share a mutual pride and sense of achievement from the successful installation of the OBR Acro system.

Sincerely yours,

William R. Jackson

William R. Jackson
Sales Manager

Enclosures

Letter 19

and need not be entered. CRT units permit you to offer better service to your customers by permitting your personnel to make online inquiries while talking to customers in-house or on the telephone.

Order processing is an area where a savings in manpower can be achieved while simplifying the overall task. We are recommending that a one-digit suffix be added to your present aisle numbers to aid pickers filling orders. A digit 1 refers to aisles to the left of your main aisle, and a digit 2 refers to aisles to the right of the main aisle. Sorting ordered items into the same sequence as your bin locations and printing a pick list in this sequence while using the additional one-digit code to further indicate where an item is located can result in considerable savings in the cost of processing orders.

Open-item invoicing and a full range of accounts receivable reporting, including an age analysis of delinquent accounts, are all part of the Acro system. Information necessary for an efficient collection program and information upon which decisions concerning whether marginal customers should be dropped are both vital parts of this system.

Inventory management and control can be greatly improved through the use of Acro. Automatic exception reports listing items that are out of stock or below established minimums, suggested order quantities to aid purchasing decisions, and access to the value and status of any or all items of inventory are some of the major tools available to manage your inventory.

Flash reporting is another valuable tool of the Acro system. Management is provided with up-to-the-minute information on demand. Flash reports include sales dollar volume from incoming orders, number of orders received, orders shipped, orders billed, amount invoiced and cash received from cash sales and accounts receivable receipts.

The data we gathered during our study of your operation indicate that your present return on investment is 3.84%. To provide a reasonable goal to work toward, we have made the following assumptions: (1) a 5% increase in sales, (2) a 14% decrease in inventory, and (3) a 14% decrease in accounts receivable. We have plotted the growth pattern of Bostick Hardware and determined that the pattern of growth over the past five years is approximately 15% per year. The Acro system is designed for

534   **sales proposals**

upward compatibility within the OBR 9100 series and also larger
OBR computer systems. We at OBR would like to join with
Bostick Hardware in an even faster and more profitable growth
in the future.

Illustrations. Any visual aid that you use in your sales presentation
should be made a part of your proposal and left with the customer so that
its impact will have a continuing effect. Oversized charts and graphs
used in the sales presentation should be reduced, however, before being
included in the physical proposal.

Institutional Sales Pitch. You must convince the prospective customer
that your company can meet his requirements better than its competitors
can. Keep this part of your proposal brief—possibly even using a list of
single-**sentence** items—but effective.

WHY OBR?

Every prospective customer is vitally concerned with the support
that the manufacturer will provide for his installation. OBR is
one of the oldest and largest manufacturers of business systems.
In every OBR office, world wide, there are specialists in the
various computer applications. Knowledge gained from successful EDP installations of all types and sizes combined with continuous training in modern system concepts and procedures have
permitted OBR analysts to excel in business systems knowledge.
Consequently, OBR has the experience and the manpower to
provide the support required for a computer installation. Our
local OBR district office in Plainsboro is staffed with competent
and knowledgeable vocational teams who are prepared to assist
with the installation of your system.

Before your system is installed, OBR specialists are available to
(1) assist you in designing your system, (2) assist you with the
planning of your computer site, (3) assist you in planning,
coordinating, and implementing your installation, (4) provide
you with a complete library of hardware and software reference
manuals, (5) aid in establishing training schedules for your EDP
staff, (6) provide programmer and operator training, (7) assist
you in converting your existing programs and files, (8) assist you
in compiling and debugging your programs, and (9) assist you in
developing your operating procedures.

After your computer is installed, our local specialists are available to continue this high level of support. The professional field

engineers at our Plainsboro district office are available to maintain and assure maximum efficiency of your equipment through preventive and remedial maintenance.

Cost Analysis. This section of your proposal may have little or even no writing. Its purpose is to present the dollar figure that your proposal, if accepted, will cost the prospective customer. See the following example.

COST ANALYSIS
OBR ACRO SYSTEM
Equipment Cost
    OBR Sato 9100 Basic System
        705–10 Processor
        305–06 Disc Controller
        506–13 Disc Unit
        305–07 Common Trunk
        934–70 Line Printer
        654–03 CRT Unit
        243–40 Adapter
        138–60 Card Reader
    Acro Applied System
    Acro Accounts Payable Applied System
    Acro Payroll Applied System
    Acro General Ledger Applied System
                    $xx,xxx.xx
Tax   x,xxx.xx
Total Investment $xx,xxx.xx

Training and Support. State the amount and kind of training and support offered by your proposal, as in the following example.

CUSTOMER EDUCATION PROGRAM
OBR offers a comprehensive program of customer education. Your personnel will be trained by a professional staff of instructors in programming and operating the OBR Acro system.

Educational courses are available to each OBR user upon receipt of a purchase or rental contract. These courses are designed to cover the normal training necessary for the successful use of an OBR system. You may select qualified employees for programmer training. OBR will administer a programming aptitude test to these persons and, if approved, schedule them in the required training school.

A member of top management is invited to attend an Executive EDP Seminar. In addition, you may select three managers to attend the Middle Management EDP Seminar.

CUSTOMER SUPPORT

The following customer support will be provided to you at no charge: (1) one complete library of OBR Acro software and hardware reference manuals, (2) a ten-hour operations course, and thirty hours on-site operator training up to thirty days after certification of your systems, (3) sixteen hours of test and compile computer time, (4) software distribution, and (5) program error detection/correction of any OBR applied software requiring OBR field engineering assistance.

same

When used as a **pronoun,** *same* is awkward and outmoded.

CHANGE   Your order has been received, and we will respond to *same* next week.

TO   Your order has been received, and we will respond to *it* next week.

OR   We received your order, and we will comply with *it* next week.

schematic diagrams

The schematic diagram, which is used primarily in electronics and chemistry and in electrical and mechanical engineering, attempts to show the operation of its subject with lines and **symbols** rather than by a physical likeness of it. The schematic diagram is usually a highly abstract representation of its subject because it emphasizes the relationships among the parts at the expense of precise proportions. Because a schematic diagram is a symbolic representation of the subject, as opposed to a realistic representation of it, the schematic relies heavily on the symbols and **abbreviations** that are common to the subject. (See Figure 36.)

For guidelines on how to use illustrations and incorporate them into your text, see **illustrations.**

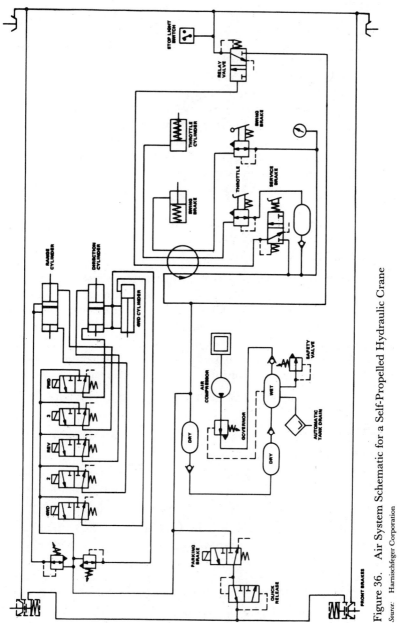

Figure 36.   Air System Schematic for a Self-Propelled Hydraulic Crane

Source: Harnischfeger Corporation

scope

If you know your **reader** and the **objective** of your writing project, you will know the type and amount of detail to include in your writing. This is *scope*, which may be defined as the depth and breadth to which you cover your subject. If you do not determine your scope of coverage in the planning stage of your writing project, you will not know how much or what kind of information to include; as a result, you are likely to invest unnecessary work during the **research** phase of your project.

Your scope should be designed to satisfy the needs of your objective and your reader. By keeping your objective and your reader's profile in mind as you work, you can easily determine those items that should and should not be included in your writing.

semicolons

The semicolon links **independent clauses** or other **sentence** elements of equal weight and grammatical rank, especially **phrases** in a series that have **commas** within them. The semicolon indicates a greater pause between clauses than a comma would, but not so great a pause as a **period** would.

When the independent clauses of a **compound sentence** are not joined by a comma and a **conjunction,** they are linked by a semicolon.

EXAMPLE   No one applied for the position; the job was too difficult.

Make sure, however, that such **clauses** balance or contrast with each other. The relationship between the two statements should be so clear that further explanation is not necessary.

EXAMPLE   It is a curious fact that there is little similarity between the chemical composition of river water and that of sea water; the various elements are present in entirely different proportions.

Do not use a semicolon between a **dependent clause** and its main clause. Remember that elements joined by semicolons must be of equal grammatical rank or weight.

CHANGE   No one applied for the position; even though it was heavily advertised.

TO   No one applied for the position, even though it was heavily advertised.

USING A SEMICOLON WITH STRONG CONNECTIVES

In complicated sentences, a semicolon may be used before transitional words or phrases *(that is, for example, namely)* that introduce examples or further explanation.

EXAMPLE The study group was aware of his position on the issue; that is, federal funds should not be used for the housing project.

A semicolon should also be used before **conjunctive adverbs** (such as *therefore, moreover, consequently, furthermore, indeed, in fact, however*) that connect independent clauses.

EXAMPLE I won't finish today; moreover, I doubt that I will finish this week.

The semicolon in this example shows that *moreover* belongs to the second clause.

USING A SEMICOLON FOR CLARITY
IN LONG AND COMPLICATED SENTENCES

Use a semicolon between two main clauses connected by a **coordinating conjunction** *(and, but, for, or, nor, yet)* if the clauses are long and contain other **punctuation.**

EXAMPLE In most cases these individuals are corporate executives, bankers, Wall Street lawyers; but they do not, as the economic determinists seem to believe, simply push the button of their economic power to affect fields remote from economics.
—Robert Lubar, "The Prime Movers," *Fortune* (February 1960), p. 98.

A semicolon may also be used if items in a series contain commas within them.

EXAMPLE Among those present were John Howard, president of the Omega Paper Company; Carol Martin, president of Alpha Corporation; and Larry Stanley, president of Stanley Papers.

Do not, however, use semicolons to enclose a parenthetical element that contains commas. Use **parentheses** or **dashes** for this purpose.

CHANGE All affected job classifications; typists, secretaries, clerk-stenographers, and word processors; will be upgraded this month.
TO All affected job classifications (typists, secretaries, clerk-stenographers, and word processors) will be upgraded this month.
OR All affected job classifications—typists, secretaries, clerk-stenographers, and word processors—will be upgraded this month.

Do not use a semicolon as a mark of anticipation or enumeration. Use a **colon** for this purpose.

CHANGE   Three decontamination methods are under consideration; a zeolite-resin system, an evaporation and resin system, and filtration followed by storage.

TO   Three decontamination methods are under consideration: a zeolite-resin system, an evaporation and resin system, and filtration followed by storage.

The semicolon always appears outside closing **quotation marks.**

EXAMPLE   The attorney said, "You must be accurate"; the client said, "I will."

sentence construction

A **sentence** is the most basic and versatile tool available to the writer. Consider the many different ways the same idea can be expressed to achieve different **emphasis.**

EXAMPLES   Every person in the department, from manager to typists, must work overtime this week to insure that we meet the deadline.

To insure that we meet the deadline, every person in the department, from manager to typists, must work overtime this week.

From manager to typists, every person in the department must work overtime this week, to insure that we meet the deadline.

To insure that we meet the deadline, every person in the department must work overtime this week, from manager to typists.

This week, every person in the department, from manager to typists, must work overtime to insure that we meet the deadline.

When shifting word order for emphasis, however, be aware that word order can make a great difference in the meaning of a sentence.

EXAMPLES   He was *only* the engineer.
He was the *only* engineer.

Except for **expletives,** all words or word groups in a sentence function as a **subject,** a **verb,** a **complement,** a **modifier,** a **connective,** an **appositive,** or an **absolute.** Subjects, verbs, and complements are the main elements of the sentence. Everything else is subordinate to them in one way or another. A sentence that progresses quickly from subject to verb to complement is clear and easy to understand. The problem is to write sentences that contain more information, and in more complicated form, with the same **clarity** and directness.

The basic sentence patterns in English are the following.

EXAMPLES   The cable snapped. (subject-verb)
Generators produce electricity. (subject-verb-direct object)
The test results gave us confidence. (subject-verb-indirect object-direct object)
Repairs made the equipment operational. (subject-verb-direct object-objective complement)
The metal was aluminum. (subject-linking verb-subjective complement)

Most sentences follow the subject-verb-object pattern. In "The company dismissed Joe," we know the subject and the object by their positions relative to the verb. The knowledge that the usual sentence order is subject-verb-object helps **readers** interpret what they read. The fact that readers tend to expect this order explains why the writer's departures from it can be effective if used sparingly for emphasis and variety, but annoying if overdone.

An inverted sentence places the elements in other than normal order.

EXAMPLES   A better job I never had. (direct object-subject-verb)
More optimistic I have never been. (subjective complement-subject-linking verb)

Inverted sentence order may be used in questions and exclamations; it may also be used to achieve emphasis.

EXAMPLES   Have you a pencil? (verb-subject-complement)
How heavy your book feels! (complement-subject-verb)
A sorry sight we presented! (complement-subject-verb)

In sentences introduced by expletives *(there, it)*, the subject generally follows the verb because the expletive precedes the verb.

EXAMPLES   *There* (expletive) *are* (verb) certain *principles* (subject) of drafting that must not be ignored.
It (expletive) *is* (verb) *difficult* (complement) *to work* (subject) in a noisy office.

Beginning sentences with expletives is often wordy.

CHANGE   *There are* certain principles of drafting that must be followed.
TO   Certain principles of drafting must be followed.

CONSTRUCTING CLEAR SENTENCES

It is always best to use uncomplicated sentences to state complex ideas. If readers must cope with a complicated sentence in addition to a complex idea, they are likely to become confused.

CHANGE   When you are purchasing parts, remember that, although an increase in the cost of aluminum forces all the vendors to increase their prices, some vendors will have a supply of aluminum purchased at the old price, and they may be willing to sell parts to you at the old price in order to get your business.

TO   Although an increase in the cost of aluminum forces all vendors to increase their prices, some vendors will have a supply of aluminum purchased at the old price. When you are purchasing aluminum parts, remember that these vendors may be willing to sell the parts to you at the old price in order to get your business.

Just as simpler sentences make complex ideas more digestible, a complex sentence construction makes a series of simple ideas more palatable.

CHANGE   The computer is a calculating device. It was once known as a mechanical brain. It has revolutionized industry.

TO   The computer, a calculating device once known as a mechanical brain, has revolutionized industry.

Do not string together in a series a number of thoughts which should be written as separate sentences or some of which should be subordinated to others. Sentences carelessly tacked together this way are monotonous and hard to read because all ideas seem to be of equal importance.

CHANGE   We started the program three years ago, there were only three members on the staff, and each member was responsible for a separate state, but it was not an efficient operation.

TO   When we started the program three years ago, there were only three members on the staff, each having responsibility for a separate state; however, that arrangement was not efficient.

It is often easy to improve sentences by eliminating trailing constructions and ineffective **repetition**.

CHANGE   We conducted a new experiment last month and learned much from it.

TO   We learned much from a new experiment last month.

CONSTRUCTING PARALLEL SENTENCES

Express coordinate ideas in similar form. The very construction of the sentence helps the reader to grasp the similarity of its components. (See also **parallel structure.**)

EXAMPLE Similarly, atoms come and go in a molecule, but the molecule remains; molecules come and go in a cell, but the cell remains; cells come and go in a body, but the body remains; persons come and go in an organization, but the organization remains.
—Kenneth Boulding, *Beyond Economics* (Ann Arbor: University of Michigan Press, 1968), p. 131.

CONSTRUCTING SENTENCES TO ACHIEVE EMPHASIS

Subordinate your minor ideas to emphasize your more important ideas. (See also **subordination.**)

CHANGE We had all arrived, and we began the meeting early.
TO Since we had all arrived, we began the meeting early.

The most emphatic positions within a sentence are at the beginning and the end. Do not waste them by tacking on **phrases** and **clauses** almost as an afterthought or by burying the main point in the middle of a sentence between less important points. For example, consider the following original and revised versions of a statement written for a company's **annual report** to its stockholders.

CHANGE Sales declined by three percent in 1981, but nevertheless the Company had the most profitable year in its history, thanks to cost savings that resulted from design improvements in several of our major products; and we expect 1982 to be even better, since further design improvements are being made.
TO Cost savings from design improvements in several major products not only offset a three-percent sales decline but made 1981 the most profitable year in the Company's history. Further design improvements now in progress promise to make 1982 even more profitable.

Reversing the normal word order is also used to achieve emphasis.

EXAMPLES I will never agree to that.
That I will never agree to.
Never will I agree to that.

BEGINNING A SENTENCE WITH A COORDINATING CONJUNCTION

There is no rule against beginning a sentence with a **coordinating conjunction;** in fact, coordinating conjunctions can be strong transitional words and at times provide emphasis.

EXAMPLE   I realize the project was more difficult than expected and that you
have also encountered personnel problems. *But* we must meet
our deadline.

Starting sentences with **conjunctions** is acceptable in all but the most formal English. But, like any other writing device, this one should be used sparingly lest it become ineffective and even annoying. (See also **sentences, sentence variety, run-on sentences, sentence fragments,** and **sentence faults**).

sentence faults

Faulty **subordination,** one of the most common sentence faults, occurs (1) when a grammatically subordinate element, such as a **dependent clause,** actually contains the main idea of the sentence or (2) when a subordinate element is so long or excessively detailed that it dominates or obscures the main idea. Avoiding the first problem (main idea expressed in a subordinate element) depends on the writer's knowing which idea is *meant* to be the main idea. Note that both of the following **sentences** appear logical; which one really is logical depends on which of two ideas the writer means to emphasize.

EXAMPLES   Although the new filing system saves money, many of the staff are
unhappy with it.
The new filing system saves money, although many of the staff
are unhappy with it.

In this example, if the writer's main point is that *the new filing system saves money,* the second sentence is correct. If the main point is that *many of the staff are unhappy,* then the first sentence is correct. It is easier than you may think to slip and put your main idea into a grammatically subordinate element (especially since people commonly do so in conversation, when they can raise their voices and make gestures to stress the main point).

The other major problem with subordination is the loading of so much detail into a subordinate element that it "crushes" the main point by its sheer size and weight.

CHANGE   Because the noise level on a typical street in New York City on a weekday is as loud as an alarm clock ringing three feet away, New Yorkers often have hearing problems.

TO   Because the noise level in New York City is so high, New Yorkers often have hearing problems.

MISCELLANEOUS SENTENCE FAULTS

The assertion made by a sentence's **predicate** about its **subject** must be logical. "Mr. Wilson's *job* is a salesman" is not logical, but "*Mr. Wilson* is a salesman" is. "Jim's *height* is six feet tall" is not logical, but "*Jim* is six feet tall" is. Do not omit a required **verb.**

CHANGE   The floor is swept and the lights out.
TO   The floor is swept and the lights *are* out.

CHANGE   I never have and probably never will write the annual report.
TO   I never have *written* and probably never will write the annual report.

Do not omit a subject.

CHANGE   He regarded price-fixing as wrong, but until abolished by law, he engaged in it as did everyone else.
TO   He regarded price-fixing as wrong, but until *it was* abolished by law, he engaged in it as did everyone else.

Avoid **compound sentences** containing **clauses** that have little or no logical relationship to each other.

CHANGE   The reactor oxidizes the harmful exhaust ingredients into harmless water vapor and carbon dioxide, and it is housed in a double-walled metal shell.
TO   The reactor oxidizes the harmful exhaust ingredients into harmless water vapor and carbon dioxide. It is housed in a double-walled metal shell.

(See also **run-on sentences** and **sentence fragments**.)

sentence fragments

A **sentence** that is missing an essential part (**subject, verb,** or **object**) is called a sentence fragment.

EXAMPLES   He quit his job. (sentence)
And quit his job. (fragment)

But having a subject and a **predicate** does not automatically make a **clause** a sentence. The clause must also make an independent statement. "I work" is a sentence; "If I work" is a fragment because the **subordinating conjunction** *if* makes the statement a **dependent clause.**

Sentence fragments are often introduced by **relative pronouns** *(who, which, that)* or **subordinating conjunctions** (such as *although, because, if, when,* and *while*). This knowledge can tip you off that what follows is a dependent clause and must be combined with a main clause.

CHANGE The new manager instituted several new procedures. *Many of which are impractical.* (The last is an adjective clause modifying *procedures* and linked to it by the relative pronoun *which*).

TO The new manager instituted several new procedures, many of which are impractical.

A sentence must contain a **finite verb; verbals** will not do the job. The following examples are sentence fragments because their verbals *(providing, to work, waiting)* cannot perform the function of a finite verb.

EXAMPLES *Providing* all employees with hospitalization insurance.
To work a forty-hour week.
The customer *waiting* to see you.

Fragments usually reflect incomplete and sometimes confused thinking. The most common type of fragment is the careless addition of an afterthought.

CHANGE These are my co-workers. *A fine group of people.*
TO These are my co-workers, a fine group of people.

The following examples are common types of sentence fragments.

CHANGE Some of our customers prefer to pay at the time of purchase. *While others find installment payments preferable.* (adverbial clause)
TO Some of our customers prefer to pay at the time of purchase, while others find installment payments preferable.

CHANGE The board approved the project. *After much discussion.* (prepositional phrase)
TO The board approved the project after much discussion.

CHANGE We reorganized the department. *Distributing the work load more evenly.* (participial phrase)
TO We reorganized the department, distributing the work load more evenly.

CHANGE   The staff decided to take a break. *It being mid-afternoon.* (absolute
          phrase)
   TO   The staff decided to take a break, it being mid-afternoon.

CHANGE   The field tests showed the prototype to be extremely rugged. *The
          most durable we've tested this year.* (appositive phrase)
   TO   The field tests showed the prototype to be extremely rugged, the
          most durable we've tested this year.

CHANGE   We have one major goal this month. *To increase the strength of the alloy
          without reducing its flexibility.* (infinitive phrase)
   TO   We have one major goal this month: to increase the strength of
          the alloy without reducing its flexibility.

A fragment usually should be part of the preceding sentence. Explanatory **phrases** beginning with *such as, for example,* and similar terms always belong with the preceding sentence.

A hopelessly snarled fragment simply has to be rewritten. This rewriting is most effectively done by pulling the main points out of the fragment, listing them in the proper sequence, and then rewriting the sentence. (See also **garbled sentences.**)

CHANGE   Removing the protection cap and the piston secured in the
          housing by means of the spring placed between the piston and
          the housing.

MAIN   1. Remove the protection cap.
POINTS   2. The piston is held in the housing by a spring.
          3. The spring is connected to the piston at one end and the
             housing at the other.
          4. To remove the piston, disconnect the spring.

   TO   Removing the protection cap enables you to remove the piston
          by disconnecting a spring that connects to the piston at one end
          and the housing at the other.

See also **sentence construction, sentence faults,** and **run-on sentences.**

sentence types

Sentences may be classified according to *structure* (simple, compound, complex, compound-complex); *intention* (declarative, interrogative, imperative, exclamatory); and *stylistic use* (loose, periodic, minor).

BY STRUCTURE

A **simple sentence** consists of one **independent clause.** At its most basic, the simple sentence contains only a **subject** and a **predicate.**

EXAMPLES   Profits (subject) rose (predicate).
The strike (subject) finally ended (predicate).

A **compound sentence** consists of two or more independent clauses connected by a **comma** and a **coordinating conjunction,** by a **semicolon,** or by a semicolon and a **conjunctive adverb.**

EXAMPLES   Drilling is the only way to collect samples of the layers of sediment below the ocean floor, but it is by no means the only way to gather information about these strata. (comma and coordinating conjunction)
—Bruce C. Heezen and Ian D. MacGregor, "The Evolution of the Pacific," *Scientific American* (November 1973), p. 103.

There is little similarity between the chemical composition of sea water and that of river water; the various elements are present in entirely different proportions. (semicolon)

It was 500 miles to the site; therefore, we made arrangements to fly. (semicolon and conjunctive adverb)

The **complex sentence** provides a means of subordinating one thought to another. A complex sentence contains one independent clause and at least one **dependent clause** that expresses a subordinate idea.

EXAMPLE   The generator will shut off automatically (independent clause) if the temperature rises above a specified point (dependent clause).

A **compound-complex sentence** consists of two or more independent clauses plus at least one dependent clause.

EXAMPLE   Productivity is central to controlling inflation (independent clause), for when productivity rises (dependent clause), employers can raise wages without raising prices (independent clause).

BY INTENTION

By intention, a sentence may be declarative, interrogative, imperative, or exclamatory.

A declarative sentence conveys information or makes a factual statement.

EXAMPLE   This motor powers the conveyor belt.

An interrogative sentence asks a direct question.

EXAMPLE   Does the conveyor belt run constantly?

An imperative sentence issues a command.

EXAMPLE   Start the generator.

An exclamatory sentence is an emphatic expression of feeling, fact, or opinion. It is a declarative sentence that is stated with great feeling.

EXAMPLE   The heater exploded!

BY STYLISTIC USE

A *loose* sentence is one that makes its major point at the beginning and then adds subordinate **phrases** and **clauses** that develop the major point. It is the pattern in which we express ourselves most naturally and easily. A loose sentence could be ended at one or more points before it actually ends.

EXAMPLE   It went up (.), a great ball of fire about a mile in diameter (.), changing colors as it kept shooting upward (.), an elemental force freed from its bonds (.) after being chained for billions of years.

Compound sentences are generally classed as loose, since the sentence could end after the first independent clause.

EXAMPLE   Copernicus is frequently called the first modern astronomer; he was the first to develop a complete astronomical system based on the motion of the earth.

Complex sentences are loose if their subordinate clauses follow their main clause.

EXAMPLE   The installation will not be completed on schedule (.) because heavy spring rains delayed construction.

A *periodic* sentence delays its main idea until the end by presenting subordinate ideas or modifiers first, thus holding the reader's interest until the end. If skillfully handled, a periodic sentence lends force, or

emphasis, to the main point by arousing the reader's anticipation and then presenting the main point as a climax.

> EXAMPLE   During the last decade or so, the attitude of the American citizen toward automation has undergone a profound change.

Do not use periodic sentences too frequently, however, for overuse results in a **style** that is irritating to the **reader.**

A *minor* sentence is an incomplete sentence. It makes sense in its context because the missing element is clearly implied by the preceding sentence.

> EXAMPLE   In view of these facts, is automation really useful? *Or economical?* There is no question that it has made our country the most wealthy and technologically advanced nation the world has ever known.

In short, minor sentences are elliptical expressions that are equivalent to complete sentences because the missing words are clearly understood without being stated.

> EXAMPLES   Why not?
> How much?
> Ten dollars.
> At last!
> This way, please.
> So much for that idea.

Minor sentences are common in advertising copy and fictional dialogue; they are not normally appropriate to technical writing.

> CHANGE   You can use the one-minute long-distance rate any time after eleven at night all the way until eight in the morning. *Any night of the week.*
>
> TO   You can use the one-minute long-distance rate any time after eleven at night all the way until eight in the morning, any night of the week.

See also **sentence construction, sentence variety,** and **sentence faults.**

sentence variety

Sentences may be long or short; they may be loose or periodic; they may be **simple, compound, complex,** or **compound-complex;** they may be declarative, interrogative, exclamatory, or imperative—they may even be elliptical. There is never a legitimate excuse for letting your sentences become tiresomely alike. However, sentence variety is best achieved during **revision:** do not let it concern you when you are **writing the draft.**

SENTENCE LENGTH

Varying sentence length makes writing more interesting to the **reader** because a long series of sentences of the same length is monotonous. For example, avoid stringing together a number of short **independent clauses.** Either connect them with subordinating **connectives,** thereby making some independent, or make some **clauses** into separate sentences.

CHANGE   This river is 60 miles long, *and* it averages 50 yards in width, *and* its depth averages 8 feet.

TO   This river, *which* is 60 miles long and averages 50 yards in width, has an average depth of 8 feet.

OR   This river is 60 miles long. It averages 50 yards in width and 8 feet in depth.

Short sentences can often be effectively combined by converting **verbs** to **adjectives.**

CHANGE   The steeplejack *fainted.* He collapsed on the scaffolding.

TO   The *fainting* steeplejack collapsed on the scaffolding.

Although too many short sentences make your writing sound choppy and immature, a short sentence can be effective at the end of a passage of long ones.

EXAMPLE   During the past two decades, many changes have occurred in American life, the extent, durability, and significance of which no one has yet measured. *No one can.*

In general terms, short sentences are good for emphatic, memorable statements. Long sentences are good for detailed explanations and support. There is nothing inherently wrong with a long sentence, or even with a complicated one, as long as its meaning is clear and direct. Sentence length becomes an element of **style** when varied for **emphasis** or contrast; a conspicuously short or long sentence can be used to good effect.

WORD ORDER

When a series of sentences all begin in exactly the same way, the result is likely to be monotonous. You can make your sentences more interesting by occasionally starting with a modifying word, **phrase,** or clause. This could be a single adjective, **adverb, participle,** or **infinitive;** it could be a **prepositional phrase,** a **participial phrase,** or an **infinitive phrase;** or it could be a subordinate clause. But overdoing this technique can be monotonous; use it with moderation.

EXAMPLES   *Exhausted,* the project director slumped into a chair. (adjective)
Recently, sales have been good. (adverb)
Smiling, he extended his hand. (participle)
To advance, one must work hard. (infinitive)
In the morning, we will finish the report. (prepositional phrase)
Reading the report, she found several errors. (participial phrase)
To reach the top job, she presented constructive alternatives when current policies failed to produce results. (infinitive phrase)
Because we now know the result of the survey, we may proceed with certainty. (adverb clause)

Inverted sentence order can be an effective way to achieve variety.

EXAMPLES   Then occurred the event that gained us the contract.
Never have sales been so good.

Too many inverted sentences, however, can lead to the kind of criticism that was aimed at *Time* magazine's writing style many years ago: "Backwards ran the sentences until reeled the mind." Use inverted sentence order sparingly.

Be careful in your sentences to avoid the unnecessary separation of **subject** and verb, **preposition** and **object,** and parts of a **verb phrase.**

CHANGE   Electrical equipment can, if not carefully handled, cause serious accidents. (parts of verb phrase separated)
TO   Electrical equipment can cause serious accidents if not carefully handled.

This is not to say, however, that subject and verb should never be separated by a modifying phrase or clause.

EXAMPLE   John Stoddard, who founded the firm in 1943, it still an active partner.

Vary the position of **modifiers** in your sentences to achieve variety as well as precision. The following examples illustrate four different ways the same sentence could be written by varying the position of its modifiers.

EXAMPLES Gently, with the square end up, slip the blasting cap down over the time fuse.

With the square end up, gently slip the blasting cap down over the time fuse.

With the square end up, slip the blasting cap gently down over the time fuse.

With the square end up, slip the blasting cap down over the time fuse gently.

LOOSE/PERIODIC/INSERTION SENTENCES

A loose sentence makes its major point at the beginning and then adds subordinate phrases and clauses that develop or modify the point. A periodic sentence delays its main idea until the end by presenting modifiers or subordinate ideas first, thus holding the reader's interest until the end.

EXAMPLES The attitude of the American citizen toward automation has undergone a profound change during the last decade or so. (loose)

During the last decade or so, the attitude of the American citizen toward automation has undergone a profound change. (periodic)

Experiment in your own writing, especially during revision, with shifts from loose sentences to periodic sentences. Avoid the sing-song monotony of a long series of loose sentences, particularly a series containing coordinate clauses joined by **conjunctions.** Subordinating some thoughts to others makes your sentences more interesting.

CHANGE The auditorium was filled to capacity, *and* the chairman of the board came onto the stage. The meeting started at eight o'clock, *and* the president made his report of the company's operations during the past year. The audience of stockholders was obviously unhappy, *but* the members of the board of directors were all re-elected.

TO By eight o'clock, *when* the chairman of the board came onto the stage and the meeting began, the auditorium was filled to capacity. *Although* the audience of stockholders was obviously unhappy with the president's report of the company's operations during the past year, the members of the board of directors were all re-elected.

For variety, you may also alter the normal sentence order with an inserted phrase or clause.

> EXAMPLE Titanium fills the gap, both in weight and strength, between aluminum and steel.

The technique of inserting such a phrase or clause is good for **emphasis,** for providing detail, for breaking monotony, and for regulating **pace.** (See also **sentences, sentence construction,** and **sentence faults.**)

sequential method of development

The sequential, or step-by-step, method of development is especially effective for **explaining a process** or describing a mechanism in operation. It would also be the logical method for writing **instructions.**

The main advantage of the sequential method of development is that it is easy to follow because the steps correspond to the elements of the process or operation being described.

The disadvantages of the sequential method are that it can become monotonous and that it does not lend itself very well to achieving **emphasis.**

Practically all **methods of development** have elements of sequence within them. The **chronological method of development,** for example, is also sequential: to describe a trip chronologically, from beginning to end, is also to describe it sequentially.

THE PROPER WAY TO CLEAN YOUR TEETH

Proper tooth cleaning begins with the proper equipment. Choose a brush stiff enough to remove particles from between your teeth, yet soft enough to massage your gums without making them bleed. A toothbrush labeled ''medium'' will usually accomplish both objectives.

Squeeze out a bead of toothpaste just long enough to cover the bristles of your brush. Too much paste will prevent sufficient contact between the bristles and your teeth and gums.

Brush your teeth with an up-and-down motion. The American Dental Association's Council on Dental Therapeutics recommends this technique to remove food particles from between the teeth and to give gums adequate stimulation.

Rinse your mouth after brushing. Then use dental floss to remove food particles between your teeth that the toothbrush may have missed. Begin by wrapping the floss securely around your forefinger. Using a gentle back-and-forth motion, insert the floss between your teeth and move it gently up and down. Be sure to floss in the area where the teeth and gums meet.

To maintain good oral hygiene, brush your teeth within one hour after each meal. If brushing is not possible after meals, rinse your mouth thoroughly instead.

service

When used as a **verb,** *service* means "keep up or maintain" as well as "repair."

EXAMPLE   Our company will *service your equipment.*

To mean providing a more general benefit, use *serves.*

CHANGE   Our company *services* the northwest area of the state.
TO   Our company *serves* the northwest area of the state.

set/sit

Sit is an **intransitive verb;** it does not, therefore, require an **object.** Its past tense is *sat.*

EXAMPLES   I *sit* by a window in the office.
We *sat* around the conference table.

Set is usually a **transitive verb,** meaning "put or place," "establish," or "harden" (something). Its past tense is *set.*

EXAMPLES   Please *set* the trophy on the shelf.
The jeweler *set* the stone beautifully.
Can we *set* a date for the tests?
The high temperature *sets* the epoxy quickly.

Set is occasionally intransitive.

EXAMPLES   The sun *sets* a little earlier each day.
The glue *sets* in 45 minutes.

shall/will

It was at one time fairly common to use *shall* for first-person constructions and *will* for second- and third-person constructions.

EXAMPLES   I *shall* go.
You *will* go.
He *will* go.

This is an unnecessary distinction, however, because no one could be confused by "I will go" to express an action in the near future. *Shall* is often used in all **persons,** nonetheless, to emphasize determination that something will occur.

EXAMPLES   I *shall* go.
We *shall* go.
They *shall* go.

sic

Latin for "thus," *sic* is often used in **quotations** to indicate that the writer has quoted the material exactly as it appeared in the original source. It is most often used when the original material contains an obvious error or might in some other way be questioned. *Sic* is placed within **brackets.**

EXAMPLE   In *Basic Astronomy,* Professor Jones notes that the "earth does not revolve around the son [sic] at a constant rate."

simile

A simile is a direct **comparison** of two essentially unlike things, making the comparison with the word *like* or *as.* Similes state that A is *like* B.

EXAMPLE   Constructing the frame is *like piecing together the borders of a jigsaw puzzle.*

Like **metaphors,** similes can help illuminate difficult or obscure ideas.

EXAMPLE   The odds against having your plant or business destroyed by fire are at least as great *as the odds against hitting the jackpot on a Las Vegas slot machine.* Because of the long odds, top management may consider sophisticated fire protection and alarm systems more costly than the risk merits and refuse to approve the expenditure. The plant fire protection chief is free to recommend whatever system he likes, but unless the system is required by fire codes or ordered by the insurance company, chances are it will never win top management's approval.

And yet, plants are destroyed by fire, *just as gamblers sometimes hit the jackpot on a one-armed bandit.* Losing big if fire strikes the plant, or winning big on a slot machine both depend on the chance alignment of three events or factors. . . .

—Jay R. Asher and Daniel T. Williams, Jr., "How Not to Lose the Fire Protection Gamble," *Occupational Hazards* (September 1974), p. 130.

simple sentences

A simple sentence has one **clause.** At its most basic, the simple sentence contains only a **subject** and a **predicate.**

EXAMPLES Profits rose.
The strike ended.

Both the subject and the predicate may be compounded without changing the basic structure of the simple sentence.

EXAMPLES *Bulldozers and road graders* have blades. (compound subject)
Bulldozers *strip, ditch, and backfill.* (compound predicate)

Although **modifiers** may lengthen a simple sentence, they do not alter its basic structure.

EXAMPLE The *recently introduced* procedure works *very well.*

Inverting subject and predicate does not alter the basic structure of the simple sentence.

EXAMPLE A better job I never had.

A simple sentence may contain **noun phrases** and **prepositional phrases** in either the subject or the predicate.

EXAMPLES The man *in the blue suit* is my boss. (prepositional phrase in the subject)
I wrote the report *in a day.* (prepositional phrase in the predicate)

A simple sentence may include modifying **phrases** in addition to the **independent clause.** This fact causes most of the confusion about simple sentences. The following sentence, for example, is a simple sentence because the introductory phrase is an adjective phrase and not a **dependent clause.**

EXAMPLE *Hard at work in my office,* I did not realize how late it was.

-size/-sized

As **modifiers,** the **suffixes** *-size* and *-sized* are more common to advertising copy than to general writing.

EXAMPLES   a *king-sized* bed, an *economy-sized* carton

However, they are usually redundant unless they are part of the name of a product and should generally not be used.

CHANGE   It was a *small-size* desk.
TO   It was a *small* desk.

slashes

Although not always considered a mark of **punctuation,** the slash performs punctuating duties by separating and showing omission. The slash is called a variety of names, including slant line, virgule, bar, shilling sign.

The slash is often used to separate parts of addresses in continuous writing.

EXAMPLE   The return address on the envelope was Ms. Rose Howard/62 W. Pacific Court/Dalton/Ontario/Canada.

The slash often indicates omitted words and letters.

EXAMPLES   miles/hour for "miles per hour"
c/o for "in care of"
w/o for "without"

In fractions the slash separates the numerator from the denominator.

EXAMPLES   2/3 (2 of 3 parts); 3/4 (3 of 4 parts); 27/32 (27 of 32 parts)

The slash is sometimes used to indicate **brackets** on a typewriter that has no bracket key.

EXAMPLE   Dr. Smith wrote that the earth "does not revolve around the son /sic/ at a constant rate."

In informal writing, the slash is also used to separate day from month and month from year in **dates.**

EXAMPLE   12/29/74

so/such

So is often vague and should be avoided if another word would be more exact.

CHANGE   She writes faster, *so* she finished before I did.
TO   *Because* she writes faster, she finished before I did.

Another problem occurs with the **phrase** *so that,* which should never be replaced with *so* or *such that.*

CHANGE   The report should be written *so* it can be copied.
TO   The report should be written *so that* it can be copied.

CHANGE   The report should be written *such that* it can be copied.
TO   The report should be written *so that* it can be copied.

Such, an **adjective** meaning "of this or that kind," should never be used as a **pronoun.**

CHANGE   Our company does not need computers and we do not anticipate using *such.*
TO   Our company does not need computers and we do not anticipate using *any.*

some

When *some* functions as an **indefinite pronoun** for plural **count nouns,** or as an **indefinite adjective** modifying plural count nouns, use a plural **verb.**

EXAMPLES   *Some* people *are* kinder than others.
Some of us *are* prepared to leave.

Some is singular, however, when used with **mass nouns.**

EXAMPLES   *Some* sand *has* trickled through the crack.
Some oil *was* spilled on the highway.
Some stationery *is* missing from the supply cabinet.
Most of the water evaporated, but *some* remains.

some/somewhat

Some, an **adjective** or **pronoun** meaning "an undetermined quantity" or "certain unspecified persons," should not replace the **adverb** *somewhat*, which means "to some extent."

> CHANGE   His writing has improved *some.*
> TO   His writing has improved *somewhat.*
> OR   His writing is *somewhat* improved.

some time/sometime/sometimes

Some time refers to a duration of time.

> EXAMPLE   We waited for *some time* before calling the customer.

Sometime refers to an unknown or unspecified time.

> EXAMPLE   We will visit you *sometime.*

Sometimes refers to occasional occurrences at unspecified times.

> EXAMPLE   He *sometimes* visits the branch offices.

spatial method of development

In a spatial sequence, you describe an object or a process according to the physical arrangement of its features. Depending upon the subject, you may describe the features from top to bottom, from side to side, from east to west (or west to east), from inside to outside, and so on. Descriptions of this kind rely mainly on dimension (height, width, length), direction (up, down, north, south), shape (rectangular, square, semicircular), and proportion (one-half, two-thirds). Features are described in relation to one another.

> EXAMPLE   One end is raised six to eight inches higher than the other end to permit the rain to run off.

Features are also described in relation to their surroundings.

> EXAMPLE   The lot is located on the east bank of the Kingman River.

The spatial method of development is commonly used in **descriptions** of laboratory equipment, **proposals** for landscape work, construction-site

progress reports, and, in combination with a step-by-step sequence, in many types of **instructions.**

The following instructions, which explain how a two-person security team should conduct a methodical room search, make use of spatial sequencing:

Conducting a Methodical Room Search

First, look around the room to decide how the room should be divided for the search and to what height the first searching sweep should extend. The first sweep should include all items resting on the floor up to the selected height.

As nearly as possible, divide the room into two equal parts. Base the division on the number and type of objects in the room, not on the size of the room. Divide the room with an imaginary line that extends from one object to another—for example, the window on the north wall to the floor lamp next to the south wall.

Next, select a search height for the first sweep. Base this height on the average height of the majority of objects resting on the floor. In a typical room, this height is established by such objects as table and desk tops, chair backs, and so on. As a rule, the first sweep will be made hip high and below.

After dividing the room and establishing the first search height, go to one end of the agreed upon room division. This point will be the starting point for the first and all subsequent search sweeps. Beginning back to back, work your way around the room along the walls toward the other team member at the other end of the imaginary dividing line. Check all items resting on the floor adjacent to the walls; be sure to check the floor under the rugs. When you meet your partner at the other end of the room, return to the starting point and search all items in the middle of the room up to the first search height. Include all items mounted in or on the walls in the first search sweep, such as air conditioning ducts, baseboards, heaters, built-in storage units, and so on. During this and all subsequent searches, use an electronic or a medical stethoscope.

Then determine the search height for the second search sweep. This height is usually set at the chin or at the top of the head of the searchers. Return to the starting point and repeat the searching technique up to the second search height. This sweep typically covers objects hanging on walls, built-in storage units, tall standing items on the floor, and the like.

Next, establish the third search height. This area usually includes everything in the room from the top of the searcher's head to the ceiling. In this sweep, features like hanging light fixtures and mounted air conditioning units are examined.

Finally, if the room has a false or suspended ceiling, perform a fourth sweep. Check flush or ceiling light fixtures, air conditioning or ventilation ducts, speaker systems, structural frames, and so forth.

specie/species

Specie means "coined money" or "in coin."

EXAMPLE   Paper currency was virtually worthless, and creditors began to demand payment in *specie*.

Species means a category of animals, plants, or things having some of the same characteristics or qualities. *Species* is the correct **spelling** for both singular and plural.

EXAMPLE   The wolf is a member of the canine *species*.
Many animal *species* are represented in the Arctic.

specific-to-general method of development

This **method of development** begins with a specific statement and builds to a general conclusion. It is somewhat like the **increasing-order-of-importance method of development** in that it carefully builds its case, often with examples and **analogies** in addition to facts or statistics, and does not actually make its point until the end. For example, if your subject were highway safety, you might begin with a specific highway accident and then go on to generalize about how details of the accident were common enough to many similar accidents that recommendations could be made to reduce the probability of such accidents. Following is one example of specific-to-general development.

The Facts About Seat Belts

Statistic   Recently the Highway Safety Foundation studied the use of seat belts in 4,500 accidents involving nearly 13,000 people. Nearly all these accidents occurred on routes which had a speed limit of

at least 40 mph. Only 20 percent of all the vehicle occupants were
Statistic wearing any kind of seat belt.

The shoulder-type belts were even more unpopular than the
lap belts, and only 4 percent of the occupants who had shoulder
belts were wearing them.

In this study, as in other studies, it was found that vehicle
Statistic passengers not using seat belts were more than 4 times as likely to
be killed as those using them. The driver in some cases escaped a
more serious injury by being thrown against the steering wheel.

General A conservative estimate is that 40 percent of the front-seat
Conclusion passenger car deaths could be prevented if everyone used the seat
belts, which the law requires the manufacturer to put into each
automobile. If you are in an accident, your chances of survival
are far greater if you are using your seat belt.

—*The Safe Driving Handbook* (New York: Grosset & Dunlap, 1970), pp. 84–85.

specifications

By definition, a specification is "a detailed and exact statement of
particulars; especially a statement prescribing materials, dimensions,
and workmanship for something to be built, installed, or manufactured." The most significant words in this definition are "detailed" and
"exact." Although there are two broad categories of specifications—
government specifications and industrial specifications—both require
the writer of the specification to achieve a high degree of accuracy and
exact technical detail. A specification must be written so clearly that no
one could misinterpret any statement contained in it; therefore, do not
imply or suggest—state explicitly what is needed. **Ambiguity** in a
specification not only can waste money but it can result in a lawsuit.
Because of the stringent requirements for completeness and exactness of
detail in writing specifications, careful **research** and **preparation** are
especially important before you begin to write, as is careful **revision** after
you have completed the draft of your specification.

GOVERNMENT SPECIFICATIONS

Government agencies are required by law to contract for equipment
strictly according to definitions provided in formal specifications. Government specifications are contractual documents that protect both the
procuring government agency and the contractor.

A government specification is a precise definition of exactly what the contractor is to provide for the money he is paid. In addition to a technical description of the device to be purchased, the specification normally includes an estimated cost, an estimated delivery date, and the standards for the design, manufacture, workmanship, testing, training of government personnel, governing codes, inspection, and delivery of the item to be purchased. Government specifications are often used to prescribe the content and deadline for a **government proposal** submitted by a vendor or a company that wishes to bid on a project.

Government specifications are engineering and contractual documents with rules, **formats,** and peculiarities often known only to people trained in this specialty. What follows is a general description of the contents of a government specification. Detailed information on military and nonmilitary government specifications can be found in the following publications.

MILITARY *Specifications, Types and Forms.* Mil-S-83490. October 1968.
Standardization Policies, Procedures and Instructions. DOD-4120.3-M. January 1972.

Both are available from the Naval Publications and Forms Center, 5801 Tabor Avenue, Philadelphia, Pennsylvania 19120.

NONMILITARY *Index of Federal Specifications, Standards, and Commercial Descriptions.* FPMR101-29. (updated annually)

This publication is available from the U.S. General Services Administration Specification and Consumer Information Distribution Section, Building 197, Navy Yard Annex, Washington, D.C. 20407.

Government specifications contain details on (1) the scope of the project, (2) the documents that the contractor is required to furnish with the purchased device, (3) the required product characteristics and functional performance of the purchased device, (4) the required tests, test equipment, and test procedures, (5) the required preparations for delivery, (6) notes, and (7) an **appendix.**

The "characteristics and performance" section of the specification (item three in the previous paragraph) must precisely define every product characteristic not covered by the engineering drawings, and it must precisely define every functional performance requirement the device must meet when it is operational. The "test" section of the specification (item four in the previous paragraph) must specify how the device is to be tested to verify that it meets all requirements. The test section includes the precise tests that are to be performed, the procedure to be used in conducting the tests, and the test equipment to be used.

The following example is the beginning of a very long government specification for a computer installation.

Specification for JC/80
Computerized Building Automation System

1700. PREFACE

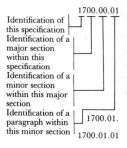

Identification of this specification
Identification of a major section within this specification
Identification of a minor section within this major section
Identification of a paragraph within this minor section

1700.00.01 The building automation system as herein specified shall be provided in its entirety by the building automation contractor. The building automation contractor shall base his bid on the system as specified. Alternate techniques, modifications, or changes to any aspect of these specifications shall be submitted as a voluntary alternate.

1700.01. VOLUNTARY ALTERNATES

1700.01.01 Voluntary alternates shall be fully documented and submitted to the consulting engineer ten working days prior to bid date for permission to bid the voluntary alternate. Permission to bid a voluntary alternate in no way implies that the alternate will be acceptable. Each voluntary alternate will be itemized as an add or deduct to base bid.

1700.01.02 No voluntary alternate shall be considered from any bidder who has failed to submit a base quotation in full compliance with the building automation system specified herein.

1701. General Requirements

1701.01. GENERAL

1701.01.01 The "computerized building automation system" herein specified shall be fully integrated and installed as a complete package by the building automation contractor. The system shall include all computer software and hardware, sensors, transmission equipment, required wiring, piping, preassembled control consoles, local panels, and labor supervision. Adjustment and calibration shall be provided as a prerequisite in the service contract specified hereafter.

1701.01.02   All input/output for central processing unit operation shall be ASCII (American Standard Code for Information Interchange) coded with standard EIA (Electronic Industries Association) interface hardware. Future incorporation of marketplace equipment prohibits any deviations from the standard ASCII-coded input/output or the standard EIA interface hardware.

1701.01.03   The building automation system operator shall have the capability to make several on-line adjustments to various system parameters and response to his requests shall occur immediately. The system shall be a combination of hardware and software to permit simultaneous data processing, output printing, and operator communications.

1701.02.   CONTRACTOR

1701.02.01   The building automation contractor shall have a local office within a 75-mile radius of the job site, staffed with factory-trained engineers fully capable of providing instructions and routine emergency maintenance service on all system components.

1701.03.   EXPERIENCE RECORD

1701.01.01   The building automation contractor shall have a 5-year successful history of the design and installation of fully computerized building systems similar in performance to that specified herein and shall be prepared to evidence this history as condition of acceptance and approval prior to bidding.

1701.04.   INSTALLATION

1701.04.01   The installation shall include computer programming, drawings, supervision, adjusting, validating, and checkout necessary for an operational system.

1701.04.02   The building automation contractor shall provide all other wiring necessary for system operation, including tie-ins from building automation system relays into motor starting circuits.

1701.04.03   All wiring performed by the building automation contractor shall be installed in accordance with all applicable local and national electrical codes.

1701.04.04   It shall be the responsibility of the building automation contractor to make a complete job survey prior to bid date to ascertain job conditions.

1702. WORK PROGRESS

1702.00.01 To assure on-time completion of the project, the building automation contractor shall use a computerized quantitative weekly reporting system. Each week a progress report will be submitted to the owner's representative. The report shall be of actual construction progress, within plus or minus 1% accuracy, and shall be submitted within one week after work has been performed at the job site. Subjective reports and estimates of project completion are not acceptable.

1702.00.02 All weekly reports shall relate to manpower analysis of the project, giving dates the work is planned for completion and the number of calendar days actual job progress varies from the original plan.

1703. CONTRACT COMPLETION, GUARANTEE, AND SERVICE

1703.00.01 All components, parts, and assemblies shall be guaranteed against defects in workmanship and materials for a period of one year after acceptance. In addition, the building automation contractor shall provide operator instruction and, if desired by the owner, system maintenance training as described hereinafter for the primary system as well as the subsystems.

1703.00.02 The following procedures shall govern the guarantee period. Within thirty days after the owner is receiving beneficial use of the building automation system, the contractor shall initiate the guarantee period by formally transmitting to the owner a 12-month service and maintenance contract marked "Paid in full." The contract shall become effective upon being dated with countersignature by the owner or his authorized representative. This service contract shall be a formal service agreement of the contractor, signed by an authorized employee, and shall include the monthly cost of the services to be provided.

1703.01. SERVICE CONTRACT

1703.01.01 The service contract shall include these minimum provisions:
01. Provide regularly scheduled maintenance and service of at least one man-day per month by factory-trained service representatives of the building automation contractor.
02. Replace all defective parts and components as required.
03. Make available, upon request, emergency maintenance service.

04. Unless canceled by the owner 45 days prior to termination of the service agreement, the building automation contractor shall agree to continue the service on a monthly basis for an additional one-year period at the maintenance contract rates set forth in the agreement. . . .

INDUSTRIAL SPECIFICATIONS

The industrial specification is used in areas like computer software where there are no engineering drawings or other means of documentation. It is a permanent document, the purpose of which is twofold: (1) to document the item being implemented so that it can be maintained if the person who designed it is promoted, transferred, or leaves the company and (2) to provide detailed technical information on the item being implemented to all those in the company who need it (this includes other engineers and technicians, technical writers, technical instructors, and possibly salespeople and purchasing agents).

The industrial specification describes (1) a planned project, (2) a newly completed project, or (3) an old project. The specification for a planned project describes how it is going to be implemented, the specification for a newly completed project describes how the newly completed project was implemented, and the specification for an old project describes the project as it finally exists after it has been operational long enough for all the problems to have been discovered and corrected. All three types of industrial specifications include a detailed technical description of all aspects of the item being described, including what was done, how it was done, what is required to use the item, how it is used, what its function is, who would use it, and so on. In addition, the specification for a planned project often includes the estimated time to complete the project (in man-hours or man-months in addition to calendar months) and estimated costs.

The industrial specification differs from the **technical manual** that is given to the customer for the same project in that the specification contains very detailed information that the customer does not need.

The following example is a specification for a computer utility routine that reads a deck of punched cards and prints in English the contents of each card. The name of the routine is CDPRNT. GAC is the name of another utility routine, and NEAT/3 is the name of a computer programming language.

2.8.8 CARD TO PRINTER

The purpose of this routine is to enable the user quickly and easily to verify and list a Hollerith Code punched card deck.

2.8.8.1 *Scope*

Identification of this specification
Identification of a major section within this specification
Identification of a minor section within this major section
Identification of a paragraph within this minor section

The CDPRNT routine will be of valuable use to any site that uses cards either as Monitor Control String records, NEAT/3 source lines, and/or data records. It also can be used in conjunction with other utility routines. For example, a deck of cards that has just been punched may be printed for easy visual validation. As another example, if (during an GAC run, using a card deck as the source file) an error or errors occur on the reading and translating of the cards, a simple run of the deck through the CDPRNT routine will list the cards and also verify the punch configurations against the set being used. If any of the punches are in error, the columns containing the errors are printed behind the card image in the listing.

2.8.8.2 *User Options*

The user is provided with a few simple but useful options. These are as follows:

1. *SET TYPE* The user may choose either the extended A-set, H-set, or Site Code for translation of his cards.

2. *SLEW* The user may slew either 1 or 2 lines between each card image on the printer.

3. *FILENAME* An optional filename of ten characters may be entered, which will be printed at the top of each page.

4. *SPREAD* The spread function, also optional, will differentiate between NEAT/3 Source Card types (C,D,F, etc.) and list them in a format similar to the compiler. If no entry is made on the parameter for the spread option, the cards will be printed in a straight 80-character format.

2.8.8.3 *Parameter Card*

The parameter card is designed as follows:

	ⅺ	ⅺ			ⅺ		ⅺ	
PAGE	LINE			REFERENCE		OPERATION		OPERANDS
1 2 3	4 5 6	7	8 9 10 11 12 13 14 15 16 17		18 19 20 21 22 23		24 25 26 27 28 29 30 31 32 33 34 35 36 37 38 39 40 41 42 43 44	

```
     |   |  C | S P E C S U T I L | C D P R N T | S U D . S . S . F I L E N A M E D O . S .
     |   |  C |                   |             |                   E   L                 P
     |   |  C |                   |             |                   T   E                 R
     |   |  C |                   |             |                   W                     E
     |   |  C |                   |             |                                         A
     |   |  C |                   |             |                                         D
     |   |  C |
```

The explanation of the parameter is:

Column 7–C indicates a control card.
    8–16–SPEC$UTIL, identifies a utility routine.
    18–21–CDPRNT, routine name.
    24–26–SUD, symbolic unit designator of the card reader to be used as source.
    28–SET, card set to be used.
        H = Extended H-set
        A = Extended A-set
        S = SITECODE of Alternate Systems Disc
    30–SLEW, signals printer number of lines to slew.
        1 = Single space
        2 = Double space
    32–41–FILENAME, optional entry, allows the user to give a name to card file being printed. Ten-character name and version number.
    43–SPREAD, an *S* entered in this column will cause the cards to be printed in NEAT/3 compiler type format. No entry in this column gives 80-character format.
        S = SPREAD
        □ = 80-character image

The parameter card is entered in the normal utility fashion.

There should be only one parameter card for each run.

2.8.8.4   *Card Verification*

As the card file is being read and translated, the CDPRNT routine detects any characters that cannot be translated and notes the column of the card that is in error. The column or columns in error are then printed out immediately following the card image. The routine checks up to 10 errors in any single card. Upon reaching the 10th, the routine moves a message "BAD CARD" under the CHECK COLUMN heading on the print page. If such a message appears, it should inform the user that this particular card needs to be completely re-created as it has at least 10 errors in it.

2.8.8.5   *Translation Using Site Code*

By using the Side Code, the CDPRNT routine will be able to handle a nonstandard card code. The routine will expect the user to have the pack with the nonstandard code (which he has built by using SITECD routine) on the Alternate Systems Disc unit. CDPRNT will access the Site Code from the Alternate Systems Disc. The reason for specifying the alternate disc is that the program may be loaded in the standard A-set from the current systems disc, while the card file can still be translated in any code desired by the user.

spelling

Learning to spell requires a systematic effort. The following system will help you learn to spell correctly.

1. Keep your dictionary handy, and use it regularly. If you are unsure about the spelling of a word, don't rely on memory or guesswork— consult the dictionary. When you look up a word, focus on both its spelling and its meaning.
2. After you have looked in the dictionary for the spelling of the word, write the word from memory several times. Then check the accuracy of your spelling. If you have misspelled the word, repeat this step. If you do not follow through by writing the word from memory, you lose the chance of retaining it for future use. Practice is essential.
3. Keep a list of the words you commonly misspell, and work regularly at whittling it down. Do not load the list with exotic words; many of us would stumble over *asphyxiation* or *pterodactyl*. Concentrate instead on words like *calendar, maintenance,* and *unnecessary.* These and other frequently used words should remain on your list until you have learned to spell them.
4. Check all your writing for misspellings by **proofreading.**

The American Heritage Dictionary of the English Language
Twenty-Thousand Words Spelled and Divided for Quick Reference
Webster's New Collegiate Dictionary
Webster's New World Dictionary of the American Language
Webster's Third International Dictionary of the English Language
Random House Dictionary of the English Language

Any of these dictionaries should serve you well.

spin-off

In technical usage, *spin-off* refers to benefits that come about in one area (for example, housing materials) as the result of achievements in another area (for example, space technology research). Because it is **jargon,** do not use the term unless you are certain that all your **readers** understand it.

> EXAMPLE   The Teflon coating on cookware is a *spin-off* from the space program.

standard English

Standard English, unlike nonstandard English, is distinguished by its adherence to the conventional grammatical features of the language, by correctly spelled and carefully chosen words, and by **punctuation** that adheres to conventions established by the editorial policies of newspapers, magazines, publishing houses, dictionaries, usage handbooks, teachers, and others concerned with how our language is used. Because of mass education, travel, and the various media, standard English is the most common variety we encounter. Indeed, its chief advantage is that because it is commonly shared it promotes efficient understanding—a fact that explains its appropriateness to technical writing. It is, in fact, the language of business, industry, government, education, and the professions. (See also **English, varieties of.**)

strata

Strata is the plural form of *stratum,* meaning a "layer of material."

EXAMPLES   The land's *strata are* exposed by erosion.
Each *stratum is* clearly visible in the cliff.

style

A dictionary definition of *style* is "the way in which something is said or done, as distinguished from its substance." A writer's style is determined by the way he thinks and by the way he transfers his thoughts to paper—uses words, **sentences,** images, **figures of speech, description, methods of development,** and so on.

A writer's style is the way his language functions in particular situations. For example, a letter to a friend would be relaxed, even chatty in **tone,** while a job application letter would be more restrained and deliberate. Obviously, the style appropriate to the one letter would not be appropriate to the other. In both letters, the audience and subject determine the manner or style the writer adopts.

Standard English can be divided into two broad categories of style—formal and informal—according to how it functions in certain situations. Understanding the distinction between a **formal writing style** and an **informal writing style** helps us to use the appropriate style in the appropriate place. We must recognize, however, that no clearcut line divides the two catgories and that some writing may call for a combination of the two styles.

When we consciously attempt to create a "style," we usually defeat our purpose. One writer may attempt to impress the **reader** with a flashy writing style, which can become **affectation.** Another may attempt to impress the **reader** with scientific objectivity and produce a style that is dull and lifeless. Technical writing need be neither affected nor dull. It can be simple, clear, direct, and interesting—the key is to master the basic writing skills and to keep your reader in mind always. What will be both informative and interesting to your reader? When this question is uppermost in your mind, as you carefully apply the steps of the writing process, you will achieve an interesting and informative writing style. (See the Checklist of the Writing Process.)

The following guidelines will help you produce a brisk, interesting style. Concentrate on them as you revise your rough draft.

1. Use the active **voice**—not exclusively, but as much as possible without becoming awkward or illogical.
2. Use **parallel structure** whenever a sentence presents two or more thoughts that are equal in importance.
3. Avoid the monotony of a sing-song style by varying your sentence types and the length of your sentences.
4. Avoid stating positive thoughts in negative terms (write *40% responded* instead of *60% failed to respond*).
5. Concentrate on achieving the proper balance between **emphasis** and **subordination.**

Beyond an individual's personal style, there are various kinds of writing that have distinct stylistic traits, such as **technical writing style.**

subjective complements

A subjective complement is a **noun** or **adjective** in the **predicate** of a **sentence;** it completes the meaning of a **linking verb** by describing or renaming the subject of the **verb.**

EXAMPLES   The project director seems *confident.* (adjective)
Acme Corp. is *our major competitor.* (noun phrase)
His excuse was *that he had been sick.* (noun clause)
He is an *engineer.* (noun)

The subjective complement is also known as a **predicate nominative** (noun) or a **predicate adjective** (adjective).

subjects of sentences

The subject of a sentence is a word or group of words about which the **sentence** or **clause** makes a statement. It indicates the topic of the sentence, telling what the **predicate** is about. It may appear anywhere in a sentence, but most often appears at the beginning.

EXAMPLES *To increase sales* is our goal.
The wiring is defective.
We often work late.
That he will be fired is now doubtful.

The simple subject is a **substantive;** the complete subject is the simple subject and its **modifiers.** In the following sentence, the simple subject is *procedures;* the complete subject is *the procedures that you instituted.*

EXAMPLE The procedures that you instituted have increased efficiency.

Grammatically, a subject must agree with its **verb** in **number.**

EXAMPLES The *departments have* much in common.
The *department has* several advantages.

The subject is the actor in active **voice** sentences.

EXAMPLE The *aerosol bomb* propels the liquid as a mist.

A compound subject has two or more elements as the subject of one verb.

EXAMPLE *The president* and *the treasurer* agreed to withhold the information.

Alternative subjects are joined by *or* and *nor.*

EXAMPLE Either *cash* or a *check* is acceptable.

Be careful not to shift subjects in a sentence; doing so may confuse your **reader.**

CHANGE *Radio amateurs* stay on duty during emergencies, and *sending messages* to and from disaster areas is their particular job.
TO *Radio amateurs,* who stay on duty during emergencies, are responsible for sending messages to and from disaster areas.
OR *Radio amateurs* stay on duty during emergencies, sending messages to and from disaster areas.

subordinating conjunctions

Subordinating conjunctions connect **sentence** elements of varying importance, normally **independent clauses** and **dependent clauses;** they usually introduce the dependent clause. The most frequently used subordinating conjunctions are *so, although, after, because, if, where, than, since, as, unless, before, though, when,* and *whereas.* They are distinguished from **coordinating conjunctions,** which connect elements of equal importance.

> EXAMPLES   He finished his report *and* he left the office. (coordinating conjunction)
> He left the office *after* he finished his report. (subordinating conjunction)

A clause introduced with a subordinating conjunction is a dependent clause.

> EXAMPLE   *Because we did not oil the machine,* the engine was ruined.

The word *because* is a subordinating conjunction. Do not use the coordinating conjunction *and* as a substitute for it.

> CHANGE   We didn't add oil to the crankcase, *and* it ruined the engine.
> TO   *Because* we didn't add oil to the crankcase, the engine was ruined.

Be aware that words may serve multiple functions. *When, where, how,* and *why* serve both as **interrogative adverbs** and as subordinating conjunctions.

> EXAMPLES   *When* will we go? (interrogative adverb)
> We will go *when* he arrives. (subordinating conjunction)

Since, until, before, and *after* are used not only as subordinating conjunctions but also as **prepositions.**

> EXAMPLES   *Since* we had all arrived, we decided to begin the meeting early. (subordinating conjunction)
> I have worked on this project *since* May. (preposition)

subordination

Subordination is a technique used by writers to show, in the structure of a **sentence,** the appropriate relationship between ideas of unequal importance by subordinating the less important ideas to the more important ideas. The skillful use of subordination is a mark of mature writing.

CHANGE   Beta Corp. now employs 500 people. It was founded just three years ago.

TO   Beta Corp., which now employs 500 people, was founded just three years ago.

OR   Beta Corp., which was founded just three years ago, now employs 500 people.

Effective subordination can be used to achieve **sentence variety, conciseness,** and **emphasis.** For example, take the sentence "The city manager's report was carefully illustrated, and it covered five typed pages." See how it might be rewritten, using subordination, in any of the following ways.

EXAMPLES   The city manager's report, *which covered five typed pages,* was carefully illustrated. (adjectival clause)

The city manager's report, *covering five typed pages,* was carefully illustrated. (participial phrase)

The carefully illustrated report of the city manager covered five typed pages. (participial phrase)

The city manager's *five-page* report was carefully illustrated. (single modifier)

The city manager's report, *five typed pages,* was carefully illustrated. (appositive phrase)

We sometimes use a **coordinating conjunction** to concede that an opposite or balancing fact is true; however, a subordinating **connective** can often make the point more smoothly.

CHANGE   Their bank has a lower interest rate on loans, *but* ours provides a fuller range of essential services.

TO   *Although* their bank has a lower interest rate on loans, ours provides a fuller range of essential services.

The relationship between a conditional statement and a statement of consequences is clearer if the condition is expressed as a subordinate **clause.**

> CHANGE   The bill was incorrect, *and* the customer was angry.
>
> TO   The customer was angry *because* the bill was incorrect.

Subordinating connectives (such as *because, if, while, when, though*) and **relative pronouns** *(who, whom, which, that)* achieve subordination effectively when the main clause states a major point and the **dependent clause** establishes a relationship of time, place, or logic with the main clause.

> EXAMPLE   A build-up of deposits is impossible *because* the apex seals are constantly sweeping the inside chrome surface of the rotor housing.

Relative pronouns *(who, whom, which, that)* can be used effectively to combine related ideas that would be stated less smoothly as independent clauses or sentences.

> CHANGE   The generator is the most common source of electric current. It uses mechanical energy to produce electricity.
>
> TO   The generator, *which* is the most common source of electric current, uses mechanical energy to produce electricity.

Avoid overlapping subordinate constructions, with each depending upon the last. Often the relationship between a relative pronoun and its antecedent will not be clear in such a construction.

> CHANGE   Shock, *which* often accompanies severe injuries, severe infections, hemorrhages, burns, heat exhaustion, heart attacks, food or chemical poisoning, and some strokes, is a failure of the circulation, *which* is marked by a fall in blood pressure *that* initially affects the skin (*which* explains pallor) and later the vital organs of kidneys and brain; there is a marked fall in blood pressure.
>
> TO   Shock often accompanies severe injuries, severe infections, hemorrhages, burns, heat exhaustion, heart attacks, food or chemical poisoning, and some strokes. It is a failure of the circulation, initially to the skin (this explains pallor) and later to the vital organs of kidneys and brain; there is a marked fall in blood pressure.

substantives

A substantive is a word, or a group of words, that functions in its **sentence** as a **noun.** It may be a noun, a **pronoun,** or a **verbal (gerund** or **infinitive),** or it may be a **phrase** or even a **clause** that is used as a noun.

EXAMPLES   The *report* is due today. (noun)
We must finish the project on schedule. (pronoun)
Several college graduates applied for the job. (noun phrase)
Drilling is expensive. (gerund)
To succeed will require hard work. (infinitive)
What I think is unimportant. (noun clause)

suffixes

A suffix is a letter or letters added to the end of a word to change its meaning in some way. Suffixes can change the **part of speech** of a word.

EXAMPLES   The *wire* is on the truck. (noun)
The *wirelike* tubing is on the truck. (adjective)

What is the *length* of the unit? (noun)
Move the unit *lengthwise* through the conveyor. (adverb)

sweeping generalizations

When the scope of an opinion is unlimited, the opinion is called a sweeping generalization. Such statements, though at times tempting to make, should be qualified during **revision.** Consider the following:

EXAMPLES   Anyone who succeeds in business today is dishonest.
Engineers are poor writers.

These statements ignore any possibility that someone may have succeeded in business without being dishonest or that there might be an engineer who writes superbly. Moreover, one person's definition of "success" or "dishonesty" or "poor writing" may be different from another's. No matter how certain you are of the general applicability of an opinion, use such all-inclusive terms as *anyone, everyone, no one, all,* and *in all cases* with caution. Otherwise, your opinions are likely to be judged irresponsible. (See also **logic.**)

syllabication

When a word is too long to fit on a line, the proper place to divide it is often a troublesome question. The most general rule is to divide words between syllables. The following guidelines may be useful.

1. One-syllable words should not be divided.
2. Words with less than six letters should not normally be divided.
3. Fewer than three letters should not be left at the end of a line or carried over to begin a new line.
4. Words with **suffixes** or **prefixes** should be divided at, rather than within, the suffix or prefix.

   CHANGE su-permarket
       TO super-market

5. A hyphenated word should not be divided at the **hyphen** if the hyphen is essential to the meaning of the word. For example, *re-form* (to change the shape of something) and *reform* (to improve something) have different meanings. Thus it could be confusing to divide *re-form* at the hyphen.
6. **Abbreviations** and **contractions** should never be divided.
7. Proper names and company titles should not be divided.
8. The last word of a **paragraph** should not be divided.
9. The last word on a page of a typewritten manuscript should not be divided.

Secretarial handbooks with **spellings** and word divisions and **dictionaries** that indicate syllable breaks provide authoritative help. A handy guide is *20,000 Words for Stenographers, Students, Authors & Proofreaders,* compiled by Louis A. Leslie (New York: The McGraw-Hill Book Co., 1977).

symbols

From highway signs to mathematical equations, people communicate in written symbols. When a symbol seems appropriate in your writing, either be certain that your **reader** understands its meaning or place an explanation in **parentheses** following the symbol the first time it appears. However, never use a symbol when your reader would more readily understand the full term. Following is a list of symbols and their appropriate uses.

Symbol	Meaning and Use
£	pound (basic unit of currency in the United Kingdom)
$	dollar (basic unit of currency in the United States)
O	oxygen (For a listing of all symbols for chemical elements, see a periodic table of elements in a **dictionary** or handbook.)
+	plus
−	minus
±	plus or minus
∓	minus or plus
×	multiplied by
÷	divided by
=	equal to
≠ or ≒	not equal to
≈ or ≑	approximately (or nearly equal to)
≡	identical with
≢	not identical with
>	greater than
≯	not greater than
<	less than
≮	not less than
:	is to (or ratio)
≐	approaches (but does not reach equality with)
∥	parallel
⊥	perpendicular
√	square root
∛	cube root
∞	infinity
π	*pi*
∴	therefore (in **mathematical equations**)
∵	**because** (in mathematical equations)
( )	parentheses (See also **punctuation.**)
[ ]	brackets (See also **punctuation.**)
{ }	braces (used to group two or more lines of writing, to group figures in **tables** and to enclose figures in mathematical equations)

Symbol	Meaning and Use
°F or °C	degree (Fahrenheit or Celsius)
'	minute *or* foot
"	second *or* inch
#	number
*	asterisk (used to indicate a **footnote** when there are very few)
&	**ampersand**
♂	**male**
♀	**female**
©	**copyright**
%	**percent**
ᶜ/ₒ	in care of
a/o	account of
@	at (used in **tables,** but never in writing)
´	acute (accent mark in French and other languages)
`	grave (accent mark in French and other languages)
^	circumflex (accent mark in French and other languages)
~	tilde (**diacritical mark** identifying the palatal nasal in Spanish and Portuguese)
⁻	macron (marks a long phonetic sound, as in *cāke*)
ᵙ	breve (marks a short phonetic sound, as in *brăcket*)
¨	dieresis or umlaut (mark placed over the second of two consecutive vowels indicating that the sound is to be pronounced—*coöperate*)
¸	cedilla (mark placed beneath the letter *c* in French, Portuguese, and Spanish to indicate the letter is pronounced as *s*—*garçon*)
∧	caret (a proofreader's mark used to indicate inserted material)
FR	franc (basic unit of currency in France)
Mex $	peso (basic unit of currency in Mexico)
$	peso (Philippine peso)
R	ruble (basic monetary unit of the U.S.S.R.)
¥	yen (basic unit of currency in Japan)

See also **abbreviations** and **proofreader's marks.**

synonyms

A synonym is a word that means nearly the same thing as another word.

EXAMPLES   purchase, acquire, buy
seller, vendor, supplier

The **dictionary** definitions of synonyms are usually identical; the **connotations** of such words, however, may differ. (A *seller* may be the same thing as a *supplier* but the term *supplier* does not suggest a commercial transaction as strongly as *seller.*)

Do not try to impress your **reader** by finding fancy or obscure synonyms in a **thesaurus;** the result is likely to be **affectation.** (See also **connotation/denotation** and **antonyms.**)

syntax

Syntax refers to the way words, **phrases,** and **clauses** are put together to form an orderly and logical **sentence.** Because English sentences depend heavily on word order for meaning, the position a word occupies can change the meaning of the sentence.

EXAMPLES   He was *only* the engineer.
He was the *only* engineer.

T

table of contents

A table of contents is a list of chapters or sections in a book or **report.** It lists them in their order of appearance and cites their page numbers. Appearing at the front of a work, a table of contents permits the researcher to preview what is in the work and assess the work's value to him. It also aids the **reader** who may want to read only certain sections.

The length of your report should determine whether it needs a table of contents. If it does, use the major **heads** and subheads of your **outline** to create your table of contents, as in the following example.

Table of Contents

For punctuating a table of contents, see "leaders" in **periods.**

tables

A table is useful for showing large numbers of specific, related facts or statistics in a brief space. A table can present **data** in a more concise form than the text can, and a table is more accurate than graphic presentations because it provides numerous facts that a **graph** cannot convey. A table makes **comparisons** between figures easy because of the arrangement of the figures into rows and columns, although overall trends about the information are more easily seen in charts and graphs. (See Table 1.)

Table 1   Recreational fresh-water angling by water-body type and geographical region*

Geographical Regions	Reservoirs	Man-Made Ponds	Natural Lakes & Ponds	Rivers & Streams	Farm Ponds
New England	130	40	570	410	410
Middle Atlantic	710	290	780	1200	630
East North Central	1200	760	3100	1600	1300
West North Central	810	550	1200	970	980
South Atlantic	1100	760	640	1500	1600
East South Central	890	630	190	670	1200
West South Central	1700	610	430	880	1300
Mountain	820	50	280	600	230
Pacific	950	200	820	1400	470
Totals	8300	3900	8000	9200	7800

*In thousands of anglers. Anglers who fished in more than one water body or region are represented in more than one category.

Source: U.S. Department of the Interior

Caption

Column captions

Body

Boxhead

Stub

Rule

Footnote

Source line

GUIDELINES FOR CREATING TABLES

Table Number. If you are using several tables, assign each a specific number; center the number and title above the table. The numbers are usually Arabic, and they should be assigned sequentially to the tables throughout the text. Tables should be referred to in the text by table number rather than by direction ("Table 4" rather than "the above table"). If there are more than five tables in your **report** or paper, give them a separate heading ("List of Tables") and list them by title and page number, together with any figure numbers, on a separate page immediately after the **Table of Contents.**

Title. The title, which is placed just above the table, should describe concisely what the table represents.

Boxhead. The boxhead carries the column headings. These should be kept concise but descriptive. Units of measurement, where necessary, should be specified either as part of the heading or enclosed in **parentheses** beneath the heading. Avoid vertical lettering where possible.

Stub. The left-hand vertical column of a table is the stub. It lists the items about which information is given in the body of the table.

Body. The body comprises the data below the boxhead and to the right of the stub. Within the body, columns should be arranged so that the terms to be compared appear in adjacent rows and columns. Leaders are sometimes used between figures to aid the eye in following data from column to column. Where no information exists for a specific item, leave an empty space to acknowledge the gap.

Rules. These are the lines that separate the table into its various parts. Horizontal lines are placed below the title, below the body of the table, and between the column headings and the body of the table. They should not be closed at the sides. The columns within the table may be separated by vertical lines if they aid clarity.

Footnotes. **Footnotes** are used for explanations of individual items in the table. Symbols (*, #) or lower-case letters, rather than numbers, are ordinarily used to key table footnotes because numbers might be mistaken for the data in a numerical table.

Source Line. The source line, which identifies where the data were obtained, appears below any footnotes (when a source line is appropriate.)

Continued Lines. When a table must be divided so that it can be continued on another page, repeat the boxhead and give the table number at the head of each new page with a "continued" label ("Table 3, continued").

technical information letters

The letter of technical information is, in effect, a small technical **report.** It may take the form of a **memorandum** for use within your firm, or it may be a letter or a report to another firm.

Since this letter or memo is an abbreviated report, follow the appropriate steps of the writing process as you compose it. It is most helpful to your **reader** if you restate the problem in the letter or memo. Above all, be as clear and to the point as possible.

Consider Letter 20. It addresses an updating problem peculiar to a specific computer. The writer spells out the problem by way of introduction, gives the background of the problem, and then goes on to suggest two possible solutions to it.

technical manuals

Technical manuals are provided to customers and the manufacturer's technical personnel to enable the technical specialists and customers to use and maintain the manufacturer's product. Technical manuals are normally written by professional technical writers, although in smaller companies engineers and technicians may be asked to write a technical manual.

The customers are normally businesses or industrial concerns. The equipment is normally complicated mechanical, electrical, hydraulic, or pneumatic devices. The manuals may include operating instructions, assembly and disassembly instructions, troubleshooting guides, theory of operation, parts lists, engineering drawings, and so on. They may be either bound or loose-leaf, although they are more often loose-leaf to provide for easy updating.

In writing a technical manual, adhere to the Checklist of the Writing Process. **Preparation** and **organization** are especially important because of the complexity of the project; give particular care to **outlining** and review the entries on **instructions, explaining a process,** and **technical writing style.**

technical writing style

Technical writing is standard **exposition** in which the **tone** is objective, with the author's voice taking a back seat to the subject matter. Because the focus is on an object or a process, the language is utilitarian,

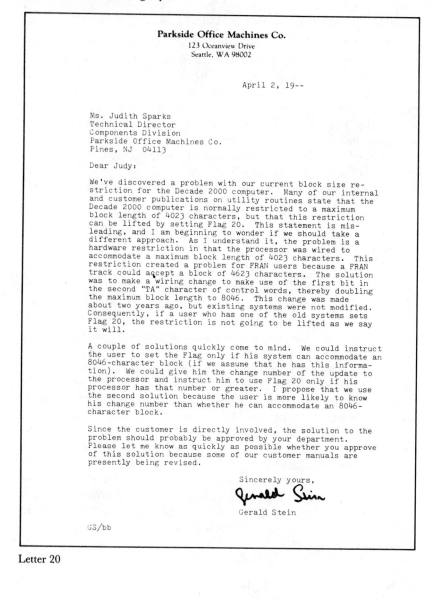

Parkside Office Machines Co.
123 Oceanview Drive
Seattle, WA 98002

April 2, 19--

Ms. Judith Sparks
Technical Director
Components Division
Parkside Office Machines Co.
Pines, NJ 04113

Dear Judy:

We've discovered a problem with our current block size restriction for the Decade 2000 computer. Many of our internal
and customer publications on utility routines state that the
Decade 2000 computer is normally restricted to a maximum
block length of 4023 characters, but that this restriction
can be lifted by setting Flag 20. This statement is misleading, and I am beginning to wonder if we should take a
different approach. As I understand it, the problem is a
hardware restriction in that the processor was wired to
accommodate a maximum block length of 4023 characters. This
restriction created a problem for FRAN users because a FRAN
track could accept a block of 4623 characters. The solution
was to make a wiring change to make use of the first bit in
the second "TA" character of control words, thereby doubling
the maximum block length to 8046. This change was made
about two years ago, but existing systems were not modified.
Consequently, if a user who has one of the old systems sets
Flag 20, the restriction is not going to be lifted as we say
it will.

A couple of solutions quickly come to mind. We could instruct
the user to set the Flag only if his system can accommodate an
8046-character block (if we assume that he has this information). We could give him the change number of the update to
the processor and instruct him to use Flag 20 only if his
processor has that number or greater. I propose that we use
the second solution because the user is more likely to know
his change number than whether he can accommodate an 8046-
character block.

Since the customer is directly involved, the solution to the
problem should probably be approved by your department.
Please let me know as quickly as possible whether you approve
of this solution because some of our customer manuals are
presently being revised.

Sincerely yours,

Gerald Stein

Gerald Stein

GS/bb

Letter 20

emphasizing exactness rather than elegance. Thus the writing is usually not adorned with figurative language, except where a **figure of speech** would genuinely promote understanding. Technical writing requires an effective **introduction** or **opening,** good **organization,** and **sentence variety,** and benefits greatly from the effective use of **heads,** and **illustrations.**

Effective technical writing avoids overuse of the passive **voice.** Its vocabulary is appropriately technical, although the general word is preferable to the technical word when the material allows the general word. (A jargony shorthand is a poor substitute for clear and direct writing; do not use a big or technical word merely because you know it— make certain that it belongs. See also **affectation.**)

Technical writing is direct and to the point as the following excerpt from a customer manual prepared by a computer manufacturer illustrates.

> The executive software controls all program-related functions, such as handling input and output operations, interpreting and executing macro instructions, providing format and translation capabilities, and handling error conditions. To conserve valuable memory space, only the pertinent software overlays are memory-resident during program execution; all other software overlays reside on disc and are called into memory only when needed.
>
> The software overlays that reside in memory during program execution are collectively called the executive software. Program-related software overlays (such as arithmetic and logic routines, the program overlay caller, the GET and PUT routines, the format and translation routines, etc.) are included in the user's object program by the compiler and are, therefore, called into memory with the program. Other executive routines and certain hardware control information are read into dedicated memory locations (called the Resident Executive Area) by Monitor at the start of the day and between programs; this group of routines and the control information comprise the Resident Executive. When necessary, the Resident Executive's Special I/O routine calls disc-resident software overlays into the Primary Software Overlay Area (PSOA), which is a 512-byte area of memory located beyond the user's program area.
>
> —*NCR Century Operating System Manual,* (Dayton: NCR Corporation, 1969), p. 14.

The **format** in this type of writing is usually clear-cut. The chief advantage of a well-defined format is that it gives the writer a mechanical way of organizing his material while helping the **reader** keep track of where the writer is going. In writing that is concerned with descriptions, processes, and sequences, a well-defined format grows out of the subject matter itself. (See also **style** and **formal writing style.**)

telegraphic style

Telegraphic style condenses writing by omitting **articles, pronouns, conjunctions,** and transitional expressions. Although **conciseness** is important in writing, writers sometimes make their **sentences** too brief by omitting these words. Telegraphic style forces the **reader** to mentally supply the missing words. Compare the following passages; notice how much easier the revised version reads (the added words are italicized).

> CHANGE   Take following action when treating serious burn. Remove loose clothing on or near burn. Cover injury with clean dressing and wash area around burn. Secure dressing with tape. Separate fingers/toes with gauze/cloth to prevent sticking. Do not apply medication unless doctor prescribes.

> TO   Take *the* following action when treating *a* serious burn. Remove *any* loose clothing on or near *the* burn. Cover *the* injury with *a* clean dressing and wash *the* area around *the* burn. *Then* secure *the* dressing with tape. Separate fingers *or* toes with gauze *or* cloth to prevent *them from* sticking *together.* Do not apply medication unless *a* doctor prescribes *it.*

Telegraphic style can also produce **ambiguity,** as the following example demonstrates.

> CHANGE   Grasp knob and adjust lever before raising the boom.

Does this sentence mean that the reader should *adjust the lever* or *grasp an adjust lever?*

> TO   Grasp the knob and the adjust lever before raising the boom.
> OR   Grasp the knob and adjust the lever before raising the boom.

Some writers excuse telegraphic style with the claim that their readers will "understand." Maybe so, but they will probably have to work hard to understand—and may therefore even misunderstand. Though telegraphic writing can save space, it never really saves time. As a writer, remember that although you may save yourself work by writing telegraphically, your readers will have to work that much harder to read—or decipher—your writing. (See also **transitions.**)

tenant/tenet

A *tenant* is one who holds or temporarily occupies a property owned by another person.

EXAMPLE   The *tenants* were upset by the increased rent.

A *tenet* is an opinion or belief held by a person or an organization.

EXAMPLE   The idea that competition will produce adequate goods and services for a society is a central *tenet* of capitalism.

tense

Tense is the grammatical term for **verb** forms that indicate time distinctions. There are six basic tenses in English: past, past perfect, present, present perfect, future, and future perfect. Each of these has a corresponding progressive form.

Tense	Basic	Progressive
Past	I began	I was beginning
Past Perfect	I had begun	I had been beginning
Present	I begin	I am beginning
Present Perfect	I have begun	I have been beginning
Future	I will begin	I will be beginning
Future Perfect	I will have begun	I will have been beginning

Perfect tenses allow you to express a prior action or condition that continues in a present, past, or future time.

EXAMPLES   I *have begun* to write the annual report, and I will work on it for the rest of the month. (present perfect)

I *had begun* to read the manual when the lights went out. (past perfect)

I *will have begun* this project by the time funds are allocated. (future perfect)

Note from the table that the progressive forms are created by combining the **helping verb** *be,* in the appropriate tenses, with the present **participle** *(-ing)* form of the main verb.

PAST TENSE

The simple past tense indicates that an action took place in its entirety in the past. The past tense is usually formed by adding *-d* or *-ed* to the root form of the verb.

EXAMPLE   We *closed* the office early yesterday.

PAST PERFECT TENSE

The past perfect tense indicates that one past event preceded another. It is formed by combining the helping verb *had* with the past participle form of the main verb.

EXAMPLE   He *had finished* by the time I arrived.

PRESENT TENSE

The simple present tense represents action occurring in the present, without any indication of time duration.

EXAMPLE   I *use* the beaker.

A general truth is always expressed in the present tense.

EXAMPLE   He learned that the saying "time *heals* all wounds" is true.

The present tense can be used to present actions or conditions that have no time restrictions.

EXAMPLE   Water *boils* at 212°F.

The present tense can be used to indicate habitual action.

EXAMPLE   I *pass* the paint shop on the way to my department every day.

The present tense can be used as the "historical present" to make things that occurred in the past more vivid.

EXAMPLE   He *asks* for more information on production statistics and *receives* a detailed report on every product manufactured by the company. Then he *asks*, "Is each department manned at full strength?" In his office, surrounded by his staff, he *goes* over the figures and *plans* for the coming year.

PRESENT PERFECT TENSE

The present perfect tense describes something from the recent past that has a bearing on the present—a period of time before the present but after the simple past. The present perfect tense is formed by combining a form of the helping verb *have* with the past participle form of the main verb.

EXAMPLES He *has retired,* but he visits the office frequently.
We *have finished* the draft and are ready to begin revising it.

FUTURE TENSE

The simple future tense indicates a time that will occur after the present. It uses the helping verb *will* (or *shall*) plus the main verb.

EXAMPLE I *will finish* the job tomorrow.

FUTURE PERFECT TENSE

The future perfect tense indicates action that will have been completed at a future time. It is formed by linking the helping verbs *will have* to the past participle form of the main verb.

EXAMPLE He *will have driven* the test car 40 miles by the time he returns.

TENSE AGREEMENT OF VERBS

The verb of a subordinate **clause** should usually agree in tense with the verb of the main clause. (See also **agreement.**)

EXAMPLES When the supervisor *presses* the starter button, the assembly line
begins to move.
When the supervisor *pressed* the starter button, the assembly line
began to move.

Shift in Tense. Be consistent. The only legitimate shift in tense records a real change in time. When you choose a tense in telling a story or discussing an idea, stay with that tense. Illogical shifts in tense will confuse your **reader.**

CHANGE Before he *installed* the printed circuit, the technician *cleans* the
contacts.

TO Before he *installed* the printed circuit, the technician *cleaned* the
contacts.

test reports

The *test report* differs from the more formal **laboratory report** in both size and **scope.** Considerably smaller and less formal than the laboratory report, the test report can be a **memorandum** or a formal business letter,

=== **Biospherics, Inc.**
                                                       4928 Wyaconda Road
                                                       Rockville, MD 20852

                                        March 14, 19--

Mr. John Sebastiani, General Manager
Midtown Development Corporation
114 West Jefferson Street
Milwaukee, WI   53201

Subject: Results of Analysis of Soil Samples for Arsenic

Dear Mr. Taylor:

     Following are the results of the analysis of 22 soil samples for
arsenic. The arsenic values listed are based on a wet-weight determination.
The moisture content of the soil is also given to allow conversion of the
results to a dry-weight basis if desired.

Hole Number	Depth	Percent of Moisture	Arsenic Total As ppm
1	12"	19.0	312.0
2	Surface	11.2	737.0
3	12"	12.7	9.5
4	12"	10.8	865.0
5	12"	17.1	4.1
6	12"	14.2	6.1
7	12"	24.2	2540.0
8	Surface	13.6	460.0

     I noticed that some of the samples contained large amounts of
metallic iron coated with rust. Arsenic tends to be absorbed into soils
high in iron, aluminum, and calcium oxides. The large amount of iron
present in some of these soil samples is probably responsible for
retaining high levels of arsenic. The soils highest in iron, aluminum,
and calcium oxides should also show the highest levels of arsenic,
provided the soils have had approximately equal levels of arsenic exposure.

     If I can be of further assistance, please do not hesitate to contact
me.

                              Yours truly,

                              Gunther Gottfried

                              Gunther Gottfried
                              Chemist

GG/jrm

Figure 37

Biospherics, Inc.
4928 Wyaconda Road
Rockville, MD 20852

September 9, 19--

Mr. Leon Hite, Administrator
The Angle Company, Inc.
1869 Slauson Boulevard
Waynesville, VA 23927

Dear Mr. Hite:

On Tuesday, 30 August, Biospherics, Inc., performed asbestos-in-air monitoring at your Route 66 construction site, near Front Royal, Virginia. Six persons and three construction areas were monitored.

All monitoring and analyses were performed in accordance with "Occupational Exposure to Asbestos," U.S. Department of Health, Education, and Welfare, Public Health Service, National Institute for Occupational Safety and Health, 1972. Each worker or area was fitted with a battery-powered personal sampler pump operating at a flow rate of approximately one liter per minute. The airborne asbestos was collected on a 37-mm Millipore type AA filter mounted in an open-face filter holder. Samples were collected over an 8-hour period.

A wedge-shaped piece of each filter was mounted on a microscope slide with a drop of 1:1 solution of dimethyl phthalate and diethyl oxalate and then covered with a cover clip. Samples were counted within 24 hours after mounting, using a microscope with phase contrast option.

In all cases, the workers and areas monitored were exposed to levels of asbestos fibers well below the NIOSH standard. The highest exposure found was that of a driller who was exposed to 0.21 fibers per cubic centimeter. The driller's sample was analyzed by scanning electron microscopy followed by energy dispersive X-ray techniques which identify the chemical nature of each fiber, thereby verifying the fibers as asbestos or identifying them as other fiber types. Results from these analyses show that the fibers present are tremolite asbestos. No nonasbestos fibers were found.

Yours truly,

Gary Willis

Gary Willis
Chemist

GW/jrm
Enclosures

Figure 38

depending on its recipient. Either way, the report should have a subject line at the beginning to identify the test under discussion.

The **opening** of a test report should state the test's purpose, unless it is completely obvious. The body of the report presents the data. A report on the tensile strength of metal, for example, includes the readings from the test equipment. If the procedure used to conduct the test would interest the **reader,** it should be described. The results of the test should be stated and, if necessary, interpreted. Often there is reason to discuss their significance. The report should conclude with any recommendations made as a result of the test.

The test report in Figure 37 does not explain how the tests were performed because such an explanation is unnecessary. The example in Figure 38 does explain how the tests were conducted.

that

Avoid unnecessary repetition of *that*.

CHANGE   I think *that* when this project is finished *that* you should write the results.

TO   I think *that* you should write the results when the project is finished.

OR   Write the results when the project in finished.

CHANGE   You will note *that* as you assume greater responsibility and as your years of service with the company increase, *that* your benefits increase accordingly.

TO   You will note *that* as you assume greater responsibility and as your years of service with the company increase, your benefits increase accordingly.

OR   You will note *that* your benefits increase as you assume greater responsibility and as your years of service with the company increase.

However, do not delete *that* from a **sentence** in which it is necessary for the **reader's** understanding.

CHANGE   Quarreling means trying to show the other person is in the wrong.

TO   Quarreling means trying to show *that* the other person is in the wrong.

CHANGE   Some engineers fail to recognize sufficiently the human beings
         who operate the equipment constitute an important safety
         system.
TO   Some engineers fail to recognize sufficiently *that* the human
     beings who operate the equipment constitute an important
     safety system.

See also **conciseness/wordiness.**

that/which/who

Who refers to persons, whereas *that* and *which* refer to animals and things.

EXAMPLES   John Brown, *who* is retiring tomorrow, has worked for the com-
           pany for twenty years.
           The test animal *that* was given the injection responded exactly as
           anticipated.
           The jet stream, *which* is approximately 8 miles above the earth,
           blows at an average of 64 miles per hour from the west.

That is often overused (see **that**). However, do not eliminate it if to do so
would cause **ambiguity** or problems with **pace.**

CHANGE   On the file specifications input to the compiler for any chained
         file, the user must ensure the number of sectors per main file
         section is a multiple of the number of sectors per bucket.
TO   On the file specifications input to the compiler for any chained
     file, the user must ensure *that* the number of sectors per main
     file section is a multiple of the number of sectors per bucket.

Which, rather than *that*, should be used with nonrestrictive **clauses**
(clauses that do not change the meaning of the basic **sentence**).

EXAMPLES   After John left the restaurant, *which* is one of the best in New
           York, he came directly to my office.
           A company *that* diversifies often succeeds. (restrictive)

See also **who/whom, relative pronoun,** and **restrictive and nonrestrictive elements.**

there/their/they're

There, their, and *they're* are often confused because they sound alike. *There*
is an **expletive** or an **adverb.**

EXAMPLES　*There* were more than 1,500 people at the conference. (expletive)
More than 1,500 people were *there*. (adverb)

Their is the possessive form of *they.*

EXAMPLE　Our employees are expected to keep *their* desks neat.

They're is a **contraction** of *they are.*

EXAMPLE　If *they're* right, we should change the design.

thesaurus

A thesaurus is a book of words with their **synonyms** and **antonyms,**
arranged by categories. Thoughtfully used, it can help you with **word
choice** during the **revision** phase of the writing process. However, this
variety of words may tempt you to choose inappropriate or obscure
synonyms; use a thesaurus to clarify your meaning, not to impress your
reader. Never use a word unless you are sure of its meanings; **connotations** of the word that might be unknown to you could mislead your
reader. See ''synonyms'' in **reference books** for names of thesauruses.
(See also **affectation** and **defining terms.**)

thus/thusly

Thus is an **adverb** meaning ''in this manner'' or ''therefore.'' The *-ly* in
thusly is superfluous and should be omitted.

CHANGE　The committee's work is done. *Thusly* we should have the report
by the twenty-fifth.

TO　The committee's work is done. *Thus* we should have the report by
the twenty-fifth.

'til/until

'Til is a nonstandard **spelling** of *until,* meaning ''up to the time of.'' Use
until.

CHANGE　We worked *'til* eight o'clock.

TO　We worked *until* eight o'clock.

titles

Since the title of your writing project is the first thing your **reader** sees, it should perform several functions: (1) it should indicate the specific **topic,** (2) it should suggest the project's **scope** and **objective,** (3) it should attract the reader's interest, and (4) it should possibly even reflect the **tone** of the writing.

Do not create titles that are too vague, such as "Technical Publications," or titles that are too broad, such as "A Survey of Technical Publications." Titles are often important to researchers because they may have only a title on which to decide whether to read a particular article.

to/too/two

To, too, and *two* are confused only because they sound alike. *To* is used as a **preposition** or to mark an **infinitive.**

EXAMPLES   Send the report *to* the district manager. (preposition)
I wish *to* go. (mark of the infinitive)

Too is an **adverb** meaning "excessively" or "also."

EXAMPLES   The price was *too* high. ("excessively")
I, *too,* thought it was high. ("also")

Two is a number.

EXAMPLE   Only *two* buildings have been built this fiscal year.

tone

In writing, tone is the attitude a writer takes toward his subject and his **readers.** Tone may be casual or serious, enthusiastic or skeptical, friendly or hostile. In technical writing, tone can be especially important; just as it can help you to gain your reader's sympathy, so it can rub him or her the wrong way. When you write you must always consider your objective; in addition, you must then maintain an appropriate tone.

Your tone will be set by many factors in your writing. Obviously, **formal writing style** will normally have a different tone than **informal writing style. Word choice** and sentence structure are especially important in establishing tone. The **title, introduction** or **opening,** even the

method of development you follow all contribute to the overall tone. For instance, a title such as "Some Observations on the Diminishing Oil Reserves in Wyoming" clearly sets a tone quite different from "What Happens When We've Pumped Wyoming Dry?"

The tone used in **reports** should normally be objective and impersonal, much like the first title in the previous **paragraph.** In some writing (such as **house organ articles**), the tone may be more personal and biased, as in the second title. The important thing is to make sure that your tone is the one best suited to your objective. To make sure that it is, always keep your reader in mind.

For business **correspondence,** tone is particularly crucial because the correspondence represents a direct communication between one person and another. Furthermore, good business letters establish a rapport between your organization and the public. A tone that is positive and considerate is essential to achieving this goal.

topics

On the job, the topic of a writing project is usually determined by need. In a college writing course, on the other hand, you may have to select your own topic. If you do, keep the following points in mind.

1. Select a topic that interests you.
2. Select a topic that you can **research** adequately with the facilities available to you.
3. Limit your topic so that its **scope** is small enough to handle within the time you are given. A topic like "Air Pollution," for example, would be too broad. On the other hand, keep the topic broad enough so that you will have enough to write about.
4. Select a topic for which you can make adequate **preparation** to ensure a good final **report.**

topic sentences

A topic sentence states the controlling idea of a **paragraph;** the rest of the paragraph supports and develops that statement with carefully related details. The topic sentence may appear anywhere in the paragraph—a fact that permits the writer variety in **style**—and a topic statement may also be more than one sentence if necessary.

The topic sentence is often the first sentence because it states the subject, as in the following example.

The arithmetic of searching for oil is stark. For all his scientific methods of detection, the only way the oilman can actually know for sure that there is oil in the ground is to drill a well. The average cost of drilling an oil well is over $300,000, and drilling a single well may cost over $8,000,000! And once the well is drilled, the odds against its containing any oil at all are 8 to 1! Even after a field has been discovered, one out of every four holes drilled in developing the field is a dry hole because of the uncertainty of defining the limits of the producing formation. The oilman can never know what Mark Twain once called ''the calm confidence of a Christian with four aces in his hand.''

On rare occasions, the topic sentence logically falls in the middle of a paragraph.

It is perhaps natural that psychologists should awaken only slowly to the possibility that behavioral processes may be directly observed, or that they should only gradually put the older statistical and theoretical techniques in their proper perspective. But it is time to insist that science does not progress by carefully designed steps called ''experiments,'' each of which has a well-defined beginning and end. *Science is a continuous and often a disorderly and accidental process.* We shall not do the young psychologist any favor if we agree to reconstruct our practices to fit the pattern demanded by current scientific methodology. What the statistician means by the design of experiments is design which yields the kind of data to which *his* techniques are applicable. He does not mean the behavior of the scientist in his laboratory devising research for his own immediate and possibly inscrutable purposes.

—B. F. Skinner, ''A Case History in Scientific Method,'' *The American Psychologist,* II, No. 5 (May 1956), p. 232.

Although the topic sentence is usually most effective early in the paragraph, a paragraph can effectively be made to lead up to the topic sentence; this is sometimes done to achieve **emphasis.** When a topic sentence concludes a paragraph, it can also serve as a summary or **conclusion,** based on the details that were designed to lead up to it.

Energy does far more than simply make our daily lives more comfortable and convenient. Suppose you wanted to stop—and reverse—the economic progress of this nation. What would be the surest and quickest way to do it? Find a way to cut off the nation's oil resources! Industrial plants would shut down; public utilities would stand idle; all forms of transportation would halt. The economy would plummet into the abyss of national economic ruin. *Our economy, in short, is energy-based.*

—*The Baker World* (Los Angeles: Baker Oil Tools, 1964), p. 5.

Occasionally, the controlling idea of a paragraph—what would normally be its topic sentence—is not explicitly stated but only implied. Nevertheless, if the paragraph has **unity,** a sentence stating the controlling idea *could* be written and all the sentences in the paragraph would be found to be related directly to it. However, people writing in technology and industry are well advised to make their topic sentences *explicit.*

tortuous/torturous

Tortuous means "marked by twisting and winding."

> EXAMPLE   The highway through the mountain was *tortuous.*

Torturous means "causing pain, as to punish."

> EXAMPLE   The prisoner's experience was *torturous.*

toward/towards

Both *toward* and *towards* are acceptable variant spellings of the **preposition** meaning "in the direction of"; *toward* is more common in the United States, and *towards* is more common in Great Britain.

> EXAMPLES   We walked *toward* the Empire State Building.
> Lady Diana walked *towards* the Thames.

transition

Transition is the means of achieving a smooth flow of ideas from **sentence** to sentence and from **paragraph** to paragraph. Transition is a two-way indicator of what has been said and what will be said; that is, it provides a means of linking ideas so that the relationship between them is clear. It may be accomplished by a word, a **phrase,** or even a paragraph. Without the guideposts of transition, **readers** can lose their way.

Certain words and phrases are inherently transitional. Consider the following terms and their functions.

Result: *therefore, as a result, consequently, thus, hence*
Example: *for example, for instance, specifically, as an illustration*
Comparison: *similarly, likewise*
Contrast: *but, yet, still, however, nevertheless, on the other hand*
Addition: *moreover, furthermore, also, too, besides, in addition*
Time: *now, later, meanwhile, since then, after that, before that time*
Sequence: *first, second, third, then, next, finally*

Within a paragraph, such transitional expressions make for clarity and smoothness in the movement from idea to idea. Conversely, the lack of transitional devices can make the going bumpy for the reader. Consider first the following passage, which lacks adequate transition.

EXAMPLE   People had always hoped to fly, but until 1903 it was only a dream. It was thought by some that human beings were not meant to fly. The Wright brothers launched the world's first heavier-than-air flying machine. The airplane has become a part of our everyday life.

Now read the same passage with words and phrases of transition added (in italics), and notice how much more smoothly the thoughts flow.

EXAMPLE   People had always hoped to fly, but until 1903 it was only a dream. *Before,* it was thought by some that human beings were not meant to fly. *In 1903* the Wright brothers launched the world's first heavier-than-air flying machine. *Now* the airplane has become a part of our everyday life.

And finally, read the same passage with stronger transition provided.

EXAMPLE   People had always hoped to fly, but until 1903 it was only a dream. *Before that time,* it was thought by some that human beings were not meant to fly. *However,* in 1903 the Wright brothers launched the world's first heavier-than-air flying machine. *Since then* the airplane has become a part of our everyday life.

If your **organization** and **outline** are good, your transitional needs will be less difficult to satisfy (although they must nonetheless *be* satisfied). Just as the outline is a road map for the writer, transition is a road map for the reader.

TRANSITION BETWEEN SENTENCES

In addition to using transitional words and phrases such as those shown above, the writer may achieve effective transition between sentences by repeating key words or ideas from preceding sentences, by using **pronouns** that refer to antecedents in previous sentences, and by using **parallel structure**—that is, by repeating the pattern of a phrase or **clause**. Consider the following short paragraph, in which all these means are employed. (A few examples are indicated by italics; however, to mark them all would be confusing.)

EXAMPLE   Representative of many American university towns is Millville. *This midwestern town,* formerly *a sleepy farming community,* is today the home of a large and bustling *academic community.* Attracting students from all over the Midwest, *this university* has grown very rapidly in the last ten years. *This same decade* has seen a physical expansion of *the campus.* The state, recognizing *this expansion,* has provided additional funds for the acquisition of land adjacent to the *university.* The *university* has become *Millville's* major industry, generating most of *the town's* income—and, of course, many of *its* problems, too.

Another device for achieving transition is enumeration:

EXAMPLE   The recommendation rests upon *three conditions. First,* the department staff must be expanded to a sufficient size to handle the increased work load. *Second,* sufficient time must be provided for the training of the new members of the staff. *Third,* a sufficient number of qualified applicants must be available within the allotted time.

TRANSITION BETWEEN PARAGRAPHS

All the means discussed above for achieving transition between sentences—and especially the device of repetition of key words or ideas—may also be effective for transition between paragraphs. For paragraphs, however, longer transitional elements are often required. One technique is to use an opening sentence that summarizes the preceding paragraph and then moves ahead to the business of the new paragraph.

EXAMPLE   One property of material considered for manufacturing processes is hardness. Hardness is the internal resistance of the material to the forcing apart or closing together of its molecules. Another property is ductility, the characteristic of material that permits it to be drawn into a wire. The smaller the diameter of the wire into which the material can be drawn, the greater the ductility. Material also may possess malleability, the property that makes it capable of being rolled or hammered into thin sheets of various shapes. Engineers, in selecting materials to employ in manufacturing, must consider these properties before deciding on the most desirable for use in production.

The requirements of hardness, ductility, and malleability accounts for the high cost of such materials . . .

Ask a question at the end of one paragraph and answer it at the beginning of the next.

EXAMPLE   Automation has become an ugly word in the American vocabulary because it has at times displaced some jobs. But the all-important fact that is so often overlooked is that it invariably creates many more jobs that it eliminates. (The vast number of people employed in the great American automobile industry as compared with the number of people that had been employed in the harness-and-carriage-making business is a classic example.) Almost always, the jobs that have been eliminated by automation have been menial, unskilled jobs, and those who have been displaced have been forced to increase their skills, which resulted in better and higher-paying jobs for them. *In view of these facts, is automation really bad?*

Certainly automation has made our country the most wealthy and technologically advanced nation the world has even known . . .

A purely transitional paragraph may be inserted to aid readability.

EXAMPLE   . . . that marred the progress of the company.

There were two other setbacks to the company's fortunes that year which also marked the turning of the tide: the loss of many skilled workers through the Early Retirement Program and the intensification of the devastating rate of inflation.

The Early Retirement Program . . .

CHECKING FOR TRANSITION DURING REVISION

Check for **unity** and **coherence** to determine whether transition is effective. If a paragraph has unity and coherence, the sentences and ideas are tied together and contribute directly to the subject of the paragraph. Look for places where transition is missing and provide it. Look for places where it is weak and strengthen it.

transitive verbs

A transitive verb is one that requires a **direct object,** with or without an **indirect object,** to complete its meaning.

The direct object is normally a **noun** or **pronoun** that names the thing being acted upon by the **verb.** A direct object answers the questions "what?" or "whom?"

> EXAMPLES   Ms. Jones wrote the *memorandum.*
> Ms. Jones hired *John.*

An indirect object answers the questions "*to* whom or what?" or "*for* whom or what?" That is, it normally names the person or thing that receives the direct object.

> EXAMPLES   John gave *Ms. Jones* the printout. (*Ms. Jones,* the indirect object, tells "to whom"; *gave* is the transitive verb, and *printout* is the direct object.)
> Ms. Jones gave *the printout* careful study. (*The printout,* the indirect object, tells "to what"; *gave* is the transitive verb, and *study* is the direct object.)

See also **complements.**

transmittal letters

The transmittal letter, also known as a cover letter, identifies an item being sent, the person to whom it is being sent, and the reason for sending it. A transmittal letter provides a permanent record of the transmittal for both the writer and the **reader.**

Keep your remarks brief in a transmittal letter. Open with a **paragraph** explaining what is being sent and why. In an optional second paragraph, you might want to include a summary of any information you're sending. A letter accompanying a **proposal,** for example, might point out any specific sections in the proposal of particular interest to the reader. The letter could then go on to present evidence that the writer's firm is the best one to do the job. This paragraph could also mention any specific conditions under which the material was prepared, such as limitations of time or budget. The closing paragraph should contain any acknowledgments, offer additional assistance, or express the hope that the material will fulfill its purpose.

The first of the following examples (Letter 21) is brief and to the point. The second example (Letter 22) is a bit more detailed, for it touches on the manner in which the information was gathered.

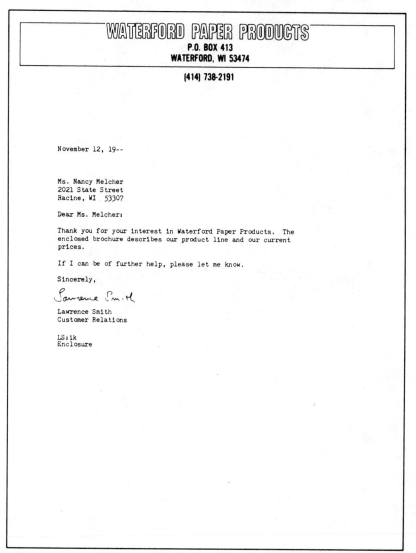

WATERFORD PAPER PRODUCTS
P.O. BOX 413
WATERFORD, WI 53474

(414) 738-2191

November 12, 19--

Ms. Nancy Melcher
2021 State Street
Racine, WI 53307

Dear Ms. Melcher:

Thank you for your interest in Waterford Paper Products. The
enclosed brochure describes our product line and our current
prices.

If I can be of further help, please let me know.

Sincerely,

Lawrence Smith

Lawrence Smith
Customer Relations

LS:lk
Enclosure

Letter 21

WATERFORD PAPER PRODUCTS
P.O. BOX 413
WATERFORD, WI 53474

(414) 738-2191

January 16, 19--

Mr. Roger Hammersmith
Ecology Systems, Inc.
1015 Clarke Street
Chicago, IL 60615

Dear Mr. Hammersmith:

Enclosed is the report estimating our power consumption for the year as
requested by John Brenan, Vice President, on September 4.

The report is a result of several meetings with the Manager of Plant
Operations and her staff and an extensive survey of all our employees.
The survey was delayed by the temporary layoff of key personnel in
Building "A" from October 1 to December 5. We believe, however, that
the report will provide the information you need to furnish us with a
cost estimate for the installation of your Mark II Energy Saving System.

We would like to thank Diana Biel of ESI for her assistance in preparing
the survey. If you need any more information, please let me know.

Sincerely,

James G. Evans
New Projects Office

JGE/fst
Enclosure

Letter 22

trip reports

Many companies require or encourage reports of the business trips their employees take. A trip report not only provides a permanent record of a business trip and its accomplishments but also enables many employees to benefit from the information one employee has gained.

A trip report should normally be in the **format** of a **memorandum,** addressed to your immediate superior. On the subject line give the destination and dates of the trip. The body of the **report** will explain why you made the trip, who you visited, and what you accomplished. The report should devote a brief section to each major event and may include a **head** for each section (you needn't give equal space to each event, but instead, elaborate on the more important events). Follow the body of the report with any appropriate **conclusions** and recommendations. The following trip report is typical.

MEMORANDUM

February 6, 19--

TO: J. L. Watson
FROM: T. R. Santow
SUBJECT: Trip to Milwaukee the Week of January 17, 19--.

The purpose of this trip was to gather information on (1) operating the new disc in a single-spindle environment, (2) system scheduling, (3) data base, (4) indexed sequential, (5) the small inquiry system, (6) random filing system, and (7) the roll-in/roll-out software.

DISC

I talked with Kevin Hutch, Janet Martin, and Gerald Mac-Dougal about the need for a separate card control string to bring the new disc up in a single-spindle mode. I also discussed the possibility of altering the software so that it will search for a program on the current system disc and then on the library disc in the single-spindle mode, and the possibility of identifying a library disc on the PAL printout.

I observed the disc in operation when Ralph Stevens ran some test routines. It appeared to run faster than the same routines run on the old disc. It soon became apparent, as I watched the tests being run, that the speed of the actuator and the rotational speed of the disc make the operating environment considerably different from what anyone had anticipated. Timing tests will have to be run on the relationship between file positions and throughput before we can write our file placement document.

SYSTEM SCHEDULING

I talked to Bill Wilson about S7 System Scheduling. At the conclusion of our discussion, he agreed to send someone to Dayton to work with us when all the necessary information is available. I also talked to Bill about the interim publications/ pilot site relationship; he felt that we should have our interim publications, as well as the writer, as the key pilot sites so that documentation can be tested as well as the software and the hardware.

DATA BASE

I discussed Data Base with Wayne Sewell and Ellen Golden to get a feel for the current status of the project, as well as for its progress, to determine when we might expect to become involved; however, the project does not seem to be far enough along yet for us to be concerned.

INDEXED SEQUENTIAL

I discussed Indexed Sequential with Margaret Miller; she had the information I needed, but informed me that Indexed Sequential will not be her responsibility in the future. On the basis of that, I tried to determine new assignments of responsibility in Milwaukee. Things are shifting so fast that it is difficult to keep up.

SMALL INQUIRY SYSTEM AND RANDOM FILING SYSTEM

I discussed the Small Inquiry System with Wally Bevins. He feels that the software is firm and that we should begin work on it at our earliest opportunity. I also discussed the Random Filing System with him; we must add the new disc to this publication and do some additional updating.

ROLL-IN/ROLL-OUT SOFTWARE

I talked to Janet Martin about Roll-In/Roll-Out specifications and found that they are as complete as we can expect them to be. She feels that the specifications are firm and that we should begin work on this project soon.

trite language

Trite language is made up of words, **phrases,** or ideas that have been used so often that they are stale.

CHANGE   *It may interest you to know* that all the folks in the branch office are *hale and hearty.* I should finish my report *quick as a wink,* and we should *clean up* on it.

TO   Everyone here at the branch office is well. I should finish my project within a week, and I'm sure it will prove profitable for us.

Trite language shows that the writer either is thoughtless in his **word choice** or is not thinking carefully about his topic; he is relying solely on what others have thought and said before him. (See also **affectation.**)

trouble reports

The trouble report is used to report an accident, an equipment failure, an unplanned work stoppage, and so on. The **report** enables the management of an organization to determine the cause of the problem and to make any changes necessary to prevent its recurrence. The trouble report normally follows a simple **format** (and is often a **memorandum**) since it is an internal document and is not large enough in either size or **scope** to require the format of a **formal report.**

In the subject line of the memorandum, state the precise problem you are reporting. Then begin your description of the problem in the body of your report. What happened? Where did it occur? When did it occur? Was anybody hurt? Was there any property damage? Was there a work stoppage? Since insurance claims, worker's compensation awards, and, in some instances, lawsuits may hinge on the information contained in a trouble report, be sure to include precise data on times, dates, locations, treatment of injuries, names of any witnesses, and any other crucial information. Give a detailed analysis of what caused the problem. Be thorough and accurate in your analysis, and support any judgments or conclusions with facts. Be careful about your **tone;** avoid any condemnation or blame. If you speculate about the cause of the problem, make it very clear to your **reader** that you are speculating. In your **conclusion,** state what has been done, what is being done, or what will be done to correct the conditions that led to the problem. This may include training in safety practices, better or improved equipment, protective clothing (for example shoes or goggles), and so on.

The following report about an accident involving personal injury was written by the foreman of a group of punch press operators for the plant's safety officer, at the safety officer's request. Since there were no witnesses, the foreman obtained the information for the report by talking to the plant nurse, to hospital personnel, and to the victim—and by inspecting the equipment used by the victim at the time of the accident.

MEMORANDUM

To: James K. Arburg, Safety Officer
From: Lawrence T. Baker, Foreman of Section A-40
Date: November 30, 19--

Subject: Personal-Injury Accident in Section A-40
          October 10, 19--

On October 10, 19--, at 10:15 P.M., Jim Hollander, operating punch press #16, accidentally brushed the knee switch of his punch press with his right knee as he swung a metal sheet over the punching surface. The switch activated the punching unit, which severed Hollander's left thumb between the first and second joints as his hand passed through the punch station. While an ambulance was being summoned, Margaret Wilson, R.N., administered first aid at the plant dispensary. There were no witnesses to the accident.

The ambulance arrived from Mercy Hospital at 10:45 P.M., and Hollander was admitted to the emergency room at the hospital at 11:00 P.M. He was treated and kept overnight for observation, then released the next morning.

Hollander returned to work one week later, on October 17. He has been given temporary duties in the tool room until his injury heals.

<u>Conclusions About the Cause of the Accident</u>

The Maxwell punch press on which Hollander was working has two switches, a hand switch and a knee switch, and <u>both</u> must be pressed to activate the punch mechanism. The hand switch must be pressed first, and then the knee switch, to trip the punch mechanism. The purpose of the knee switch is to leave the operator's hands free to hold the panel being punched. The hand switch, in contrast, is a safety feature. Because the knee switch cannot activate the press until the hand switch has been pressed, the operator cannot trip the punching mechanism by touching the knee switch accidentally.

Inspection of the punch press that Hollander was operating at the time of the accident made it clear that Hollander had taped the hand switch of his machine in the ON position, effectively eliminating its safety function. He could then pick up a panel, swing it onto the machine's punching surface, press the knee switch, stack the newly punched panel, and grab the next unpunched panel, all in one continuous motion--eliminating the need to let go of the panel, after placing it

on the punching surface, in order to press the hand
switch.
    To prevent a recurrence of this accident, I
have conducted a brief safety session with all punch
press operators, at which I described Hollander's
experience and cautioned them against tampering with
the safety features of their machines.

try and

The **phrase** *try and* is colloquial for *try to*. Unless you are writing a casual
personal letter, it is better to use *try to*.

CHANGE   Please *try and* finish the report on time.

TO   Please *try to* finish the report on time.

U

unity

Unity is singleness of purpose and treatment, the cohesive element that holds a piece of writing together; it means that everything in an article or paper is essentially about one thing or idea.

To achieve unity, the writer must select one **topic** and then treat it with singleness of purpose—without digressing into unrelated paths. The prime contributors to unity are a good outline and effective **transition.** Transition dovetails **sentences** and **paragraphs** together like the joints of a well-made drawer. Notice, for example, how neatly the sentences in the following paragraph are made to fit together by the italicized words and **phrases** of transition.

EXAMPLE   Any company which operates internationally today faces a host of difficulties. Inflation is worldwide. Most countries are struggling with other economic problems *as well. In addition,* many monetary uncertainties and growing economic nationalism are working against multinational companies. *Yet* ample business is available in most developed countries if you have the right products, services, and marketing organization. To maintain the growth Data Corporation has achieved overseas, we recently restructured our international operations into four major trading areas. *This* reorganization will improve the services and support which the corporation can provide to its subsidiaries around the world. *At the same time,* the reorganization establishes firm management control, ensuring consistent policies around the world. *So* you might say the problems of doing business abroad will be more difficult this year, but we are better organized to meet those problems.

The logical sequence provided by a good outline is essential to achieving unity. An outline enables the writer to lay out the most direct route from **introduction** to **conclusion** without digressing into side issues that are not related, or that are only loosely related, to the subject. Without establishing and following such a direct route, the writer's work is not likely to be unified.

up

Adding the word *up* to **verbs** often creates a redundant **phrase.**

CHANGE   Next open *up* the exhaust valve.
TO   Next open the exhaust valve.

usage

Usage describes the choices we make among the various words and constructions available in our language. The line between **standard English** and **nonstandard English,** or between formal and informal English, is determined by these choices. Your guideline in any situation involving such choices should be appropriateness: is the word or expression you use appropriate to your **reader** and subject? When it is, you are practicing good usage.

This book has been designed to help you sort out the appropriate from the inappropriate: just look up the item in question in the index or consult the List of Usage Entries at the front of the book. A good **dictionary** is also an invaluable aid in your selection of the right word.

utilize

Utilize should not be used as a **long variant** of *use,* which is the general word for "employ for some purpose."

CHANGE   You can *utilize* the fourth elevator to reach the 50th floor.
TO   You can *use* the fourth elevator to reach the 50th floor.

V

vague words

A vague word is one that is imprecise in the context in which it is used. Some words encompass such a broad range of meanings that there is no focus for their definition. Words such as *real, nice, important, good, bad, thing,* and *fine* are often called "omnibus words" because they can mean everything to everybody. In speech we sometimes use words that are less than precise, but our vocal inflections and the context of our conversation make their meanings clear. Since writing cannot rely upon vocal inflections, avoid the use of vague words. Be concrete and specific. (See also **abstract words/concrete words.**)

CHANGE It was a *meaningful* meeting, and we got *a lot* done.
TO The meeting resolved three questions: pay scales, fringe benefits, and work loads.

verbs

A verb is a word, or a group of words, that describes an action (The antelope *bolted* at the sight of the hunters), states the way in which something or someone is affected by an action (He *was saddened* by the death of his friend), or affirms a state of existence (He *is* a wealthy man now).

TYPES OF VERBS

Verbs may be described as being either **transitive verbs** or **intransitive verbs.**

Transitive Verbs. A transitive verb is a verb that requires a **direct object** to complete its meaning.

EXAMPLES They *laid* the foundation on October 24. ("foundation" is the direct object of the transitive verb *laid*)
George Anderson *wrote* the treasurer a letter. ("letter" is the direct object of the transitive verb *wrote*)

Intransitive Verbs. An intransitive verb is a verb that does not require an object to complete its meaning. It is able to make a full assertion about the **subject** without assistance (although it may have modifiers).

EXAMPLES   The water *boiled.*
The water *boiled* rapidly.
The engine *ran.*
The engine *ran* smoothly and quietly.

Although intransitive verbs do not have an **object,** certain intransitive verbs may take a **complement.** These verbs are called **linking verbs** because they link the complement to the subject. When the complement is a **noun** (or **pronoun**), it *refers* to the same person or thing as the noun (or pronoun) that is the subject.

EXAMPLES   The conference table *is* an antique.
Mary *remains* the director.

When the complement is an **adjective,** it *modifies* the subject.

EXAMPLES   The study *was* thorough.
The report *seems* complete.

Such intransitive verbs as *be, become, seem,* and *appear* are almost always linking verbs. A number of others, such as *look, sound, taste, smell,* and *feel,* may function either as linking verbs or as simple intransitive verbs. If you are unsure about whether one of these verbs is a linking verb, try substituting *seem;* if the sentence still makes sense, the verb is probably a linking verb.

EXAMPLES   Their antennae *feel* delicately. (simple intransitive verb)
Their antennae *feel* delicate. (linking verb)

FORMS OF VERBS

By form, verbs may be described as being either finite or infinite.
   Finite Verbs. A **finite verb** is the main verb of a **clause** or **sentence.** It makes an assertion about its subject and it can serve as the only verb in its clause or sentence. Finite verbs may be either transitive or intransitive (including linking) verbs. They are subject to changes in form to reflect **person** (I *see,* he *sees*), **tense** (I *go,* I *went*), and **number** (he *writes,* they *write*).

EXAMPLE   The telephone *rang* and the secretary *answered* it.

A **helping verb** (sometimes called an *auxiliary verb*) is used in a **verb phrase** to help indicate **mood,** tense, and **voice.**

EXAMPLES The work *had* begun.
I *am* going.
I *was* going.
I *will* go.
I *should have* gone.
I *must* go.

The most commonly used helping verbs are the various forms of *have* (*has, had*), *be* (*is, are, was*, etc.), *do* (*did, does*), and *can, may, might, must, shall, will, would, should,* and *could.* **Phrases** that function as helping verbs are often made up of combinations with the sign of the infinitive, *to:* for example, *am going to* and *is about to* (compare *will*), *has to* (compare *must*), and *ought to* (compare *should*).

The helping verb always precedes the main verb, although other words may intervene.

EXAMPLE Machines *will* never completely *replace* people.

Nonfinite Verbs. Nonfinite verbs are **verbals,** which, although they are derived from verbs, actually function as nouns, adjectives, or **adverbs.** When the *-ing* form of a verb is used as a noun, it is called a **gerund.**

EXAMPLE *Seeing* is *believing.*

An **infinitive,** which is the root form of a verb (usually preceded by *to*), can be used as a noun, an adverb, or an adjective.

EXAMPLES He hates *to complain.* (noun, direct object of *hates*)
The valve closes *to stop* the flow. (adverb, modifies *closes*)
This is the proposal *to select.* (adjective, modifies *proposal*)

A **participle** is a verb form used as an adjective.

EXAMPLES His *closing* statement was very *convincing.*
The *rejected* proposal was ours.

PROPERTIES OF VERBS

Person is the grammatical term for the form of a **personal pronoun** that indicates whether the pronoun refers to the speaker, the person spoken to, or the person (or thing) spoken about. Verbs change their forms to agree in person with their subjects.

EXAMPLES I *see* (first person) a yellow tint, but he *sees* (third person) a yellow-green hue.
I *am* (first person) convinced, and you *are* (second person) convinced; but unfortunately he *is* (third person) not convinced.

Voice refers to the two forms of a verb that indicate whether the subject of the verb acts or receives the action. If the subject of the verb acts, the verb is in the active voice; if it receives the action, the verb is in the passive voice.

EXAMPLES   The aerosol bomb *propels* the liquid as a mist. (active)
The liquid *is propelled* as a mist by the aerosol bomb. (passive)

In your writing, the active voice provides force and momentum, while the passive voice lacks these qualities. The reason is not difficult to find. In the active voice, the verb plays the key role in identifying what the subject is doing. The passive voice, on the other hand, consists of a form of the verb *to be* and a past participle of another verb. In the passive voice, the **emphasis** is on what is being done to the subject.

EXAMPLES   Things are seen by the normal human eye in three dimensions: length, width, and depth. (*Things* take precedence over the eye's function.)
The normal human eye sees things in three dimensions: length, width, and depth. (Here the eye's function—which is what the sentence is about—receives the emphasis.)

Number refers to the two forms of a verb that indicate whether the subject of a verb is singular or plural.

EXAMPLES   The machine *was* in good operating condition. (singular)
The machines *were* in good operating condition. (plural)

Tense refers to verb forms that indicate time distinctions. There are six tenses: present, past, future, present perfect, past perfect, and future perfect. All verb tenses are derived from the three principal parts of the verb: the *infinitive* (without *to,* which normally precedes it); the *past* form; and the *past participle.* Following are the six tenses of *write* (infinitive), whose other two principal parts are *wrote* (past form) and *written* (past participle).

EXAMPLES   I *write* (present, based on infinitive)
I *wrote* (past, based on past form)
I *will write* (future, based on infinitive)
I *have written* (present perfect, based on past participle)
I *had written* (past perfect, based on past participle)
I *will have written* (future perfect, based on past participle)

Note that *write* is an irregular verb; that is, its past and past participle forms are produced by internal changes: *write, wrote, written.* Regular verbs, on the other hand, form the past and past participle by the addition of the suffixes *-ed, -d,* or *-t: talk, talked; believe, believed; spend, spent.*

The tenses of regular verbs, like those of irregular verbs, are derived from the three principal parts.

EXAMPLES   I *believe* (present, based on infinitive)
I *believed* (past, based on past form)
I *will believe* (future, based on infinitive)
I *have believed* (present perfect, based on past participle)
I *had believed* (past perfect, based on past participle)
I *will have believed* (future perfect, based on past participle)

Each of the six tenses also has a progressive form. The progressive form is created by adding the appropriate tense of *be* to the present participle (*-ing*) form of the verb.

EXAMPLES   I *am writing.* (present progressive)
I *was writing.* (past progressive)
I *will be writing.* (future progressive)
I *have been writing.* (present perfect progressive)
I *had been writing.* (past perfect progressive)
I *will have been writing.* (future perfect progressive)

CONJUGATION OF VERBS

The conjugation of a verb arranges all forms of the verb so that the differences caused by the changing of the tense, number, person, and voice are readily apparent. Following is a conjugation of the verb *drive*.

Tense	*Number*	*Person*	*Active Voice*	*Passive Voice*
Present	Singular	1st	I drive	I am driven
		2nd	You drive	You are driven
		3rd	He drives	He is driven
	Plural	1st	We drive	We are driven
		2nd	You drive	You are driven
		3rd	They drive	They are driven
Progressive Present	Singular	1st	I am driving	I am being driven
		2nd	You are driving	You are being driven
		3rd	He is driving	He is being driven
	Plural	1st	We are driving	We are being driven
		2nd	You are driving	You are being driven
		3rd	They are driving	They are being driven
Past	Singular	1st	I drove	I was driven
		2nd	You drove	You were driven
		3rd	He drove	He was driven
	Plural	1st	We drove	We were driven
		2nd	You drove	You were driven
		3rd	They drove	They were driven

Tense	Number	Person	Active Voice	Passive Voice
Progressive Past	Singular	1st	I was driving	I was being driven
		2nd	You were driving	You were being driven
		3rd	He was driving	He was being driven
	Plural	1st	We were driving	We were being driven
		2nd	You were driving	You were being driven
		3rd	They were driving	They were being driven
Future	Singular	1st	I will drive	I will be driven
		2nd	You will drive	You will be driven
		3rd	He will drive	He will be driven
	Plural	1st	We will drive	We will be driven
		2nd	You will drive	You will be driven
		3rd	They will drive	They will be driven
Progressive Future	Singular	1st	I will be driving	I will have been driven
		2nd	You will be driving	You will have been driven
		3rd	He will be driving	He will have been driven
	Plural	1st	We will be driving	We will have been driven
		2nd	You will be driving	You will have been driven
		3rd	They will be driving	They will have been driven
Present Perfect	Singular	1st	I have driven	I have been driven
		2nd	You have driven	You have been driven
		3rd	He has driven	He has been driven
	Plural	1st	We have driven	We have been driven
		2nd	You have driven	You have been driven
		3rd	They have driven	They have been driven
Past Perfect	Singular	1st	I had driven	I had been driven
		2nd	You had driven	You had been driven
		3rd	He had driven	He had been driven
	Plural	1st	We had driven	We had been driven
		2nd	You had driven	You had been driven
		3rd	They had driven	They had been driven
Future Perfect	Singular	1st	I will have driven	I will have been driven
		2nd	You will have driven	You will have been driven
		3rd	He will have driven	He will have been driven
	Plural	1st	We will have driven	We will have been driven
		2nd	You will have driven	You will have been driven
		3rd	They will have driven	They will have been driven

verb phrases

A verb phrase is a group of words that functions as a single **verb.** It consists of a main verb preceded by one or more **helping verbs.**

EXAMPLE   He *is* (helping verb) *working* (main verb) hard this summer.

The main verb is always the last verb in a verb phrase.

EXAMPLES   You *will file* your tax return on time if you begin early.
You *will have filed* your tax return on time if you begin early.
You *are* not *filing* your tax return too early if you begin now.

Questions often begin with a verb phrase.

EXAMPLE   *Will* he *audit* their account soon?

Words can appear between the helping verb and the main verb of a verb phrase.

EXAMPLE   He *is* always *working.*

The **adverb** *not* may be appended to a helping verb in a verb phrase.

EXAMPLES   He *cannot work* today.
He *did not work* today.

verbals

Verbals, which are derived from **verbs,** function as **nouns, adjectives,** and **adverbs.** There are three types of verbals: gerunds, infinitives, and participles.

When the *-ing* form of a verb is used as a noun, it is called a **gerund.**

EXAMPLE   *Seeing* is *believing.*

An **infinitive** is the root form of a verb, usually preceded by *to (to analyze, to build).* Infinitives can be used as nouns, adjectives, or adverbs.

EXAMPLES   He likes *to jog.* (noun)
That is the main problem *to be solved.* (adjective)
The valve acts *to control* the flow of liquid. (adverb)

When a verb is used as an adjective, it is called a **participle;** present participles commonly end in *-ing,* past participles in *-ed, -en, -d,* or *-t.*

EXAMPLES   A *burning* candle consumes stale tobacco smoke.
The *published* report contains up-to-date information.

very

The temptation to overuse **intensifiers** like *very* is great. Evaluate your use of them carefully. Where you use them, go on to clarify their meaning.

EXAMPLE   Bicycle manufacturers had a *very* good year: sales across the country were up 43 % over the previous year.

In many sentences, however, the word can simply be deleted.

CHANGE   The board was *very* angry about the newspaper report.
TO   The board was angry about the newspaper report.

via

Via is Latin for "by way of."

EXAMPLE   The equipment is being shipped to Los Angeles *via* Chicago.

The term should be used only in routing instructions.

CHANGE   His project was funded *via* the recent legislation.
TO   His project was funded *through* the recent legislation.
OR   His project was funded *as the result of* the recent legislation.

vogue words

Vogue words are words that suddenly become popular and, because of an intense period of overuse, lose their freshness and preciseness. They may become popular through their association with science, technology, or even sports. We include them in our vocabulary because they seem to give force and vitality to our language. Ordinarily, this language sounds pretentious in our day-to-day writing.

EXAMPLES   super, interface, bottom line input, mode, variable, ballgame, parameter, communication, feedback, environment, and many words ending in *-wise.*

Obviously, some of these terms are appropriate in the right context. It is when they are used outside that context that imprecision becomes a problem. This book discusses some vogue terms individually; use the index to find the word in question. For those not included, check a current **dictionary.**

voice

In grammar, voice indicates the relation of the subject to the action of the verb. When the **verb** is in the active voice, the **subject** acts; when it is in the passive voice, the subject is acted upon.

EXAMPLES   David Cohen *wrote* the advertising copy. (active)
The advertising copy *was written* by David Cohen. (passive)

Both of the sentences say the same thing, but each has a different **emphasis**: In the first sentence emphasis is on the subject, David Cohen, whereas in the second sentence the focus is on the object, the advertising copy. Notice how much stronger and more forceful the active sentence is.

One of the rules of good writing is always to use the active voice unless there is good reason to use the passive. Because they are wordy and convoluted, passive sentences are hard for the **reader** to understand. Notice how much clearer the following passive sentence becomes when it is rewritten in the active voice.

CHANGE   Things *are seen* by the normal human eye in three dimensions: length, width, and depth.
TO   The human eye *sees* things in three dimensions: length, width, and depth.

One difficulty with passive sentences is that they can bury the subject, or performer of the action, in **expletives** and prepositional phrases.

CHANGE   It *was reported* by Engineering that the new relay is defective.
TO   Engineering *reported* that the new relay is defective.

Sometimes writers using the passive voice even forget to name the performer—information that might be missed.

CHANGE   The problem *was discovered* yesterday.
TO   The Engineering Department *discovered* the problem yesterday.

Very often the passive voice can be just plain confusing, especially when used in **instructions.**

CHANGE   Plates B and C should be marked for revision. (Are they already marked?)
TO   You *should mark* plates B and C for revision.
OR   *Mark* plates B and C for revision.

Another problem that sometimes occurs with the passive voice is **dangling modifiers.**

CHANGE Hurrying to complete the work, the wires *were connected* improperly. (Who was hurrying, the wires?)

TO Hurrying to complete the work, the technician improperly *connected* the wires. (Here, "hurrying to complete the work" properly modifies "technician.")

There are, however, certain instances when the passive voice is effective or even necessary. When the performer of the action is either unknown or unimportant, use the passive voice.

EXAMPLES The copper mine *was discovered* in 1929.
Fifty-six barrels *were processed* in two hours.

Or, when the performer of the action is less important than the receiver of that action, the passive voice is sometimes more appropriate.

EXAMPLE Ann Bryant *was presented* with an award by the president.

Whether you use the passive or the active voice, however, be careful not to *shift* voices in a sentence.

CHANGE Ms. McDonald *corrected* the malfunction as soon as it *was identified* by the technician.

TO Ms. McDonald *corrected* the malfunction as soon as the technician *identified* it.

W

wait for/wait on

Wait on should be restricted in writing to the activities of waitresses and waiters—otherwise, use *wait for.* (See also **idioms.**)

EXAMPLES   Be sure to *wait for* Ms. Sturgess to finish reading the report before asking her to make a decision.
Joe's feet ached from *waiting on* tables at the diner.

when and if

When and if (or *if and when*) is a colloquial expression that should not be used in writing.

CHANGE   *When and if* your new position is approved, I will see that you receive adequate staff help.

TO   *If* your new position is approved, I will see that you receive adequate staff help.

OR   *When* your new position is approved, I will see that you receive adequate staff help.

where . . . at

In **phrases** using the *where . . . at* construction, *at* is unnecessary and should be omitted.

CHANGE   *Where* is his office *at?*

TO   *Where* is his office?

where/that

Do not substitute *where* for *that* to anticipate an idea or fact to follow.

CHANGE   I read in the IEEE journal *where* modules will be used in our process.

TO   I read in the IEEE journal *that* modules will be used in our process.

while

While, meaning "during an interval of time," is sometimes substituted for **connectives** like *and, but, although,* and *whereas.* Because it has its own meaning in addition to this use as a connective, *while* often causes **ambiguity.**

> CHANGE   John Evans is sales manager, *while* Joan Thomas is in charge of research.
>
> TO   John Evans is sales manager, *and* Joan Thomas in in charge of research.

Do not use *while* to mean *although* or *whereas.*

> CHANGE   *While* Ryan Patterson wants the job of chief engineer, he has not yet asked for it.
>
> TO   *Although* Ryan Patterson wants the job of chief engineer, he has not yet asked for it.

Restrict *while* to its meaning of "during the time that."

> EXAMPLE   I'll have to catch up on my reading *while* I am on vacation.

who/whom

Writers often have difficulty with the choice of *who* or *whom. Who* is the subjective **case** form, *whom* is the objective case form, and *whose* is the possessive case form. When in doubt about which form to use, try substituting a **personal pronoun** to see which one fits. If *he* or *they* fits, use *who.*

> EXAMPLE   *Who* is the congressman from the 45th district?
> *He* is the congressman from the 45th district.

If *him* or *them* fits, use *whom.*

> EXAMPLE   It depended upon *them.*
> It depended upon *whom?*

It is becoming common to use *who* for the objective case when it begins a sentence, although some still object to such an ungrammatical construction. The best advice is to know your **reader.**

The choice between *who* and *whom* is especially complex in sentences with interjected **clauses.**

CHANGE   These are the men *whom* you thought were the architects.
TO   These are the men *who* you thought were the architects.

The interjected clause "you thought" tends to attract the form *whom,* a form that appears logical as the object of the verb form *thought.* In fact, *who* is the correct form, since it is the subject of the verb form *were.* Note that the interjected clause "you thought" can be omitted without disturbing the construction of the clause.

EXAMPLE   These are the men *who* were the architects.

whose/of which

Whose should normally be used with persons; *of which* should normally be used with inanimate objects.

EXAMPLES   The man *whose* car had been towed away was angry.
The mantle clock, the parts *of which* work perfectly, is over one hundred years old.

If these uses cause a **sentence** to sound awkward, however, *whose* may be used with inanimate objects.

EXAMPLE   There are added fields, for example, *whose* totals should never be zero.

-wise

Although the **suffix** *-wise* often seems to provide a tempting shortcut, it leads more often to inept than to economical expression. It is better to rephrase the **sentence.**

CHANGE   Our department rates high efficiency-wise.
TO   Our department has a high efficiency rating.

The *-wise* suffix is appropriate, however, to **instructions** that indicate certain space or directional requirements.

EXAMPLES   Fold the paper *lengthwise.*
Turn the adjustment screw half a turn *clockwise.*

word choice

As Mark Twain once said, "The difference between the right word and almost the right word is the difference between 'lightning' and 'lightning bug.' " The most important goal in choosing the right word in technical writing is the preciseness implied by Twain's comment. **Vague words** and **abstract words** defeat preciseness because they do not convey the writer's meaning directly and clearly. Vague words are imprecise because they can mean many different things.

> CHANGE   It was a *meaningful* meeting.
>   TO   The meeting helped both sides understand each other's position.

In the first sentence, *meaningful* ironically conveys no meaning at all. See how the revised sentence says specifically what made the meeting meaningful. Although abstract words may at times be appropriate to your **topic,** their unnecessary use creates dry and lifeless writing.

> ABSTRACT   work, fast, food
> CONCRETE   sawing, 110 m.p.h., steak

Being aware of the **connotation** and denotation of words will help you anticipate the **reader's** reaction to the words you choose. Connotation is the suggested or implied meaning of a word beyond its dictionary definition. Denotation is the literal, or primary, dictionary meaning of a word.

Understanding **antonyms** and **synonyms** will increase your ability to choose the proper word. Antonyms are words with nearly the opposite meaning *(fresh/stale),* and synonyms are words with nearly the same meaning *(notorious/infamous).*

Malapropisms and **trite language** are likely to irritate your readers, either by confusing them or by boring them. Malapropisms are words that sound like the intended word but have different, sometimes even humorous meanings.

> CHANGE   The printer and the card reader are in *convention* with each other
>   for the processor's attention.
>   TO   The printer and the card reader are in *contention* with each other
>   for the processor's attention.

Trite language consists of stale and worn phrases, frequently **clichés.**

> CHANGE   We will finish the project *quick as a flash.*
>   TO   We will finish the project *quickly.*

Avoid the use of **jargon** unless you are certain that all of your readers understand the terms you use from this specialized vocabulary. Avoid choosing words with the objective of impressing your reader, which is called **affectation;** also avoid the use of **long variants,** which are elongated forms of words clearly used only to impress (*utilize* for *use, telephonic communication* for *telephone call, analyzation* for *analysis*). Using a **euphemism** (an inoffensive substitute for a word that is distasteful or offensive) may help you avoid embarrassment, but overuse of euphemisms becomes affectation. Using more words than necessary (*in the neighborhood of* for *about, for the reason that* for *because, in the event that* for *if*) is certain to interfere with **clarity.**

Understanding the use of **clipped forms of words, compound words, blend words, idioms,** and **foreign words in English** will help you find the appropriate word for a specific use. Clipped forms of words are created by dropping the beginning or ending of a word (*phone* for *telephone, ad* for *advertisement*). Compound words are made from two or more words that are either hyphenated or written as one word (*mother-in-law, nevertheless*). Blend words are created by using two words to produce a third word (*smoke* + *fog* = *smog*).

A key to choosing the correct and precise word is to keep current in your reading and to be aware of new words in your profession and in the language. Be aware also, in your quest for the right word, that there is no substitute for a good **dictionary.**

writing the draft

If you have established your **objective, reader,** and **scope,** and if you have completed adequate **research** and created a good outline, you will find writing the rough draft fairly easy. Writing a rough draft is simply the process of transcribing and expanding the notes from your outline into **paragraphs** without worrying about refinements of language or such mechanical aspects of writing as **transition.** Refinement will come with **revision.**

It is usually best to write the rough draft quickly, concentrating entirely upon converting your outline to **sentences** and paragraphs. To achieve directness of communication, write as though you were explaining your subject to someone across the desk from you. Don't worry about a good **opening.** Just start. There is no need in the rough draft to be concerned about an **introduction** or about transition unless they come easily—concentrate on *ideas.* Don't attempt to polish or revise. Writing and revising are different activities. Keep writing quickly to achieve **unity** and proportion.

Try not to let grammatical rules get in your way; they are of little value in writing the rough draft. Learning all the grammatical rules in the world will not help you write a better rough draft, yet some people are so concerned about the proper use of **grammar** even at this early stage that it diverts energy and attention from the draft.

Don't affect a fancy writing **style.** Your function as a writer is to communicate certain information to your readers, not to impress them with your fancy footwork. Write in plain, direct, concise language. Write the draft in the style that is comfortable and natural for you.

Remember that the first rule of good writing is to help the reader. As you write the rough draft, keep your reader's level of technical knowledge in mind. This will not only help you write directly to your reader, it will also tell you which terms you must define.

When you are trying to write quickly and come to something difficult to explain, try to relate the new concept to something with which the reader is already familiar. Although **figures of speech** are not used extensively in technical writing, they can be very useful in explaining a complex process or piece of equipment. In the rough draft, a figure of speech might be just the tool you need to keep moving when you encounter a complex concept that must be explained or described.

Be consistent in the way you address your readers (see **point of view**). Don't switch needlessly from one form of address to another. For example, if you write ''*You* will notice that . . .'' one time, don't write ''*One* should note that . . .'' the next time. This is simply a matter of keeping your readers in mind and being considerate of them. It will not be a problem if you establish the habit of writing as if you were explaining your subject to someone across the desk from you (but in a **formal writing style** and using **standard English,** keeping in mind that your readers cannot see your gestures and facial expressions).

Most of your outline notes should become the **topic sentences** for **paragraphs** in your draft. As an example of this, notice how items III.A. and III.B. in the topic outline on page 632 become the topic sentences of the two paragraphs that follow. (And notice, too, how the subordinate items in the outline become sentences within the two paragraphs.)

OUTLINE

III. Advantages of Chicago as location for new plant
    A. *Outstanding Transport Facilities*
        1. Rail
        2. Air
        3. Truck
        4. Sea (except in winter)
    B. *Ample Labor Supply*
        1. Engineering and scientific personnel
            a. Many similar companies in the area
            b. Several major universities
        2. Technical and manufacturing personnel
            a. Existing programs in community colleges
            b. Possible special programs designed for us

RESULTING PARAGRAPHS

Probably the greatest advantage of Chicago as a location for our new plant is its excellent transport facilities. The city is served by three major railroads. Both domestic and international air cargo service is available at O'Hare International Airport. Chicago is a major hub of the trucking industry, and most of the nation's large freight carriers have terminals there. Finally, except in the winter months when the Great Lakes are frozen, Chicago is a seaport, accessible through the St. Lawrence Seaway.

    A second advantage of Chicago is that it offers an abundant labor force. An ample supply of engineering and scientific personnel is assured not only by the presence of many companies engaged in activities similar to ours but also by the presence of several major universities in the metropolitan area. Similarly, technicians and manufacturing personnel are in abundant supply. The seven colleges in the Chicago City College system, as well as half a dozen other two year colleges in the outlying areas, produce graduates with Associate Degrees in a wide variety of technical specialties appropriate to our needs. Moreover, three of the outlying colleges have expressed an interest in establishing special courses attuned specifically to our requirements.

Don't let writing an introduction stop you from getting started. In fact, many good writers prefer to write the introduction last. If writing the introduction bothers you, choose the section of your outline that interests you most and start there (this is possible only if you have a good outline).

Whatever is most comfortable for you is what you should do to get started: typing, longhand—just get started! Waiting for inspiration to come along and give you a lift is simply an excuse for avoiding work. If you have made adequate **preparation,** you are ready to begin. The actual writing should be the easiest part of the job. (See also the Checklist of the Writing Process on page xiii.)

X

X-ray

X-ray, usually capitalized, is always hyphenated as an **adjective** and **verb** and usually hyphenated as a **noun.** If you hyphenate the noun, be sure to do so consistently.

EXAMPLES   portable *X-ray* unit (adjective)
The technician *X-rayed* the ankle. (verb)
X-rays and gamma rays (noun)

Z

zeugma

A zeugma is a construction in which an adjective or verb is forced to apply to two words even though it is grammatically or logically correct only for one of them. Poets and occasionally public speakers, use zeugma deliberately for rhythmic effect; but it is imprecise and should therefore be avoided in technical writing.

CHANGE   The walls of the office *are* green, the rug blue.
TO   The walls of the office *are* green; the rug *is* blue.

In this case *are* (a plural **verb**) is used for both subjects, *walls* (a plural **noun**) and *rug* (a singular noun), which is grammatically and logically incorrect.

Bibliography

Alred, Gerald J., Diana C. Reep, and Mohan R. Limaye. *Business and Technical Writing: An Annotated Bibliography of Books, 1880–1980.* Metuchen, New Jersey, and London: The Scarecrow Press, 1981.

Anderson, Virgil A. *Training the Speaking Voice*, 3rd ed. New York: Oxford University Press, 1977.

Bernstein, Theodore M. *The Careful Writer: A Guide to English Usage.* New York: Atheneum, 1967.

_____. *Watch Your Language.* Manhasset, New York: Channel Press, 1968.

Corbett, Edward P. J. *Classical Rhetoric for the Modern Student.* 2nd ed. New York: Oxford University Press, 1971.

Crawford, Tad. *The Writer's Legal Guide.* New York: Hawthorn Books, Inc., 1977.

Crowell, Thomas Lee, Jr. *Index to Modern English.* New York: McGraw-Hill, 1964.

Davis, Richard M. *Thesis Projects in Science and Engineering.* New York: St. Martin's Press, 1980.

Eckersley-Johnson, Anna L., ed. *Webster's Secretarial Handbook.* Springfield, Mass.: G & C Merriam Company, 1976.

Elsbree, Langdon, and Frederick Bracher. *Heath's College Handbook of Composition.* 9th ed. Boston: D. C. Heath and Co., 1976.

_____. *Heath's Brief Handbook of Usage.* 7th ed. Boston: D. C. Heath and Co., 1977.

Fear, David E. *Technical Communication.* Glenview, Ill.: Scott, Foresman and Company, 1977.

Freeman, Joanna M. *Basic Technical and Business Writing.* Ames, Iowa: Iowa State University Press, 1979.

Gammage, Allen Z. *Basic Police Report Writing.* 2nd ed. Springfield, Illinois: Charles C. Thomas, Publisher, 1978.

Gorrell, Robert M., and Charlton Laird. *Modern English Handbook.* 6th ed. Englewood Cliffs, New Jersey: Prentice-Hall, Inc., 1976.

Hart, Andrew W., and James A. Reinking. *Writing for Career-Education Students.* New York: St. Martin's Press, 1977.

Hodges, John C., and Mary E. Whitten. *Harbrace College Handbook.* 8th ed. New York: Harcourt Brace Jovanovich, Inc., 1977.

Houp, Kenneth W., and Thomas E. Pearsall. *Reporting Technical Information.* 3rd ed. Beverly Hills, Calif.: Glencoe Press, 1977.

Irmscher, William F. *The Holt Guide to English: A Contemporary Handbook of Rhetoric, Language, and Literature.* 2nd ed. New York: Holt, Rinehart and Winston, Inc., 1976.

Mambert, W. A. *Presenting Technical Ideas.* New York: John Wiley & Sons, Inc., 1968.

A Manual of Style. 12th ed., rev. Chicago: The University of Chicago Press, 1974.

Mills, Gordon H., and John A. Walter. *Technical Writing.* 4th ed. New York: Holt, Rinehart and Winston, Inc., 1978.

Monroe, Alan H., and Douglas Erlinger. *Principles and Types of Speech.* 8th ed. Chicago: Scott, Foresman and Co., 1978.

Moore, Robert H. *Handbook of Effective Writing.* 2nd ed. New York: Holt, Rinehart and Winston, Inc., 1971.

Nist, John. *Speaking Into Writing.* New York: St. Martin's Press, 1969.

Oliu, Walter E., Charles T. Brusaw, and Gerald J. Alred. *Writing That Works.* New York: St. Martin's Press, 1980.

Patterson, Frank M., and Patrick D. Smith. *A Manual of Police Report Writing.* Springfield, Illinois: Charles C. Thomas, Publisher, 1977.

Pauley, Steven E. *Technical Report Writing Today.* 2nd ed. Boston: Houghton Mifflin Company, 1979.

Perrin, Porter G. *Writer's Guide and Index to English.* 4th ed. revised by Karl W. Dykema and Wilma R. Ebbitt. Chicago: Scott, Foresman and Co., 1965.

————, and George H. Smith. *The Perrin-Smith Handbook of Current English.* 2nd ed. Chicago: Scott, Foresman and Co., 1962.

Pickett, Nell A., and Ann A. Laster. *Handbook for Student Writing.* San Francisco: Canfield Press, 1972.

Quirk, Randolph, and Sidney Greenbaum. *A Concise Grammar of Contemporary English.* New York: Harcourt Brace Jovanovich, 1973.

Robert, Henry M. *Robert's Rules of Order.* Rev. Chicago: Scott, Foresman and Co., 1970.

Rorabacher, Louise E. *A Concise Guide to Composition.* 3rd ed. New York: Harper & Row, Publishers, 1976.

Stevens, Martin, and Charles H. Kegel. *A Glossary for College English.* New York: McGraw-Hill, 1966.

Strunk, William, Jr., and E. B. White. *The Elements of Style.* 3rd ed. Toronto: The Macmillan Co. of Canada, Ltd., 1978.

Tichy, H. J. *Effective Writing for Engineers, Managers, Scientists.* New York: John Wiley & Sons, Inc., 1966.

Ulman, Joseph N., Jr., and Jay R. Gould. *Technical Reporting.* 3rd ed. New York: Holt, Rinehart and Winston, Inc., 1972.

Watkins, Floyd C., William B. Dillingham, and Edwin T. Martin. *Practical English Handbook.* 5th ed. Boston: Houghton Mifflin Company, 1977.

Weismann, Herman M. *Basic Technical Writing.* 3rd ed. Columbus, Ohio: Charles E. Merrill Publishing Co., 1974.

Wilcox, Roger P. *Oral Reporting in Business and Industry.* Englewood Cliffs, New Jersey: Prentice-Hall, Inc., 1967.

Words Into Type. 3rd ed. Englewood Cliffs, New Jersey: Prentice-Hall, Inc., 1974.

Index

Correction Chart

The following symbols are commonly used in correcting written work. Most of the symbols are keyed to the pages in the text where the problem is discussed.

Symbol	Problem	Page
ab	abbreviation	2–6
adj	adjective	15–18
adv	adverb	22–25
agr	agreement	29–35
amb	ambiguity	38–39
ap	apostrophe	46–48
awk	awkwardness	57–58
[ ]	brackets	67
cap	capital letter	70–74
lc	no capital letter (lower case)	70–74, 366
case	case (grammatical)	74–80
coh	coherence (*see also* unity)	90–91, 614
:/	colon	91–94
⌃	comma	94–104
cs	comma splice	105
dm	dangling modifiers	147–148
dash	dash	148–149
def	define term	153–155
⌁	delete (omit)	
div	word division (syllabication)	580
el	ellipsis (ellipses)	182–184
emph	emphasis	184–187
excl	exclamation mark	192–193
fn	footnote	216–221
frag	sentence fragment	545–547
h	hyphen	274–276
ital	italics	322–324

Symbol	Problem	Page
jarg	jargon	325–326
mm	misplaced modifier	380–381
org	organization	424–425
¶	paragraph	432–438
//	parallel structure	438–442
( )	parentheses	443–444
⊙	period	449–451
pv	point of view	459–462
p	punctuation	482
ques	question mark	483–484
quot	quotation marks	488–490
ref	pronoun reference	474–476
rep	repetition	513
r-o	run-on (fused) sentence	528
;/	semicolon	538–540
s	sentence structure	547–550
sp	word misspelled	571–572
sub	subordination	577–578
t	tense	591–593
trans	transition	602–605
trite	trite language; cliché	610–611, 88–89
v	verb, verb form	616–622
voice	active or passive voice	624–625
wc	word choice (diction)	629–630
wordy	wordiness	117–120
X	obvious error	
?	meaning unclear	
∧	something missing	